CZECHOSLOVAK ACADEMY OF SCIENCES

Contributions to Statistics

Jaroslav Hájek Memorial Volume

Scientific Editor Dr. Zbyněk Šidák, DrSc.

Reviewer Doc. Dr. Václav Dupač, CSc.

Contributions to Statistics

Jaroslav Hájek Memorial Volume

Editor

Dr. Jana Jurečková, CSc.

Springer-Science+Business Media, B.V.

Library of Congress Cataloging in Publication Data CIP

Main entry under title:

Contributions to statistics.
Includes indexes.
1. Mathematical statistics—Addresses, essays, lectures. 2. Hájek, Jaroslav.
I. Hájek, Jaroslav. II. Jureckova, Jana, 1940—
QA276.16.C55 519.5 79—13758

ISBN 978-94-009-9364-8 ISBN 978-94-009-9362-4 (eBook)
DOI 10.1007/978-94-009-9362-4

Originally published by D. Reidel Publishing Company in 1979
Softcover reprint of the hardcover 1st edition 1979

To the memory of
Professor Jaroslav Hájek, DrSc.

Professor Jaroslav Hájek, DrSc.
(4. 2. 1926 — 10. 6. 1974)

CONTENTS

PREFACE

The death of Jaroslav Hájek on June 10, 1974 was a heavy loss to many mathematicians all around the world. The impact of his work on mathematical statistics has been so extraordinary that it has completely changed the character of some fields of this science. Some of his ideas have become a part of the common statistical consciousness.

Hájek's contribution to statistics includes research in the theory of rank tests, parametric estimation, probability sampling, statistical inference in stochastic processes and various other specializations. His results were always of fundamental character for the corresponding field and they continue to stimulate further research and progress. For proving the results, he developed original methods which are now commonly used for the solution of many related problems.

Hájek was an enthusiastic mathematician; the secret of his success was in his love for the subject and in his great sense for practical applications; just practical problems provided him the source of excellent mathematical problems and ideas.

We see the best way of commemorating Jaroslav Hájek in the arrangement of the present volume which collects papers of authors whose work is related to Hájek's work and who were friends of him. The authors come from Czechoslovakia, Hungary, the Netherlands, Sweden, U.S.A., U.S.S.R., and West Germany.

I wish to thank most cordially all the authors and all who contributed in any way to the success of the publication.

Prague, April 1976 *Jana Jurečková*

ON INTERPOLATION OF MULTIPLE AUTOREGRESSIVE PROCESSES

by

JIŘÍ ANDĚL

1. Introduction and preliminaries

Let $\mathbf{X}_1, \ldots, \mathbf{X}_n$ be p-dimensional random vectors with zero means and with a regular variance matrix $\mathbf{B} = \text{Var}\,(\mathbf{X}_1', \ldots, \mathbf{X}_n')'$. Let $\mathbf{Y}_{n+1}, \ldots, \mathbf{Y}_N$ be p-dimensional random vectors with zero means and with the unit variance matrices. Suppose that $\text{Cov}\,(\mathbf{X}_s, \mathbf{Y}_t) = \mathbf{0}$ and $\text{Cov}\,(\mathbf{Y}_t, \mathbf{Y}_v) = \mathbf{0}$ for $1 \leqq s \leqq n < t \neq v \leqq N$. Let $\mathbf{A}_0, \ldots, \mathbf{A}_n$ be $p \times p$ matrices such that $\mathbf{A}_0$ is regular and $\mathbf{A}_n \neq \mathbf{0}$. Put

$$(1) \qquad \mathbf{U}_j = -\mathbf{A}_0^{-1}\mathbf{A}_j\,, \quad j = 1, 2, \ldots, n\,.$$

Define $\mathbf{X}_{n+1}, \ldots, \mathbf{X}_N$ recursively by

$$(2) \qquad \sum_{k=0}^{n} \mathbf{A}_k \mathbf{X}_{t-k} = \mathbf{Y}_t\,, \quad t = n+1, n+2, \ldots, N\,.$$

Formula (2) is equivalent to

$$(3) \qquad \mathbf{X}_t = \sum_{k=1}^{n} \mathbf{U}_k \mathbf{X}_{t-k} + \mathbf{A}_0^{-1}\mathbf{Y}_t\,, \quad t = n+1, n+2, \ldots, N\,.$$

It is a well-known property of autoregressive processes that the best linear extrapolation $\hat{\mathbf{X}}_{i+k}$ of the vector $\mathbf{X}_{i+k}$ based on $\mathbf{X}_1, \ldots, \mathbf{X}_i$ $(n \leqq i < i + k \leqq N)$ depends only on the n last known vectors, i.e. on $\mathbf{X}_{i-n+1}, \ldots, \mathbf{X}_i$. A similar assertion holds also for the backward extrapolation (see [3]): the best linear extrapolation $\hat{\mathbf{X}}_{i-k}$ of the vector $\mathbf{X}_{i-k}$ based on $\mathbf{X}_i, \mathbf{X}_{i+1}, \ldots, \mathbf{X}_N$ $(1 \leqq i - k < i \leqq N - n + 1)$ depends only on $\mathbf{X}_i, \mathbf{X}_{i+1}, \ldots, \mathbf{X}_{i+n-1}$. In this paper, it will be proved that the best linear interpolation uses only n previous and n subsequent known vectors $\mathbf{X}_t$. All these results are proved without the assumption of stationarity.

Let the random vectors

$$(4) \qquad \mathbf{X}_1, \mathbf{X}_2, \ldots, \mathbf{X}_{i-1}, \mathbf{X}_{i+q}, \mathbf{X}_{i+q+1}, \ldots, \mathbf{X}_N$$

be known, where i and q are such natural numbers that $n \leqq i - 1$, $i + q \leqq N - n + 1$. The problem is to find the best linear estimate $\hat{\mathbf{X}}_j$ of the vector $\mathbf{X}_j$ where $i \leqq j < i + q$.

Lemma 1.1. *Let $V_1, \ldots, V_p$ and $Z_1, \ldots, Z_s$ be random variables with finite second moments. Denote $\mathbf{V} = (V_1, \ldots, V_p)'$, $\mathbf{Z} = (Z_1, \ldots, Z_s)'$. Let H be the linear space spanned by $Z_1, \ldots, Z_s$. Define the scalar product by the covariance and denote $\hat{V}_1, \ldots, \hat{V}_p$ projections of $V_1, \ldots, V_p$ onto H. If the matrix $\mathbf{G} = \operatorname{Var} \mathbf{Z}$ is regular, then the vector $\mathbf{V} = (\hat{V}_1, \ldots, \hat{V}_p)'$ is given by*

$$\hat{\mathbf{V}} = \operatorname{Cov}(\mathbf{V}, \mathbf{Z})\, \mathbf{G}^{-1}\mathbf{Z}\,. \tag{5}$$

Proof is obvious.

Lemma 1.2. *Let*

$$\mathbf{S} = \begin{Vmatrix} \mathbf{A} & \mathbf{B} \\ \mathbf{C} & \mathbf{D} \end{Vmatrix}$$

be a regular matrix with square blocks $\mathbf{A}$ *and* $\mathbf{D}$. *Suppose that* $\mathbf{A}$ *and* $\mathbf{D} - \mathbf{C}\mathbf{A}^{-1}\mathbf{B}$ *are regular. If*

$$\mathbf{S}^{-1} = \begin{Vmatrix} \mathbf{K} & \mathbf{L} \\ \mathbf{M} & \mathbf{N} \end{Vmatrix}$$

is divided into blocks of the same types as $\mathbf{S}$ *is, then*

$$\mathbf{A}^{-1} = \mathbf{K} - \mathbf{L}\mathbf{N}^{-1}\mathbf{M}\,. \tag{6}$$

Proof. It is well-known that

$$\mathbf{K} = \mathbf{A}^{-1} - \mathbf{A}^{-1}\mathbf{B}\mathbf{M}\,, \quad \mathbf{L} = -\mathbf{A}^{-1}\mathbf{B}\mathbf{N}\,, \quad \mathbf{N} = (\mathbf{D} - \mathbf{C}\mathbf{A}^{-1}\mathbf{B})^{-1}\,.$$

This implies the formula (6).

Lemma 2 is useful when we know the inverse $\mathbf{S}^{-1}$ of a "large" matrix $\mathbf{S}$ and we need the inverse $\mathbf{A}^{-1}$ of a "small" matrix $\mathbf{A}$ which is a part of $\mathbf{S}$.

Lemma 1.3. *If* $\mathbf{B} = \operatorname{Var}(\mathbf{X}'_1, \ldots, \mathbf{X}'_n)'$ *is regular, then* $\mathbf{G} = \operatorname{Var}(\mathbf{X}'_1, \ldots, \mathbf{X}'_N)'$ *is also regular. Denote* $\mathbf{B}^{-1} = \mathbf{E} = \|\mathbf{E}_{st}\|_{s,t=1}^{n}$, $\mathbf{G}^{-1} = \mathbf{H} = \|\mathbf{H}_{st}\|_{s,t=1}^{N}$, *where* $\mathbf{E}_{st}$ *and* $\mathbf{H}_{st}$ *are* $p \times p$ *blocks. Then*

$$\mathbf{H}_{st} = \mathbf{E}_{st} + \sum_{k=n+1}^{\min(n+s,n+t,N)} \mathbf{A}'_{k-s}\mathbf{A}_{k-t} \quad \textit{for} \quad 1 \leqq s\,,\ t \leqq n\,, \tag{7}$$

$$\mathbf{H}_{st} = \sum_{k=\max(s,t)}^{\min(n+s,n+t,N)} \mathbf{A}'_{k-s}\mathbf{A}_{k-t} \quad \textit{for} \quad n < \max(s,t) \leqq N\,. \tag{8}$$

Especially,

$$\mathbf{H}_{st} = \mathbf{0} \quad \textit{for} \quad |s - t| > n\,. \tag{9}$$

Proof. See [1].

2. Interpolation in autoregressive processes

Theorem 2.1. *The best linear interpolation* $\hat{\mathbf{X}}_j$ *of* $\mathbf{X}_j$ *has the form*

$$\hat{\mathbf{X}}_j = \sum_{t=1}^{n} \mathbf{P}_{j,q,t}\mathbf{X}_{i-t} + \sum_{t=1}^{n} \mathbf{Q}_{j,q,t}\mathbf{X}_{i+q-1+t}, \tag{10}$$

where $\mathbf{P}_{j,q,t}$ *and* $\mathbf{Q}_{j,q,t}$ *are matrices of the type* $p \times p$.

Proof. Denote $\mathbf{R}_{st} = E\mathbf{X}_s\mathbf{X}_t'$ for $1 \leqq s, t \leqq N$. Let $\mathbf{I}_b$ be the unit matrix of the type $b \times b$. From $\mathbf{GH} = \mathbf{I}_N$ we get

$$\sum_{t=1}^{N} \mathbf{R}_{jt}\mathbf{H}_{tk} = \delta_{jk}\mathbf{I}_p, \tag{11}$$

where $\delta_{jk} = 0$ for $j \neq k$ and $\delta_{jk} = 1$ for $j = k$. Denoting

$$J = \{i, i+1, \dots, i+q-1\} \tag{12}$$

we have

$$\sum_{t \notin J} \mathbf{R}_{jt}\mathbf{H}_{tk} = \begin{cases} \mathbf{I}_p - \sum_{t \in J} \mathbf{R}_{jt}\mathbf{H}_{tj} & \text{for } k = j, \\ -\sum_{t \in J} \mathbf{R}_{jt}\mathbf{H}_{tj} & \text{for } k \neq j. \end{cases} \tag{13}$$

Denote

$$\mathbf{T} = \begin{Vmatrix} \mathbf{I}_{i-1} & \mathbf{0} & \mathbf{0} \\ \mathbf{0} & \mathbf{0} & \mathbf{I}_q \\ \mathbf{0} & \mathbf{I}_{N-i-q+1} & \mathbf{0} \end{Vmatrix}, \qquad \mathring{\mathbf{X}} = \begin{Vmatrix} \mathbf{X}_1 \\ \dots \\ \mathbf{X}_{i-1} \\ \mathbf{X}_{i+q} \\ \dots \\ \mathbf{X}_N \end{Vmatrix}.$$

Consider the matrix $\mathbf{S} = \mathbf{TGT}'$. Obviously, $\mathbf{S}^{-1} = \mathbf{THT}'$ in view of $\mathbf{T}^{-1} = \mathbf{T}'$. Now, the matrix $\operatorname{Var} \mathring{\mathbf{X}}$ is placed in $\mathbf{S}$ in the upper left-hand corner. In view of Lemma 1.2 we obtain

$$(\operatorname{Var} \mathring{\mathbf{X}})^{-1} = \mathbf{H}_{(J)(J)} - \mathbf{H}_{(J)J}\mathbf{H}_{JJ}^{-1}\mathbf{H}_{J(J)}, \tag{14}$$

where

$$\mathbf{H}_{(J)(J)} = \|\mathbf{H}_{st}\|_{s \notin J, t \notin J}, \quad \mathbf{H}_{JJ} = \|\mathbf{H}_{st}\|_{s \in J, t \in J},$$
$$\mathbf{H}_{(J)J} = \|\mathbf{H}_{st}\|_{s \notin J, t \in J}, \quad \mathbf{H}_{J(J)} = \|\mathbf{H}_{st}\|_{s \in J, t \notin J}.$$

Lemma 1.1 implies

$$\hat{\mathbf{X}}_j = \operatorname{Cov}(\mathbf{X}_j, \mathring{\mathbf{X}})(\operatorname{Var} \mathring{\mathbf{X}})^{-1} \mathring{\mathbf{X}}. \tag{15}$$

Obviously,

$$\operatorname{Cov}(\mathbf{X}_j, \mathring{\mathbf{X}}) = (\mathbf{R}_{j1}, \ldots, \mathbf{R}_{j,i-1}, \mathbf{R}_{j,i+q}, \ldots, \mathbf{R}_{jN}) .$$

Taking into account (13) and (9) we can write

$$\hat{\mathbf{X}}_j = -\sum_{k=1}^{n} \sum_{t\in J} \mathbf{R}_{jt}\mathbf{H}_{t,i-k}\mathbf{X}_{i-k} \tag{16}$$

$$-\sum_{k=1}^{n} \sum_{t\in J} \mathbf{R}_{jt}\mathbf{H}_{t,i+q-1+k}\mathbf{X}_{i+q-1+k} - \mathbf{V}\mathbf{H}_{JJ}^{-1}\mathbf{W} ,$$

where

$$\mathbf{V} = \Big(-\sum_{t\in J} \mathbf{R}_{jt}\mathbf{H}_{ti}, \ldots, -\sum_{t\in J} \mathbf{R}_{jt}\mathbf{H}_{t,j-1} , \tag{17}$$

$$\mathbf{I} - \sum_{t\in J} \mathbf{R}_{jt}\mathbf{H}_{tj} , \quad -\sum_{t\in J} \mathbf{R}_{jt}\mathbf{H}_{t,j+1}, \ldots, \quad -\sum_{t\in J} \mathbf{R}_{jt}\mathbf{H}_{t,i+q-1}\Big) ,$$

$$\mathbf{W} = \left\| \begin{array}{c} \sum\limits_{t=i-n}^{i-1} \mathbf{H}_{it}\mathbf{X}_t + \sum\limits_{t=i+q}^{i+n} \mathbf{H}_{it}\mathbf{X}_t \\ \sum\limits_{t=i-n+1}^{i-1} \mathbf{H}_{i+1,t}\mathbf{X}_t + \sum\limits_{t=i+q}^{i+n+1} \mathbf{H}_{i+1,t}\mathbf{X}_t \\ \cdots\cdots\cdots\cdots\cdots\cdots \\ \sum\limits_{t=i-n+q-1}^{i-1} \mathbf{H}_{i+q-1,t}\mathbf{X}_t + \sum\limits_{t=i+q}^{i+n+q-1} \mathbf{H}_{i+q-1,t}\mathbf{X}_t \end{array} \right\| .$$

From (16) and (17) it is clear, that (10) holds. The proof is finished.

3. Remarks

Corollary 3.1. *If* $q = 1$, $j = i$, *then*

$$\hat{\mathbf{X}}_i = -\mathbf{H}_{ii}^{-1}\Big[\sum_{k=1}^{n} \mathbf{H}_{i,i-k}\mathbf{X}_{i-k} + \sum_{k=1}^{n} \mathbf{H}_{i,i+k}\mathbf{X}_{i+k}\Big] . \tag{18}$$

Proof follows immediately from (16) and (17).

The application of formula (18) is particularly simple in the cases when $i > n$. Then the blocks $\mathbf{H}_{st}$ occurring in (18) are given by formula (8) and the interpolation $\hat{\mathbf{X}}_i$ does not depend on the blocks $\mathbf{E}_{st}$. Especially, for $i > n$ we have

$$\mathbf{H}_{ii} = \sum_{t=0}^{n} \mathbf{A}_t'\mathbf{A}_t , \quad \mathbf{H}_{i,i-k} = \sum_{t=0}^{n-k} \mathbf{A}_{t+k}'\mathbf{A}_t , \quad \mathbf{H}_{i,i+k} = \sum_{t=0}^{n-k} \mathbf{A}_t'\mathbf{A}_{t+k} ,$$

$k = 1, 2, \ldots, n$. We can see that the matrix standing by $\mathbf{X}_{i-k}$ is the same as that by $\mathbf{X}_{i+k}$ for $k = 1, 2, \ldots, n$ if and only if

$$\sum_{t=0}^{n-k} \mathbf{A}'_{t+k}\mathbf{A}_t = \sum_{t=0}^{n-k} \mathbf{A}'_t\mathbf{A}_{t+k}, \quad k = 1, 2, \ldots, n. \tag{19}$$

Suppose for a moment that $\mathbf{X}_1, \ldots, \mathbf{X}_N$ is a part of a stationary series $\{\mathbf{X}_t\}_{-\infty}^{\infty}$.

Such a series is called symmetric, if its covariance function $R(t)$ fulfils the condition $R(t) = R(-t)$ for any t. It is known that a stationary autoregressive process $\{\mathbf{X}_t\}$ is symmetric, if and only if (19) holds (see [2]). Thus we can see that a symmetry of formula (18) is tightly connected with a general concept of the symmetry in time series.

Generally, formulas (16) and (17) contain the covariances $\mathbf{R}_{st}$. If only the matrix $\mathbf{B} = \operatorname{Var}(\mathbf{X}'_1, \ldots, \mathbf{X}'_n)'$ and matrices $\mathbf{A}_0, \ldots, \mathbf{A}_n$ are known, the following formulas are useful for computation of $\mathbf{R}_{st}$ for $\max(s, t) > n$.

Introduce matrices

$$\mathbf{M} = \begin{Vmatrix} \mathbf{0} & \mathbf{I} & \mathbf{0} & \ldots & \mathbf{0} \\ \cdots & \cdots & \cdots & \cdots & \cdots \\ \mathbf{0} & \mathbf{0} & \mathbf{0} & \ldots & \mathbf{I} \\ \mathbf{U}_n & \mathbf{U}_{n-1} & \mathbf{U}_{n-2} & \ldots & \mathbf{U}_1 \end{Vmatrix}, \quad \mathbf{\Lambda} = \begin{Vmatrix} \mathbf{0} & \mathbf{0} & \ldots & \mathbf{0} & \mathbf{0} \\ \cdots & \cdots & \cdots & \cdots & \cdots \\ \mathbf{0} & \mathbf{0} & \ldots & \mathbf{0} & \mathbf{0} \\ \mathbf{0} & \mathbf{0} & \ldots & \mathbf{0} & (\mathbf{A}'_0\mathbf{A}_0)^{-1} \end{Vmatrix}.$$

Let j, k be integers such that $1 \leqq j \leqq k \leqq N + 1 - n$. Then

$$\begin{Vmatrix} \mathbf{R}_{jk} & \ldots & \mathbf{R}_{j,k+n-1} \\ \cdots & \cdots & \cdots \\ \mathbf{R}_{j+n-1,k} & \ldots & \mathbf{R}_{j+n-1,k+n-1} \end{Vmatrix} = \begin{Vmatrix} \mathbf{R}_{jj} & \ldots & \mathbf{R}_{j,j+n-1} \\ \cdots & \cdots & \cdots \\ \mathbf{R}_{j+n-1,j} & \ldots & \mathbf{R}_{j+n-1,j+n-1} \end{Vmatrix} \mathbf{M}'^{k-j},$$

$$\begin{Vmatrix} \mathbf{R}_{kk} & \ldots & \mathbf{R}_{k,k+n-1} \\ \cdots & \cdots & \cdots \\ \mathbf{R}_{k+n-1,k} & \ldots & \mathbf{R}_{k+n-1,k+n-1} \end{Vmatrix} = \mathbf{M}^{k-j} \begin{Vmatrix} \mathbf{R}_{jj} & \ldots & \mathbf{R}_{j,j+n-1} \\ \cdots & \cdots & \cdots \\ \mathbf{R}_{j+n-1,j} & \ldots & \mathbf{R}_{j+n-1,j+n-1} \end{Vmatrix} \mathbf{M}'^{k-j} + \sum_{i=0}^{k-j-1} \mathbf{M}^i \mathbf{\Lambda} \mathbf{M}'^i.$$

These formulas follow easily from (3).

References

[1] Anděl, J. (1971). On the multiple autoregressive series. *Ann. Math. Statist.*, **42**, 755—759.

[2] Anděl, J. (1972). Symmetric and reversed multiple stationary autoregressive series. *Ann. Math. Statist.*, **43**, 1197—1203.

[3] Anděl, J. (1974). On the backward extrapolation of nonstationary autoregressive series. *Trans. of 7th Prague Conf. on Inf. Th.*, Vol. A, 29—36.

CHARLES UNIVERSITY, PRAGUE, CZECHOSLOVAKIA

Received October 1975

A NOTE ON UMV ESTIMATES AND ANCILLARY STATISTICS*

by

R. R. BAHADUR

1. Introduction and theorems

Let X be a space of points x, $\mathscr{S}$ a field of subsets of X, and P a given set of probability measures p on X. In this framework an estimate is defined to be a real valued $\mathscr{S}$-measurable function on X. An estimate $t(x)$ is P-integrable if $E_p|t| < \infty\ \forall p$ in P. Let U denote the class of all estimates t such that t^2 is P-integrable. t_1 is a UMV estimate if $t_1 \in U$, and if the conditions $t_2 \in U$, $E_p(t_1) = E_p(t_2)\ \forall p \in P$ imply that $E_p(t_1^2) \leqq E_p(t_2^2)\ \forall p \in P$. Let T denote the class of all UMV estimates.

A set $A \subset X$ is ancillary if $p(A)$ is the same for all $p \in P$. Let u be a measurable transformation of $(X, \mathscr{S})$ into a measurable space. u is an ancillary statistic if, for every measurable set B in the range space of u, $u^{-1}(B)$ is an ancillary subset of X. Let T_a be the class of all $t \in T$ such that, for every ancillary statistic u, t and u are independent statistics for each $p \in P$.

In a recent paper Bondesson (1975) gives a sufficient condition in order that an estimate be in T_a, and raises the question whether there are examples where $T_a \neq T$. The present note uses some of the conclusions and constructions of Bahadur (1957) to obtain further information on the problem (Theorems 1, 4 and 5 of this section), and to show that in general $T_a \neq T$ (Example 2 in the following section).

The following Lemmas 1 and 2 are quoted here for convenience of reference. Let N_1 be the class of all unbiased estimates of zero, i.e. $t \in N_1$ if and only if t is P-integrable and $E_p(t) = 0\ \forall p \in P$. Let N_2 be the intersection of N_1 and U.

Lemma 1. (Lehmann and Scheffé (1950)) *If $t \in U$, then $t \in T$ if and only if $t \, . \, n \in N_1$ for all $n \in N_2$.*

Let T_b denote the class of all bounded estimates in T. Let $\mathscr{S}_0$ be the smallest field such that each $t \in T_b$ is $\mathscr{S}_0$-measurable.

* Support for this research has been provided in part by National Science Foundation Grant No. NSF MPS 72-04364 AO3 from the Division of Mathematical, Physical, and Engineering Sciences of the National Science Foundation.

Lemma 2. (Bahadur (1957)) *If $t \in U$, and t is $\mathscr{S}_0$-measurable, then $t \in T$.*

Let us say that a $t \in T$ is hereditary if, for every real valued Borel measurable function f on the real line such that g defined by

$$g(x) = f(t(x)) \tag{1}$$

is in U, we have $g \in T$. Let T_h be the set of all hereditary $t \in T$.

Theorem 1. $T_b \subset T_h \subset T_a \subset T$.

Proof. Choose $t \in T_b$ and let f be a Borel measurable function such that g defined by (1) is in U. t is $\mathscr{S}_0$-measurable; hence g is $\mathscr{S}_0$-measurable; hence $g \in T$, by Lemma 2. Since f is arbitrary, $t \in T_h$. This establishes $T_b \subset T_h$. Now choose $t \in T_h$. Let B be a linear Borel set and let $g(x) = I_B(t(x))$. Since t is hereditary and g is bounded, g is in T_b. (It follows, incidentally, that t is $\mathscr{S}_0$-measurable.) Let A be an ancillary subset of X and suppose that $p(A) = \alpha \ \forall p \in P$. Then $I_A(x) - \alpha$ is in N_2, and so $g \,.\, (I_A - \alpha)$ is in N_1, by Lemma 1. This last conclusion can be expressed as $p(t(x) \in B, x \in A) = p(t(x) \in B) \,.\, p(x \in A) \ \forall p \in P$. Since B and ancillary A are arbitrary, it follows that $t \in T_a$. This completes the proof.

The following theorem provides a justification of the term "hereditary UMV estimate".

Theorem 2. *T_h is the class of all $\mathscr{S}_0$-measurable estimates in U.*

Proof. As noted in the proof of Theorem 1, $t \in T_h$ implies that t is an $\mathscr{S}_0$-measurable function in U. To establish the converse, let t be an $\mathscr{S}_0$-measurable estimate in U and let f be a Borel measurable function such that, with g defined by (1), g is in U. Then g is $\mathscr{S}_0$-measurable and it follows from Lemma 2 that t and g are in T. Since f is arbitrary, $t \in T_h$ and this completes the proof.

A set $C \subset X$ is P-null if C is $\mathscr{S}$-measurable and $p(C) = 0 \ \forall p \in P$. Estimates t_1 and t_2 are P-equivalent if $\{x : t_1(x) \neq t_2(x)\}$ is a P-null set. It is readily seen that if t_1 is in U or T or T_a or T_h, and t_2 is an estimate P-equivalent to t_1, then t_2 is in U or T or T_a or T_h respectively.

The fact that P-equivalence conditions are absent from Theorem 2 is due to the following property of the field $\mathscr{S}_0$.

Theorem 3. *If t_1 and t_2 are P-equivalent estimates, and t_1 is $\mathscr{S}_0$-measurable, then so is t_2.*

Proof. Define $t_i^{(k)} = t_i$ if $|t_i| < k$ and $t_i^{(k)} = 0$ (say) otherwise, for $i = 1, 2$ and $k = 1, 2, \ldots$. Then $t_1^{(k)}$ is a bounded $\mathscr{S}_0$-measurable function, so $t_1^{(k)} \in T$ by Lemma 2. It is plain that $t_1^{(k)}$ and $t_2^{(k)}$ are P-equivalent, so that $t_2^{(k)}$ is also in T; since $t_2^{(k)}$ is bounded, it is $\mathscr{S}_0$-measurable. Since k is arbitrary, and since $t_2 = \lim_{k \to \infty} t_2^{(k)}$, it follows that t_2 is $\mathscr{S}_0$-measurable, and this completes the proof.

Choose and fix a $t \in T$. Theorem 7.1 of Bondesson (1975) states that if t^k is P-integrable for each $k = 1, 2, \ldots$ and if, for each $p \in P$, there is only one probability distribution on the real line which has $\{E_p(t^k) : k = 1, 2, \ldots\}$ for its moment sequence then $t \in T_a$. Similar but stronger sufficient conditions can be given which imply that in fact $t \in T_h$. For example, if $t^k n^2$ is P-integrable for each $k = 1, 2, \ldots$ and $n \in N_2$, and if for each $p \in P$ there exists $\delta_p > 0$ such that $E_p(\exp\{rt\}) < \infty$ for $-\delta_p < r < \delta_p$, then $t \in T_h$. Both the conditions stated in this paragraph are satisfied if t is bounded, but they are too strong to be necessary for their respective purposes.

The following theorem shows that the distinctions between T_h, T_a, and T are pathologies of infinity.

Theorem 4. *If X is a finite set, or P is a finite set, then*

$$T_h = T_a = T. \tag{2}$$

Proof. If X is finite we have

$$T_b = T \tag{3}$$

and (2) follows from Theorem 1. Suppose now that P is finite, say $P = \{p_1, \ldots, p_k\}$. Let μ be the probability measure on $\mathscr{S}$ defined by $\mu = (\sum_{i=1}^{k} p_i)/k$. Then for each i there exists an $\mathscr{S}$-measurable function $f_i(x)$ with $0 \leqq f_i(x) \leqq k \;\forall x \in X$ such that $dp_i = f_i(x)\, d\mu$ on $\mathscr{S}$ $(i = 1, \ldots, k)$. Note that each f_i^2 is μ-integrable, and that an estimate t is in U if and only if t^2 is μ-integrable.

Choose and fix a $t_1 \in T$. Let $t_2 \in U$ be such that $E_p(t_1) = E_p(t_2) \;\forall p \in P$. Then $E_p(t_1^2) \leqq E_p(t_2^2) \;\forall p \in P$ and hence $E_\mu(t_1^2) \leqq E_\mu(t_2^2)$. Since t_2 is arbitrary, we conclude that t_1 minimizes $E_\mu(t^2)$ in the class of all t such that $t - t_1$ is in N_2. Hence by Theorem 1 of Bahadur (1957) there exist constants $c_1, \ldots, c_k$ such that, with $t_3 = \sum_{i=1}^{k} c_i f_i(x)$, we have $t_1 = t_3$ a.e. $[\mu]$. Hence t_1 and t_3 are P-equivalent, and so $t_3 \in T$. Since t_3 is bounded, it follows that t_3 is $\mathscr{S}_0$-measurable. It now follows from the P-equivalence of t_1 and t_3 by Theorem 3 that t_1 is $\mathscr{S}_0$-measurable. Thus each $t \in T$ is $\mathscr{S}_0$-measurable, and (2) now follows from Theorems 1 and 2. This completes the proof. The proof shows that if P is a finite set then each $t \in T$ is P-equivalent to an estimate in T_b, but it is easy to see that (3) need not hold.

A field $\mathscr{S}^* \subset \mathscr{S}$ is said to be *complete* for P if every $\mathscr{S}^*$-measurable unbiased estimate of zero which is in U is P-equivalent to the estimate which is $\equiv 0$ on X. A real valued function g defined on P is said to be estimable if there exists an estimate $t \in U$ such that $E_p(t) = g(p) \;\forall p \in P$. These definitions of Bahadur (1957) are a slight modification of the original definitions of Lehmann and Scheffé (1950). Now consider the following four statements. (i) There exists a subfield of $\mathscr{S}$ which is sufficient and complete for P. (ii) $\mathscr{S}_0$ is sufficient and complete for P. (iii) T contains an unbiased estimate of every estimable g, and $T_h = T_a = T$. (iv) T contains an unbiased estimate of every estimable g.

Theorem 5. *Statements* (i), (ii) *and* (iii) *are equivalent. If P is a dominated set then all four statements are equivalent.*

Proof. We shall show that (iii) ⇒ (ii) and that (i) ⇒ (iii); this will establish the first part of the theorem. The second part will then follow from the first part by Theorem 4 of Bahadur (1957).

Suppose then that (iii) holds. Let A be an $\mathscr{S}$-measurable subset of X, and let $g(p) = E_p(I_A) = p(A)$ for $p \in P$. By the first part of (iii) there exists a $t \in T$ which is an unbiased estimate of this g. Let B be an $\mathscr{S}_0$-measurable set. Then $I_B \in T$ by Lemma 2. Since $t - I_A \in N_2$, it follows from Lemma 1 that $I_B \,.\, (t - I_A) \in N_1$. This conclusion can be expressed as

(4) $$p(A \cap B) = E_p(I_B \,.\, t) \; \forall B \in \mathscr{S}_0 \,, \quad p \in P \,.$$

It follows from (2) by Theorem 2 that t is $\mathscr{S}_0$-measurable. It now follows from (4) that, for each $p \in P$, t serves as the conditional probability of A given $\mathscr{S}_0$. Since A is arbitrary, it is established that $\mathscr{S}_0$ is sufficient for P. It follows from Lemma 2 that $\mathscr{S}_0$ is always complete, so (ii) holds.

Suppose now that (i) holds, let $\mathscr{S}^*$ be a complete and sufficient subfield, and let T^* be the class of all $\mathscr{S}^*$-measurable estimates in U. It follows from the Lehmann-Scheffé theory (or, rather, a trivial modification thereof) that (iv) holds, and also that T is essentially the same as T^*: more precisely, (α) $T^* \subset T$, and (β) corresponding to each $t \in T$ there exists a P-equivalent $t^* \in T^*$. It follows (cf. the proof of Theorem 2) from (α) that in fact (γ) $T^* \subset T_h$. Now choose $t \in T$. By (β) there exists a $t^* \in T^*$ which is P-equivalent to t. It follows from (γ) and P-equivalence that $t \in T_h$. Thus $T \subset T_h$, and Theorem 1 implies that (2) holds. This completes the proof.

It is not known at present whether the domination assumption is essential to the second part of Theorem 5.

2. Counterexamples

In both examples of this section $X = \{1, 2, \ldots, \text{ad inf.}\}$, $\mathscr{S}$ is the field of all sets of X, and $P = \{p_1, p_2, \ldots, \text{ad inf.}\}$. The sets $\{p_2, p_3, \ldots\}$ are different in Examples 1 and 2, but in each example p_1 is determined by $\{p_2, p_3, \ldots\}$ as follows:

(5) $$p_1 = \sum_{k=2}^{\infty} c_k p_k \,,$$

where

(6) $$c_k = b/k^5 \quad (k = 2, 3, \ldots) \,,$$

and $b > 0$ is such that $\sum_{k=2}^{\infty} c_k = 1$. It will be seen presently that this definition of p_1 implies that, in each example,

$$E_{p_1}(x^3) < \infty , \quad E_{p_1}(x^4) = \infty . \tag{7}$$

Let

$$n_1(x) = (-1)^x , \quad n_2(x) = (-1)^x \cdot x , \quad n_3(x) = (-1)^x \cdot x^2 \tag{8}$$

for all $x = 1, 2, \ldots$.

Example 1 ($T_h \neq T_a$). For each $k = 2, 3, \ldots$ let p_k be defined by

$$p_k(\{x\}) = \begin{cases} k(k+1)/(4k^2-2) & \text{if } x = k-1 , \\ 2(k^2-1)/(4k^2-2) & \text{if } x = k , \\ k(k-1)/(4k^2-2) & \text{if } x = k+1 , \\ 0 & \text{otherwise} . \end{cases} \tag{9}$$

This example is virtually the same as the example in Section 6 of Bahadur (1957). As in that example, it can be shown here that N_1 is the linear manifold spanned by $\{n_2, n_3\}$; consequently, N_2 is the manifold spanned by n_2. It follows hence from Lemma 1 that T is the manifold spanned by $\{1, x\}$. Hence T_h is the set of constants. Let A be an ancillary set. Then, for some constant α, $I_A - \alpha$ is a bounded function in N_2. Since n_2 is unbounded, we must have $I_A(x) - \alpha \equiv 0$. Hence $\alpha = 1$ or 0, and A is X or the empty set. It follows that any ancillary statistic is constant. Hence $T_b = T_h \neq T_a = T$.

Example 2 ($T_a \neq T$). For each $k = 2, 3, \ldots$ let p_k be defined by

$$p_k(\{x\}) = \begin{cases} 1/8 & \text{if } x = k-1 , \\ 3/8 & \text{if } x = k , \\ 3/8 & \text{if } x = k+1 , \\ 1/8 & \text{if } x = k+2 , \\ 0 & \text{otherwise} . \end{cases} \tag{10}$$

It follows from (8) and (10) that $E_{p_k}(n_i) = 0$ for $i = 1, 2, 3$; $k = 2, 3, \ldots$. Hence $E_{p_1}(n_i) = 0$ for $i = 1, 2, 3$ by (5) and (6). Since N_1 is always a linear manifold, it follows that N_1 contains the span of $\{n_1, n_2, n_3\}$. Now let $n_4 \in N_1$ be given. It is readily seen that there exist constants α, β, γ such that n_4 agrees with $n_5 \equiv \alpha n_1 + \beta n_2 + \gamma n_3$ for $x = 1, 2, 3$. Let $n_6 = n_5 - n_4$. Then $n_6 \in N_1$, so that $E_{p_k}(n_6) = 0$ for $k = 2, 3, \ldots$. Since n_6 vanishes on $\{1, 2, 3\}$, it follows from (10) that $n_6(x) = 0 \ \forall x \in X$. We conclude that N_1 is the span of $\{n_1, n_2, n_3\}$. Consequently, by (7), N_2 is the span of $\{n_1, n_2\}$.

Consider a $t \in T$. By Lemma 1, $t \cdot n_1 \in N_1$. Since $n_1^2 \equiv 1$, $t \cdot n_1 \in N_2$, so $t \cdot n_1 \equiv \alpha n_1 + \beta n_2$ for some constants α and β; hence $t \equiv \alpha + \beta x$ by (8).

Conversely, if $t \equiv \alpha + \beta x$, and $n_4 \in N_2$, say $n_4 \equiv \gamma n_1 + \delta n_2$, it is plain from (8) that $t \,.\, n_4$ is in the span of $\{n_1, n_2, n_3\}$ $(\equiv N_1)$, so $t \in T$, by Lemma 1. Thus, as in Example 1, T is the span of $\{1, x\}$.

Let Y be the set $\{2m + 1 : m = 0, 1, 2, \ldots\}$. Then (5) and (10) imply that $p(Y) = 1/2 \; \forall p \in P$, so Y is a non-trivial ancillary set. (It can be shown by the argument used in Example 1 above that Y and $X - Y$ are the only non-trivial ancillary sets of X). However, the ancillary statistic I_Y is a function of each non-constant t in T. Hence $T_b = T_h = T_a =$ the constants, and $T_a \neq T$.

References

[1] BAHADUR, R. R. (1957). On unbiased estimates of uniformly minimum variance. *Sankhyā*, **18** (parts 3 and 4), 211—224.

[2] BONDESSON, L. (1975). Uniformly minimum variance estimation in location parameter families. *Ann. Statist.*, **3**, 637—660.

[3] LEHMANN, E. L. - SCHEFFÉ, H. (1950). Completeness. similar regions, and unbiased estimation. *Sankhyā*, **10**, 305—340.

THE UNIVERSITY OF CHICAGO, CHICAGO, ILLINOIS, U.S.A.

Received December 1975

DIRICHLET PROCESSES PRODUCE DISCRETE MEASURES: AN ELEMENTARY PROOF

by

ROBERT H. BERK (1)*, AND I. RICHARD SAVAGE (2)

In memory of a friend

The situation of interest is summarized by Ferguson (1973):

"Let $\mathscr{X}$ be a space and $\mathscr{A}$ a σ-field of subsets, and let α be a finite non-null measure on $(\mathscr{X}, \mathscr{A})$. Then a stochastic process P indexed by elements A of $\mathscr{A}$, is said to be a Dirichlet process on $(\mathscr{X}, \mathscr{A})$ with parameter α if for any measurable partition $(A_1, \ldots, A_k)$ of $\mathscr{X}$, the random vector $(PA_1, \ldots, PA_k)$ has a Dirichlet distribution with parameter $(\alpha A_1, \ldots, \alpha A_k)$. P may be considered a random probability measure on $(\mathscr{X}, \mathscr{A})$. The main theorem states that if P is a Dirichlet process on $(\mathscr{X}, \mathscr{A})$ with parameter α, and if $X_1, \ldots, X_n$ is a sample from P, then the posterior distribution of P given $X_1, \ldots, X_n$ is also a Dirichlet process on $(\mathscr{X}, \mathscr{A})$ with parameter $\alpha + \sum_1^n \delta_{X_i}$ where δ_x denotes the measure giving mass one to the point x." (*p.* 209)

A distribution p, is *discrete* if and only if there exists a finite or countable set, A, such that $pA = 1$. Ferguson and others [see Kingman (1975)] have shown:

Theorem. *If P is a Dirichlet process then*

$$Pr(P \text{ is discrete}) = 1 .$$

Previous proofs have used other, deeper, properties of the Dirichlet process than those quoted above. It should be noted, however, that the proofs of the quoted properties might be considered deep. The elementary proof follows from:

Lemma. *Let X have distribution p. Then p is discrete if and only if*

$$p(p\{X\} > 0) = 1 .$$

The proof of the lemma being straightforward, we proceed to the proof of the theorem. The symbol Pr will be used as referring specifically to a Dirichlet process. We note that

$$Pr(P\{X\} > 0) = E\, Pr(P\{X\} > 0 \mid X) .$$

* Work supported by Grant MPS72-05082-A02 from the U.S. National Science Foundation.

But given X, $P\{X\}$ is a non-degenerate Beta random variable with parameters $\alpha\{X\} + 1$ and $\alpha(\mathscr{X} - \{X\})$; thus $Pr(P\{X\} > 0 \mid X) = 1$ wp 1, which implies

$$Pr(P\{X\} > 0) = 1 . \tag{1}$$

The theorem is thus established.

The above result must actually be considered heuristic for reasons which are set forth in sections 2–4 below. What is involved are technical matters related to measurability and some more fundamental difficulties stemming from the definition of a Dirichlet process. Throughout, by a "Dirichlet process" or "Dirichlet measure", we mean a probability measure obtained by Ferguson (1973, lemma 1) using the Kolmogorov extension theorem. We do *not* mean his alternative construction (op. cit., equation 7) using a Poisson process. The latter is referred to somewhat tangentially in section 5 below. Before describing the difficulties alluded to above, we discuss certain aspects of Ferguson's definition of a Dirichlet process.

1. A Dirichlet process as defined by Ferguson is a probability triple $(\mathscr{S}, \sigma(\mathscr{S}), Pr)$, where $\mathscr{S} = [0, 1]^{\mathscr{A}}$ is the product space having for each of its factors, the closed unit interval $[0, 1]$, there being as many factors as elements of $\mathscr{A}$. (In the interesting applications, $\mathscr{A}$ has uncountably many elements.) Equivalently, $\mathscr{S}$ may be viewed as the set of all set functions defined on $\mathscr{A}$ with values in $[0, 1]$. Here $\sigma(\mathscr{S})$ is the product σ-field for $\mathscr{S}$, the σ-field generated by the measurable cylinders having a finite base (a cylinder whose base is determined by restrictions on a finite number of coordinates of $\mathscr{S}$). Then $\sigma(\mathscr{S})$ is precisely the set of measurable cylinders having a countable base (for all such cylinders are in $\sigma(\mathscr{S})$ and the set of such cylinders is a σ-field). Viewing $\mathscr{S}$ as a set of set functions, each set in $\sigma(\mathscr{S})$ may be described by restrictions on a countable collection $\{pA_i, i = 1, 2, \ldots\}$, where $\{A_i\}$ is a given countable subset of $\mathscr{A}$ and p denotes an element of $\mathscr{S}$. Thus individual points of $\mathscr{S}$ are not in $\sigma(\mathscr{S})$.

2. A Dirichlet process then selects a random element $P \in \mathscr{S}$ and it is not known, a priori (so to speak) that Pr concentrates on $\mathscr{P} \subset \mathscr{S}$, the set of (countably additive) probability measures on $(\mathscr{X}, \mathscr{A})$. Indeed, a statement like $Pr(P \in \mathscr{P}) = 1$ is not meaningful, for $\mathscr{P}$ is not in $\sigma(\mathscr{S})$. (The set $\mathscr{P}$ is not determined by a countable number of restrictions if $\mathscr{A}$ is uncountable.)

3. Further, one does not know that given P, $X \sim P$ (X has distribution P). In fact, since P may be a set function that is not additive, such a statement would also appear to be meaningless. Thus, even if it were somehow apparent from Ferguson's definition that

$$Pr(X \in A \mid P) = PA , \quad A \in \mathscr{A} , \tag{2}$$

the possible lack of additivity for P prevents one from concluding, for example, that $Pr(X \in A \cup B \mid P) = PA + PB$ when A and B are disjoint sets in $\mathscr{A}$. Technically, this means that with Ferguson's definition of Pr, it is not clear that (2) defines a conditional probability distribution (a regular conditional probability) for X given P. So, while the proof of (1) may be correct, as matters now stand, interpreting this to mean that $wp1$ P is a discrete probability distribution is necessarily heuristic.

4. Some technicalities of a milder sort are glossed over in the proof of (1). For one thing, it must be shown that $P\{X\}$ is a random variable, in order to make probability statements about it. A difficulty of a related nature is that the lemma is not correct as stated unless singletons are in $\mathscr{A}$. These last difficulties disappear when $\mathscr{A}$ contains the Borel sets for a separable metric space $\mathscr{X}$. (By the Borel sets, we mean the σ-field generated by the closed subsets of $\mathscr{X}$.) We shall, in the sequel, confine attention to the case of $\mathscr{X}$ separable metric, although it is possible to get round some of these difficulties in a more general setting. That singletons are then in $\mathscr{A}$ is clear (they are closed). Showing that $P\{X\}$ is a random variable means showing that the mapping $(p, x) \to p\{x\}$ from $\mathscr{S} \times \mathscr{X}$ to $[0, 1]$ is $\sigma(\mathscr{S}) \times \mathscr{A}$ measurable. This can be argued as follows. Let D be a countable dense subset of $\mathscr{X}$ and let $\mathscr{L}_n$ be the collection of all (open) spheres centered at some x in D and having radius $1/n$. Then for $y > 0$,

$$\{(p, x) : p\{x\} \geqq y\} = \bigcap_n \bigcup_{I \in \mathscr{L}_n} \{p : pI \geqq y\} \times \{x : x \in I\} .$$

For each I, the set on the right is a measurable rectangle in $\sigma(\mathscr{S}) \times \mathscr{A}$ and only a countable number of them are involved.

5. The difficulties mentioned in sections 2 and 3 above can be circumvented in various ways. For example, both Ferguson (op. cit.) and Blackwell (1973) construct mappings from some probability space to $\mathscr{P}$, thereby obtaining a random probability distribution $P^* \in \mathscr{P}$. [See also Kingman (1967).] In either case, it may be checked that $\sigma(\mathscr{P}) = \{B \cap \mathscr{P} : B \in \sigma(\mathscr{S})\}$ is contained in the σ-field for $\mathscr{P}$ generated by the given mapping, so that one obtains a probability measure on $(\mathscr{P}, \sigma(\mathscr{P}))$. The measure so obtained may then be extended in a unique way to $\sigma(\mathscr{S}) \vee \mathscr{P}$, the σ-field generated by $\sigma(\mathscr{S}) \cup \mathscr{P}$, so that $\mathscr{P}$ has measure one. As $\sigma(\mathscr{S}) \subset \sigma(\mathscr{S}) \vee \mathscr{P}$, one then obtains by restriction a probability measure on $\sigma(\mathscr{S})$, which can be shown to be Dirichlet. (In fact, in view of the definition of $\sigma(\mathscr{P})$, these statements follow if one checks that for every measurable partition $\{A_1, \ldots, A_n\}$ of $\mathscr{X}$, the P^*A_i are random variables and that $(P^*A_1, \ldots, P^*A_n)$ has the appropriate Dirichlet distribution.) In this sense, the constructions of Ferguson and Blackwell may be said to generate a Dirichlet process. In Ferguson's case, the mapping is actually to $\mathscr{D} \subset \mathscr{P}$, the set of discrete probability distributions, so the question of discreteness for P is automatically resolved.

6. One may, however, start with the Dirichlet measure $(\mathscr{S}, \sigma(\mathscr{S}), Pr)$ and show that it induces a probability measure on $(\mathscr{P}, \sigma(\mathscr{P}))$ having all the desired properties. One way of doing this is to observe that for α countably additive (as we henceforth assume), $\mathscr{P}$ has outer measure one for Pr. (The argument for this is given in section

10 below.) Thus Pr may be transferred to $(\mathscr{P}, \sigma(\mathscr{P}))$ as follows: Pr may be extended to $\sigma(\mathscr{S}) \vee \mathscr{P}$ so that $Pr(\mathscr{P}) = 1$. (The details of such extensions are given in Halmos (1950), p. 71, exercise 2. Alternatively, it is straightforward to verify that for any set B in $\sigma(\mathscr{S}) \vee \mathscr{P}$, necessarily of the form $B = (B_1 \cap \mathscr{P}) \cup (B_2 - \mathscr{P})$, for some B_1 and B_2 in $\sigma(\mathscr{S})$, taking $PrB = PrB_1$ defines a probability measure on $\sigma(\mathscr{S}) \vee \mathscr{P}$ with $Pr\mathscr{P} = 1$.) Then on restricting to $\sigma(\mathscr{P}) \subset \sigma(\mathscr{S}) \vee \mathscr{P}$, we obtain a probability measure on $\sigma(\mathscr{P})$.

By the same token, Pr as defined by Ferguson (op. cit.) on $(\mathscr{S} \times \mathscr{X}, \sigma(\mathscr{S}) \times \mathscr{A})$ may be transferred to a probability measure on $(\mathscr{P} \times \mathscr{X}, \sigma(\mathscr{P}) \times \mathscr{A})$, which we continue to denote by Pr. (Here, $\mathscr{X}$ and $\mathscr{A}$ may be read as surrogates for $\mathscr{X}^n$ and $\mathscr{A}^n$, $n = 1, 2, \ldots, \infty$. A similar comment applies at various places in the ensuring discussion.) We indicate in the following sections (7 and 8) that Pr, thus transferred to $\mathscr{P} \times \mathscr{X}$, has all the desired properties of a prior and sample.

7. Ferguson shows that Pr on $\mathscr{S} \times \mathscr{X}$ has a system of (regular) conditional (posterior) probability distributions $Pr(P \in B \mid X)$, $B \in \sigma(\mathscr{S})$, which are again Dirichlet measures. These are seen to transfer to $\mathscr{P} \times \mathscr{X}$ and are then a system of conditional probability distributions for the latter. Thus, in transferring to $\mathscr{P} \times \mathscr{X}$, the essential property of the Dirichlet prior is preserved, as one would hope. *Inter alia*, we have met the difficulty mentioned in section 2 above: P may now be taken to be a bona fide random probability measure on $(\mathscr{X}, \mathscr{A})$.

8. It remains to show that, given P, $X \sim P$. This question can now be begged, in a sense. For, having transferred Pr to $\mathscr{P}$, we may extend it to $\mathscr{P} \times \mathscr{X}$ in the obvious way, by defining the conditional distribution of X given P to be

$$Pr(X \in A \mid P) = PA\,, \quad A \in \mathscr{A}\,. \tag{2}$$

(Note that this does define a $\sigma(\mathscr{P})$ measurable function. Relation (2) is precisely what one has in mind in speaking of a random probability distribution and is, for example, tacitly used by Blackwell (op. cit.) on constructing a random element of $\mathscr{P}$.)

It is reassuring that the direct transfer of Pr from $\mathscr{S} \times \mathscr{X}$ to $\mathscr{P} \times \mathscr{X}$ achieves the same end: (2) is also a system of conditional probability distributions for the latter. To see that (2) is a conditional probability for Pr when transferred to $\mathscr{P} \times \mathscr{X}$, recall that Ferguson extends Pr from $\mathscr{S}$ to $\mathscr{S} \times \mathscr{X}$ by defining

$$Pr(X \in A \mid PA, PA_1, \ldots, PA_n) = PA \quad wp1 \tag{3}$$

for any finite collection $\{A, A_1, \ldots, A_n\} \subset \mathscr{A}$. Alternatively, one may assert that

$$Pr(X \in A \mid P) = PA \quad wp1\,. \tag{4}$$

For then, on taking a further conditional expectation, (4) is seen to entail (3) and is thus consistent with Pr on $\mathscr{S} \times \mathscr{X}$. That is, (4) and hence also (2) give versions of the indicated conditional probability for Pr on $\mathscr{S} \times \mathscr{X}$. On transferring to $\mathscr{P} \times \mathscr{X}$, (2) remains a valid conditional probability for the latter. That it is then also a conditional probability distribution (i.e., regular) is clear.

9. Thus the transfer of Pr from $\mathscr{S} \times \mathscr{X}$ to $\mathscr{P} \times \mathscr{X}$ carries along all of the desirable properties of the Dirichlet measure and, in fact, introduces one more: One may assert that given P, $X \sim P$. This is not quite the case in Ferguson's original framework. As already noted, (2) does give a version of the conditional probability for Pr on $\mathscr{S} \times \mathscr{X}$, but this does not obviously define a conditional probability distribution. That is, for disjoint sets A and B in $\mathscr{A}$, one may assert only that $Pr(X \in A \cup \cup B \mid P) = Pr(X \in A \mid P) + Pr(X \in B \mid P)$ $wp1$, since (as shown in section 10 below), for Pr on $\mathscr{S}$, one may assert only that $P(A \cup B) = PA + PB$ $wp1$ (and the exceptional set can, of course, depend on A and B). In many cases, when $\mathscr{X}$ is complete separable metric, for example, there will exist a system of conditional probability distributions $Pr(X \in A \mid P)$ for Pr on $\mathscr{S} \times \mathscr{X}$, but it is not clear that it is given by (2). Thus in Ferguson's framework, it cannot be asserted that given P, $X \sim P$. Indeed, the temptation to do so is diminished on recalling that P is a random element of $\mathscr{S}$.

10. To complete the discussion, we indicate an argument showing that $\mathscr{P}$ has outer measure one for Pr on $\mathscr{S}$. As noted above (section 2), $\mathscr{P}$ is not in $\mathscr{S}$. (This is also true of $\mathscr{D} \subset \mathscr{P}$ and of $\mathscr{F} \supset \mathscr{P}$, the set of finitely additive set functions on $\mathscr{A}$.) However, $\sigma(\mathscr{S})$ does separate points, so that one may, at least, speak of a discrete or degenerate measure on $(\mathscr{S}, \sigma(\mathscr{S}))$, (albeit in a slightly roundabout manner). If $\mathscr{P} \subset B \in \sigma(\mathscr{S})$, the restrictions on the countable number of coordinates defining the cylinder B must perforce be satisfied by all probability measures. Hence B must contain a countable intersection of events of one of two types: (i) $\{p \in \mathscr{S} : p(A_1 \cup \cup A_2) = pA_1 + pA_2\}$, where A_1 and A_2 are disjoint sets in $\mathscr{A}$ or (ii) $\{p \in \mathscr{S} : : pA_n \downarrow 0\}$, where $\{A_n\}$ is a countable system of decreasing sets in $\mathscr{A}$ having empty intersection. That $\mathscr{P}$ has outer measure one then follows if each event of types (i) and (ii) has probability one. Ferguson (op. cit., proposition 2) shows that all events of type (ii) have probability one for Pr. Here is a corresponding argument for (i). Note that for any measurable partition $A_1, \ldots, A_n$ of $\mathscr{X}$, $Pr(PA_1 + \ldots + PA_n = 1) = 1$ (but the exceptional set can depend on the partition). This is a property of the Dirichlet distribution of $(PA_1, \ldots, PA_n)$. By considering the partitions $(A, B, \mathscr{X} - (A \cup B))$ and $(A \cup B, \mathscr{X} - (A \cup B))$, one concludes that $P(A \cup B) = PA + PB$ $wp1$. *Inter alia*, even without Ferguson's proposition 2, this shows that $\mathscr{F}$ has outer measure one for Pr (even if α is only finitely additive). Conceivably, this is what Ferguson has in mind when he asserts, on p. 213, that his lemma 1 is valid "as a description of a random finitely additive set function".

11. We make one further small clarification in the assertion of our theorem. Having transferred Pr to $\mathscr{P}$, what is proved is that $\mathscr{D}$ has inner measure one. (Properly speaking, this is Blackwell's (op. cit.) conclusion too.) Strictly, one cannot assert $Pr(P \in \mathscr{D}) = 1$ unless one first completes $(\mathscr{P}, \sigma(\mathscr{P}), Pr)$ or knows that $\mathscr{D} \in \sigma(\mathscr{P})$. When $\mathscr{A}$ is countably generated, $\mathscr{D} \in \sigma(\mathscr{P})$ [cf. Dubins and Freedman (1964), 2.13].

12. It is clear from the above that in various ways, $\sigma(\mathscr{S})$ is an inadequate σ-field for $\mathscr{S}$. Happily, the same cannot be said about $\sigma(\mathscr{P})$. Note that $\sigma(\mathscr{P})$ is the smallest σ-field making all the maps $p \to pA$ $(A \in \mathscr{A})$ measurable. This property is inherited from $\sigma(\mathscr{S})$. Then, in fact, for $f : \mathscr{X} \to R$ bounded and measurable, $p \to \to \int f \, dp$ is $\sigma(\mathscr{P})$ measurable. In particular, $\sigma(\mathscr{P})$ contains $w(\mathscr{P})$, the σ-field for $\mathscr{P}$ generated by the topology of weak convergence; the σ-field generated by all of the maps $p \to \int c \, dp$, $c : \mathscr{X} \to R$ bounded and continuous. In fact, for $\mathscr{X}$ metric, $w(\mathscr{P}) = = \sigma(\mathscr{P})$. This is asserted without proof for $\mathscr{X}$ compact metric by Dubins and Freedman (op. cit., 3.1). The more general result may be established as follows.

Let $\mathscr{B}$ be the sets in $\mathscr{A}$ which can be approximated by a sequence of continuous functions: $A \in \mathscr{B}$ if the indicator function of A is the pointwise limit of a sequence of bounded continuous functions $c_n : \mathscr{X} \to [0, 1]$. Then $\mathscr{B}$ is a field. To see this, it suffices to note that $\mathscr{X} \in \mathscr{B}$ and to check that $\mathscr{B}$ is closed under complementation and pairwise intersections. But if $A \in \mathscr{B}$ is approximated by $\{c_n\}$, then $\mathscr{X} - A$ is approximated by $\{1 - c_n\}$, so that $\mathscr{X} - A \in \mathscr{B}$. Similarly, if A and B in $\mathscr{B}$ are approximated by $\{c_n\}$ and $\{d_n\}$ respectively, then AB is approximated by $\{c_n d_n\}$ and thus is in $\mathscr{B}$. Moreover, $\mathscr{B}$ contains all closed sets. This is a consequence of Urysohn's lemma and the fact that every closed subset of a metric space is a G_δ, a countable intersection of open sets. Thus $\mathscr{B}$ contains the field generated by the closed subsets of $\mathscr{X}$.

To finish the proof, we argue as follows. Let $\mathscr{M} = \{A \in \mathscr{A} : p \to pA \textit{ is } w(\mathscr{P}) \textit{ measurable}\}$. Since every set in $\mathscr{B}$ can be approximated by a sequence of continuous functions, $\mathscr{B} \subset \mathscr{M}$: That is, for $A \in \mathscr{B}$, $pA = \lim_n \int c_n \, dp$, showing that $p \to pA$ is the limit of a sequence of $w(\mathscr{P})$ measurable mappings. Moreover, $\mathscr{M}$ is a monotone class: If $\{A_n\} \subset \mathscr{M}$ is monotonic with limit A, then $PA = \lim_n pA_n$, so that $p \to pA$ is $w(\mathscr{P})$ measurable as well. It follows that $\mathscr{M}$ contains the σ-field generated by $\mathscr{B}$ (see Loève (1963), 1.6.A) and thus $\mathscr{M} = \mathscr{A}$. Since $\sigma(\mathscr{P})$ is the smallest σ-field for which all of the maps $A \to pA$ are measurable, $\sigma(\mathscr{P}) \subset w(\mathscr{P})$.

It is a pleasure to acknowledge helpful discussions with Prof. L. D. Brown on some of the foregoing material.

References

BLACKWELL, D. (1973). Discreteness of Ferguson selections. *Ann. Statist.*, **1**, 2, 356—358.

DUBINS, L. - FREEDMAN, D. (1964). Measurable sets of measures. *Pacific Journal of Mathematics*, **14**, 4, 1211—1222.

FERGUSON, T. S. (1973). A Bayesian analysis of some nonparametric problems. *Ann. Statist.*, **1**, 2, 209—230.

HALMOS, P. R. (1950). "Measure Theory". Van Nostrand, New York.

KINGMAN, J. F. C. (1967). Completely random measures. *Pacific Journal of Mathematics*, **21**, 1, 59—78.

KINGMAN, J. F. C. (1975). Random discrete distributions, with discussion. *J. Roy. Statist. Soc.* **B**, **37**, 1, 1—22.

LOÈVE, M. (1963). "Probability Theory", 3rd edition. Van Nostrand, New York.

(1) RUTGERS UNIVERSITY, NEW BRUNSWICK, NEW JERSEY, U.S.A.

(2) YALE UNIVERSITY, NEW HAVEN-CONNECTICUT, U.S.A.

Received October 1975

DESCRIPTIVE STATISTICS FOR NONPARAMETRIC MODELS IV. SPREAD

by

P. J. BICKEL (1)*, AND E. L. LEHMANN (2)**

1. Ordering by spread

In the preceding paper of this series [1] (to which we refer as BL III) we studied the dispersion of a symmetric distribution about its center of symmetry. In the present paper we study a related aspect of dispersion, which does not require the assumption of symmetry but which even in the symmetric case does not coincide with the concept considered in BL III. Roughly speaking, instead of looking at dispersion relative to a fixed point, we now consider the spread of a random variable throughout its distribution. The difference is perhaps best explained in terms of an example.

Example 1. Let,

$$f_p(x) = \tfrac{1}{2}p \quad \text{if} \quad |x| \leqq 1\,,$$
$$\qquad = \tfrac{1}{2}(1-p) \quad \text{if} \quad 1 < |x| \leqq 2\,, \tag{1.1}$$

for $0 < p < 1$. Let X have distribution F with density f_p, Y have distribution G with density f_{1-p} for $p > \frac{1}{2}$. Then it follows from (1.2) of BL III that $|Y|$ is stochastically larger than $|X|$ and hence Y is more dispersed about 0 than X according to the definition of BL III. This corresponds to our intuitive feeling that in a global sense G is more dispersed than F since it can be obtained by pushing some of the central mass of F into the tails. Yet, locally for $1 < |X| < 2$, G is more concentrated than F since it has a higher uniform density there.

The basic definition of BL III for calling a symmetric distribution G more dispersed about its center of symmetry than a distribution F about its center of symmetry is equivalent to

$$G^{-1}(v) - G^{-1}(\tfrac{1}{2}) \gtreqless F^{-1}(v) - F^{-1}(\tfrac{1}{2}) \quad \text{as} \quad v \gtrless \tfrac{1}{2}\,. \tag{1.2}$$

* This paper was prepared with support of U.S. Office of Naval Research Contract No. N00014-69-A-0200-1038/NR042-036.

** This research was prepared with support of National Science Foundation Grant No. GP-38485.

In the present paper, we shall call an arbitrary (i.e. not necessarily symmetric) distribution G *more spread out* than a distribution F if

(1.3) $$G^{-1}(v) - G^{-1}(u) \geqq F^{-1}(v) - F^{-1}(u) \quad \text{for all} \quad 0 < u < v < 1\,,$$

where F^{-1} is defined by $F^{-1}(u) = \sup\{x : F(x) \leqq u\}$, that is, if two percentage points of G are at least as far apart as the corresponding percentage points of F. A concept which contains the essence of this definition was introduced by Brown and Tukey (1946).

Note that the definition (1.3) reflects the greater concentration of F throughout, and it of course does not require symmetry, thus fulfilling the two desiderata mentioned above. Furthermore, a comparison of (1.3) with (1.2) shows that for symmetric distributions (1.2) is satisfied whenever (1.3) holds; that (1.3) is in fact more stringent follows from Example 1.

Two properties of (1.2) noted for symmetric distributions in BL III are seen to be implied for arbitrary distributions by (1.3), namely,

(a) Any random variable is more spread out than a constant;

(b) aX is more spread out than X if $a > 1$.

Note however, that the ordering is not invariant under monotone transformations: $G^{-1}(u) - F^{-1}(u)\uparrow$ does not imply $h[G^{-1}(u)] - h[F^{-1}(u)]\uparrow$.

In BL III it was noted that if F and G are symmetric about 0 with densities f and g satisfying: $g(x)/f(x)$ is increasing for $x > 0$, then (1.2) holds. Example 1 shows that these conditions are not enough to insure (1.3).

Example 2. Let X take on the values $a < b$ with probabilities p and q $(0 < p < 1)$. Then

(1.4) $$F^{-1}(v) - F^{-1}(u) = \begin{cases} 0 & \text{if} \quad p \leqq u \quad \text{or} \quad v < p\,, \\ b - a & \text{if} \quad u < p \leqq v\,. \end{cases}$$

From (1.4) it is easily seen that no continuous strictly increasing distribution can be either more or less spread out than F in the sense of (1.3). More generally, continuous distributions and discrete distributions are not comparable. This illustrates how strong a requirement (1.3) is. We shall show below that it is nevertheless satisfied in many cases. However, it is convenient first to give some alternative expressions for condition (1.3).

Suppose that F^{-1} and G^{-1} are differentiable. Dividing both sides of (1.3) by $v - u$, it is then clear that (1.3) implies

(1.5) $$\frac{d}{du}[G^{-1}(u)] \geqq \frac{d}{du}[F^{-1}(u)] \quad \text{for all} \quad u$$

and conversely (1.5) implies (1.3). Thus, G is more spread out than F if G^{-1} is nowhere steeper than F^{-1}.

Evaluating the derivatives in (1.5), we see that another form of the condition is

$$g[G^{-1}(u)] \leqq f[F^{-1}(u)] \quad \text{for all} \quad u\,. \tag{1.6}$$

This is a so-called tail ordering condition introduced by Doksum (1969).

We want to stress that in our opinion this ordering corresponds to spread as property (b) above indicates. Tailweight should be specified by a scale-free ordering such as van Zwet's [5] or Lawrence's [4] and measured by scale-free functionals such as the kurtosis.

Lemma 2.2 of Doksum [3] states that, for distributions symmetric about 0, G ordered with respect to F in Lawrence's sense and $g(0) \leqq f(0)$ implies that G is more dispersed than F. This is consistent with our point of view. To see this note that $1/f(0)$ is a measure of scale and the condition $g(0) \leqq f(0)$ added to the scale-free ordering of Lawrence creates a new ordering of spread which possesses properties (a) and (b).

We mention finally a form which for the sake of simplicity we shall state under the additional assumption that F and G are strictly increasing.

Theorem 1. *If F and G are strictly increasing then G is more spread out than F if and only if there exists a strictly increasing function h such that*

(i) *$x < x'$ implies $h(x') - h(x) \geqq x' - x$*

and

(ii) *if X has distribution F, then $h(X)$ has distribution G.*

Proof. Suppose first that such a function exists. Then (1.3) follows from the relation $G^{-1}(u) = h[F^{-1}(u)]$. Conversely, if (1.3) holds, it is easily seen that the function $h(x) = G^{-1}[F(x)]$ has the desired property.

Theorem 1 shows that G being more spread out than F means that one can get from F to G by spreading all pairs of points further apart. Note also that Theorem 1 shows that our dispersion ordering depends on $G^{-1}F$ only.

Example 3. Let F be the uniform distribution on (0, 1). Since the right-hand side of (1.6) is then 1 for all $0 < u < 1$, the condition for G with density g to be more spread out than F is that

$$g(x) \leqq 1 \quad \text{for all} \quad x\,. \tag{1.7}$$

Similarly, G is less spread out than F if and only if its support is an interval (a, b) of length <1 and if in this interval $g(x) \geqq 1$ for all x.

Example 4. Let $F = \Phi$ be the standard normal distribution. Then G is more spread out than Φ if and only if the normal probability plot of G, $y = \Phi^{-1}\,G(x)$ has slope $\geqq 1$ at all points.

Example 5. Let F be the double exponential distribution with density $\frac{1}{2}\exp(-|x|)$. Then

$$f[F^{-1}(u)] = \begin{cases} u & \text{if } 0 < u < \frac{1}{2}, \\ 1-u & \text{if } \frac{1}{2} < u < 1 \end{cases} \tag{1.8}$$

and G is more spread out than F if

$$g(x) \leqq \begin{cases} G(x) & \text{for all } x < 0, \\ 1 - G(x) & x > 0. \end{cases} \tag{1.9}$$

Alternatively the failure rate of G must be at most 1 when x is interpreted as time running in either direction from the origin.

Example 6. Let F be the logistic distribution with density $e^{-x}(1+e^{-x})^{-2}$. Then, G is more spread out than F if and only if $g(x) \leqq G(x)(1 - G(x))$ for all x and less spread out if and only if the reverse inequality holds for all x.

From (1.9) it follows easily, for example, that a normal distribution with sufficiently small variance is less spread out than F, but that a normal distribution can never be more spread out than F no matter how large its variance. The situation is just the reverse in the case of a Cauchy distribution, which is more spread out than F if its scale is sufficiently large, but which can never be less spread out than F.

Finally, the logistic distribution provides an example of a distribution which is more spread out than F for sufficiently large scale and less spread out than F for sufficiently small scale.

An interesting connection between the dispersion ordering of BL III and the present ordering by spread is given by the following result.

Theorem 2. *If Y is more spread out than X and if Y', Y'' and X', X'' are independent copies of Y and X respectively, then $Y'' - Y'$ is more dispersed than $X'' - X'$.*

Proof. Let h be the function guaranteed by Theorem 1. Then $|Y'' - Y'| = |h(X'') - h(X')| \geqq |X'' - X'|$ and hence $|Y'' - Y'|$ is stochastically larger than $|X'' - X'|$.

That the converse of Theorem 2 does not hold is shown by the following generalization of Example 1.

Example 7. Let $f = f_p$, $g = f_{p'}$ be defined by (1.1). Then if $p > p' \geqq \frac{1}{2}$, it is easily seen that $X'' - X'$ is less dispersed than $Y'' - Y'$; on the other hand, G is not more spread out than F.

2. Measures of spread

The axioms for a measure of spread coincide with those for a measure of dispersion given in BL III except that in (1.9) of that paper dispersion ordering is replaced by the present spread ordering and that minor differences result from the dropping of the assumption of symmetry. For the sake of completeness, we shall now restate the full set of axioms.

A *measure of spread* is a functional $\Delta(F)$ (also denoted by $\Delta(X)$ where X is a random variable with distribution F) defined over a sufficiently large class of distributions which is closed under changes of location and scale. We shall require Δ to be nonnegative and to satisfy

$$\Delta(aX) = |a|\,\Delta(X) \quad \text{for} \quad a > 0\,, \tag{2.1}$$

$$\Delta(X + b) = \Delta(X) \quad \text{for all} \quad b\,, \tag{2.2}$$

and

$$\Delta(-X) = \Delta(X)\,. \tag{2.3}$$

As before, these conditions imply that

$$\Delta(c) = 0 \quad \text{for any constant} \quad c\,. \tag{2.4}$$

The converse: $\Delta(X) = 0$ implies $X = c$, is false for measures of spread as well as for measures of dispersion. Let Δ be the measure given in (2.12) below and F assign mass $\frac{1}{4}$ each to the points ± 1 and $\frac{1}{2}$ to 0. Then $\Delta(F) = 0$. A nonnegative functional satisfying (2.1), (2.2) and (2.3) will be called a measure of spread if it satisfies in addition

$$\Delta(F) \leqq \Delta(G) \quad \text{whenever } G \text{ is more spread out than} \quad F\,. \tag{2.5}$$

Note that if $\Delta(F)$ is a measure of spread, so is $\varkappa\,\Delta(F)$ for any $\varkappa > 0$.

A large and interesting class of measures of spread is obtained when the following theorem is applied to some of the results of BL III.

Theorem 3. *Let $\tau(X)$ be a measure of dispersion in the sense of* BL III *and let X', X'' be two independent copies of X. Then*

$$\Delta(F) = \tau(X'' - X') \tag{2.6}$$

is a measure of spread.

Proof. That Δ satisfies (2.1)–(2.3) is obvious from the fact that τ satisfies the corresponding conditions. To prove (2.5), suppose that G is more spread out than F. Then $X'' - X'$ and $Y'' - Y'$ are symmetric about 0 and it follows from Theorem 2 that $Y'' - Y'$ is more dispersed than $X'' - X'$ and hence from the fact that τ is a measure of dispersion that $\Delta(F) = \tau(X'' - X') \leqq \tau(Y'' - Y') = \Delta(G)$ as was to be proved.

Example 8. As a first example let $\tau(F)$ be the standard deviation of X, which was seen in BL III to be a measure of dispersion. Then

$$\Delta(F) = \{E(X'' - X')^2\}^{1/2} = \sqrt{(2)}\,[\tau(F)]\,.$$

It follows that $\Delta(F)$ is a measure of spread, and hence also that

$$\Delta(F) = SD(F) \tag{2.7}$$

is a measure of spread although no longer restricted to symmetric F.

An obvious generalization, obtained by starting with the pth power deviations considered in BL III are the measures

$$\Delta_p(F) = \{E \mid X'' - X' \mid ^p\}^{1/p}. \tag{2.8}$$

Example 9. Similarly, by starting with $\tau(F) = \text{med } |X|$, we find that

$$\Delta(F) = \text{med } |X'' - X'| \tag{2.9}$$

is a measure of spread.

Example 10. A class of examples not having the above structure is given by

$$\Delta(F) = F^{-1}(t) - F^{-1}(1-t) \quad \text{for any} \quad t > \tfrac{1}{2}, \tag{2.10}$$

which obviously satisfies (2.1)–(2.3) and (2.5). The same is true of the more general class

$$\Delta(F) = \left[\int_{1/2}^{1} [F^{-1}(t) - F^{-1}(1-t)]^\gamma \, d\Lambda(t)\right]^{1/\gamma} \tag{2.11}$$

where Λ is any finite measure on $(\frac{1}{2}, 1)$.

A case of particular interest is $t = \frac{3}{4}$, and hence

$$\Delta(F) = F^{-1}(\tfrac{3}{4}) - F^{-1}(\tfrac{1}{4}) \tag{2.12}$$

the interquartile range of F. In BL III, this was seen to be a measure of dispersion for symmetric distribution; it now follows that it is a measure of spread for arbitrary distributions. More generally, the class of measures (2.11) when restricted to symmetric distributions for suitable Λ coincides with the class of dispersion measures (1.10) of BL III. Note, however, that even for $\gamma = 2$ and Λ = Lebesgue measure, the measure (2.11) is not a multiple of the SD when F is asymmetric.

That not every measure of spread, when restricted to symmetric distributions, reduces to a measure of dispersion is shown by the following example.

Example 11. Let X take on the values $-1, 0, 1$ with probabilities $p/2$, $1-p$, $p/2$ respectively. Then

$$E\{|X'' - X'|^r\} = 2p\, 1 - p) + 2^{r-1}p^2$$

and it is easily seen that for $r < 1$, the left-hand side is not an increasing function of p. Since $|X|$ is stochastically increasing with p, this means that $[E\{|X'' - X'|^r\}]^{1/r}$ is not a measure of dispersion for symmetric distributions although it is a measure of spread.

3. Choice of measure

The choice of a measure of spread, as was the case for measures of location and of dispersion will be based largely on the accuracy with which it can be estimated, and the appropriate measure of accuracy is the same of that discussed in BL III for dispersion, the *standardized asymptotic variance,* i.e. the asymptotic variance of the estimator of $\Delta(F)$ divided by $\Delta^2(F)$. As in the earlier paper one would again try to find measures which behave satisfactorily (a) in relation to robustness, either by being robust or at least more robust than $\Delta_1(F)$; and (b) possessing good efficiency relative to $\Delta_1(F)$, ideally for all F, but if this cannot be achieved at least for the type of F likely to occur in practice.

We have not carried out this program in the present case. The work of BL III suggests that in terms of the indicated properties no completely satisfactory measure is likely to exist. It suggests further that the following two types of measures may be reasonable compromise solutions.

(i) *The pth power measures* (2.8). While not robust, for $p < 2$ these measures are presumably more robust than $\Delta(F)$, which corresponds to the case $p = 2$. The natural estimator of $\Delta(F)$ is $\hat{\Delta}(F)$ or the asymptotically equivalent statistic

$$\hat{\Delta}_p = \left\{\frac{1}{\binom{n}{2}} \sum_{i<j}\sum |X_j - X_i|^p\right\}^{1/p}. \tag{3.1}$$

Since this is a U-statistic, it follows from the work of Hoeffding that $\sqrt{(n)}\,.\,[\hat{\Delta} - \Delta(F)]$ is asymptotically normal. It seems plausible to conjecture that for values of $p \geqq 1.5$ the asymptotic efficiency of $\hat{\Delta}_p$ relative to the standard deviation $\hat{\Delta}_2$ will be reasonably high at least for typical distributions.

(ii) *Trimmed standard deviations.* There are two possible versions of trimmed standard deviations in this context.

(a) We can take $\gamma = 2$ and Λ the uniform distribution on $(\frac{1}{2}, 1 - \beta)$ in (2.11).

(b) We can consider $\tau(X - X', \alpha, \beta)$ where $\tau(X, \alpha, \beta)$ is given by (3.1) of BL III.

(As usual we let X represent its distribution.)

It is easy to see that if $0 < \alpha < 1 - \beta < 1$ both of these measures are robust. Their estimates are given by (essentially),

(a) $$[n(\tfrac{1}{2} - \beta)]^{-1/2}\Big[\sum_{k=n/2}^{n(1-\beta)} [X_{(k)} - X_{(n-k+1)}]^2\Big]^{1/2}$$

(b) $$\left[\binom{n}{2}(1 - \beta - \alpha)\right]^{-1/2}\Big[\sum_{k=\binom{n}{2}}^{\binom{n}{2}(1-\beta)} (X - X')^2_{(k)}\Big]^{1/2}$$

where $X_{(1)} \leqq \dots \leqq X_{(n)}$ are the order statistics of $X_1, \dots, X_n$ and $(X - X')_{(1)} \leqq \dots \dots \leqq (X - X')_{\binom{n}{2}}$ are the order statistics of the pseudo sample $X_i - X_j$, $i < j$. Asymptotic normality with variance of the order $1/n$ of these estimates is evident only for case (a) when $\gamma = 1$. It seems plausible that this property holds in general and that the resulting efficiencies with respect to the S.D. have the same general numerical features as we found for their analogues in studying dispersion. However, we do not pursue this. We do not know a fortiori which of the measures (a) or (b) is preferable and leave these interesting questions open.

References

[1] Bickel, P. J. - Lehman, E. L. (1976). Descriptive statistics for nonparametric models III. Dispersion. *Ann. Statist.*, **4,** 1139—1158.

[2] Brown, G. - Tukey, J. W. (1946). Some distributions of sample means. *Ann. Math. Statist.* **7,** 1—12.

[3] Doksum, K. (1969). Starshaped transformations and the power of rank tests. *Ann. Math. Statist.,* **40,** 1167—1176.

[4] Lawrence, M. J. (1975). Inequalities of *s*-ordered distributions. *Ann. Statist.*, **3,** 413—428.

[5] Zwet, van, W. R. (1964). Convex transformations of random variables. *Math. Centrum,* Amsterdam.

(1), (2) UNIVERSITY OF CALIFORNIA, BERKELEY, CALIFORNIA, U.S.A.

Received October 1975

SOME PERSONAL RECOLLECTIONS OF JAROSLAV HÁJEK

by

TORE DALENIUS

I first met Jaroslav in Stockholm in 1957. He immediately made a deep impression on me in several ways. I was struck by his strong devotion to statistical research and his firm determination to put statistics to good use in his native country. Then there was his personality. He always took the time to listen to other people's arguments, and started his discussions with them with the assumption that they had a valid point. He was very generous and always willing to help.

Naturally, I wanted to continue our acquaintance and to work with him on problems of interest to both of us, especially in the realm of survey sampling (my own main field of interest at that time). Soon after our first meeting, we cooperated – together with the late Professor S. Zubrzycki (Wroclaw) – on problems of plane sampling; the result of our endeavors appeared in the proceedings of the fourth Berkeley symposium. I also had the opportunity and privilege of discussing a variety of problems with Jaroslav by correspondence, and of getting his help as a referee of manuscripts submitted to the Review of the International Statistical Institute. Jaroslav generously shared his time with me – and indeed with many others.

In the course of these ventures, I got to know two of Jaroslav's great characteristics. First, he was never concerned about not receiving "acknowledgment of priority" when it was due him. For instance, in his paper "Representative sampling by a two-stage method" he provided a scheme for selecting a sample of n distinct units with probability proportional to the aggregate size of the units in the sample, and developed the theory necessary to evaluate the properties of this method. This paper was published in 1949 in Statistický Obzor in Czech (with a short summary in English), which may explain why his contribution did not at that time attract the kind of attention it deserved. The same result was found independently by others, who published their results in the early 1950s and whose names have come to be associated with what more properly should be referred to as "Hájek's scheme". Jaroslav never expressed any bitterness about this state of affairs; he was satisfied to see that the scheme became known and incorporated into the kit of tools available to survey statisticians.

Second, Jaroslav was interested and compentent in a wide variety of fields of statistics: he made signal contributions to the foundations of statistical theory, the methods and theories of survey sampling, and non-parametric statistics, to give but an indication of his breadth. I refer the reader to Václav Dupač's fine evaluation of Jaroslav's work in the Annals of Statistics, Vol. 3, No. 5, September 1975 – the list of Jaroslav's publications in that paper is indeed a lasting memorial of his professional life.

When I accepted an invitation to contribute to the Jaroslav Hájek Memorial Volume, I felt that I was in a situation not easy to handle. As one of Jaroslav's many friends, I wished to pay my tribute to him. But, knowing his high standards for his own work, I felt I would not be able to do justice to him by way of my contribution to the Memorial Volume. The two papers which I have contributed are intended to focus on two of Jaroslav's characteristics not explicitly discussed above. The first paper, on randomized response, is the outcome of teamwork, an approach to research which Jaroslav himself was strongly in favor of. The second paper, on a heuristic approach to a certain sampling design problem, might never have been written had I had the opportunity of discussing the problem with Jaroslav. With his outstanding command of the mathematical and other tools needed to tackle difficult problems, it would, I am sure, have been feasible to do more and better than I have been able to do.

TORE DALENIUS, NOVEMBER 24, 1975

A NEW RANDOMIZED RESPONSE DESIGN FOR ESTIMATING THE MEAN OF A DISTRIBUTION

by

TORE DALENIUS (1)*, AND RICHARD A. VITALE (2)

Introduction

The growing concern about "invasion of privacy" represents an important challenge to the applied statistician to develop new theory and methods for the collection of sensitive data. Quite naturally, a respondent may be hesitant or evasive in providing information which may indicate deviation from a social or legal norm and which he feels may be used against him at some later time. The task of the statistician is to design a survey which at once collects useful data (i.e. subject to some precise analysis) and protects the anonymity of the *individual respondent.*

An important step was taken by Warner (1965) who introduced the idea of using a randomized response technique as a mean of coping with the problem of "evasive answer bias" in interview surveys. The Warner (1965) design is as follows. A proportion π of a population of individuals has the sensitive characteristic A; the proportion $1 - \pi$ has the non-sensitive characteristic $\bar{A}$. π is to be estimated by means of an interview survey based on a simple random sample (with replacement) of n individuals. The interview is carried out as follows. Each respondent is informed that the survey concerns the characteristics A and $\bar{A}$. The respondent is given a spinner with a face marked so that the spinner points to A with probability P and to $\bar{A}$ with probability $1 - P$. The respondent is asked to spin the spinner (unobserved by the interviewer) and report whether the spinner points to the characteristic — A or $\bar{A}$ — which the respondent has. The answers are recorded:

$R = 1$ if the respondent reports that the spinner points to the characteristic which he has;

$R = 0$ if the respondent reports that the spinner points to the characteristic which he does not have.

* The research of the senior author was supported by a grant from the Bank of Sweden Tercentenary Foundation.

The interviews provide n_1 replies of the type $R = 1$ and $n_0 = n - n_1$ replies of the type $R = 0$. Then it is shown in Warner (1965) that

$$\pi^* = \frac{1}{2P - 1}\left[P - 1 + \frac{n_1}{n}\right]$$

is an unbiased estimate of π.

Since 1965, a considerable methodological development has taken place. We will be satisfied here by pointing to the following contributions:

(1) The Warner (1965) design has been extended to deal with the multiproportions case: the population is divided into k categories, of which at most $(k - 1)$ are sensitive. Two pertinent references are Abul-Ela et al. (1967) and Dalenius and Bourke (1973).

(2) Randomized response designs have been developed for estimating the proportions π_{rc} in a $r \times c$ contingency table (cf. Eriksson (1973)).

(3) Randomized response designs have been developed for obtaining quantitative data, for example "continuous data" such as amount of income withheld from taxation (Poole (1974)) and "discrete data" such as number of abortions ((Greenberg et al. (1971)).

The purpose of this paper is to present a new randomized response design which possesses two attractive properties: it is simple to use, and it affords a large measure of anonymity to the respondent. Though a quantitative estimate is the final end, the respondent is asked only to make a qualitative response.

I. Discrete case

We consider a finite population of individuals in which the characteristic X occurs with the values* $0, 1, 2, \ldots, M - 1$ with the relative frequencies $\pi_0, \pi_1, \pi_2, \ldots, \pi_{M-1}$, respectively. The mean of this characteristic,

$$\mu = \sum_{j=0}^{M-1} j\pi_j$$

is to be estimated by an interview sample survey based on simple random sampling with replacement.

The interview is carried out as follows. The individuals selected for the survey are provided with a spinner whose face is divided into equal sectors labelled 0, 1, 2, ...

* More generally, X may occur with the values $x_0, x_0 + h, x_0 + 2h, \ldots, x_0 + (M - 1)h$. In order to keep the formulas simple, we shall limit our discussion to the case presented above.

..., $M-1$. He is asked to spin the spinner and report whether the value S he sees is at least the value of his characteristic X. The replies are recorded as follows:

$$R = 0 \quad \text{if} \quad S \geqq X\,,$$
$$1 \quad \text{if} \quad S < X\,.$$

Thus, the design calls for collecting a "weaker" kind of data than called for by the design presented in Greenberg et al. (1971). This very fact may serve to make our design more attractive in the respondent's eyes. As we will soon see, this feature is achieved at the expense of a possibly large variance.

The survey yields the observations

$$R_1, R_2, \ldots, R_i, \ldots, R_n\,.$$

We estimate μ by

$$\mu^* = M\,.\,\frac{1}{n}\sum_{i=1}^{n} R_i = M\bar{R}\,.$$

It is easy to prove the following results:

$$E\mu^* = \mu\,.$$

This result follows from

$$E\mu^* = EM\frac{1}{n}\sum_{i=1}^{n} R_i = \frac{M}{n}\sum_{i=1}^{n} ER_i = M\,.\,ER_i = M\sum_{j=0}^{M-1}\frac{j}{M}\pi_j = \mu\,.$$

$$\operatorname{Var}\mu^* = \operatorname{Var} M\frac{1}{n}\sum_{j=1}^{n} R_i = \frac{M^2}{n}\operatorname{Var} R_i = \frac{M^2}{n}\frac{\mu}{M}\left(1-\frac{\mu}{M}\right) = \frac{1}{n}\mu(M-\mu)\,.$$

Since $\mu(M-\mu) \leqq M^2/4$ (achieved for $\mu = M/2$), we have the (conservative) bound

$$\operatorname{Var}\mu^* \leqq M^2/4n\,.$$

One way of reducing the variance, if necessary, is to require each respondent to spin the spinner s times, and report the outcomes

$$R_{i1}, R_{i2}, \ldots, R_{ik}, \ldots, R_{is} \quad (i = 1, \ldots, n)\,.$$

These responses will be statistically dependent. We compute

$$\bar{R}_i = \frac{1}{s}\sum_{k=1}^{s} R_{ik}$$

and estimate μ by

$$\mu_s^* = M\frac{1}{n}\sum_{i=1}^{n}\bar{R}_i\,.$$

Again μ_s^* is unbiased and we have

$$\operatorname{Var} \mu_s^* = \frac{M^2}{n} \operatorname{Var} \bar{R}_i .$$

Since in general $\operatorname{Var} \bar{R}_i < \operatorname{Var} R_i$, there will be some reduction in variance.

II. Continuous case

If the characteristic X achieves values on an interval $[0, M]$ – the more general interval $[a, a + M]$ is treated similarly – with density $f(x)$, then the procedure and analysis mimic those in the discrete case. This time the face of the spinner is labelled continuously from 0 to M and again the respondent is asked to report whether the spinner value S represents at least the value of his characteristic X. The reply is recorded

$$\begin{aligned} R = 0 \quad &\text{if} \quad S \geqq X , \\ 1 \quad &\text{if} \quad S < X \end{aligned}$$

and as in the discrete case, we form

$$\mu^* = M \frac{1}{n} \sum_{i=1}^{n} R_i \quad (i = 1, \ldots, n)$$

with

$$E\mu^* = \mu$$

and

$$\operatorname{Var} \mu^* = \frac{1}{n} \mu(M - \mu) .$$

By the same token, the variance may be reduced by requiring each respondent to spin the spinner s times.

References

ABUL-ELA, A. L. A. - GREENBERG, B. G. - HORWITZ, D. G. (1967). A multiproportions randomized response model. *Journ. Amer. Statist. Assoc.*, **62,** 990—1008.

DALENIUS, T. - BOURKE, P. D. (1973). Multi-proportions randomized response using a single sample. *Report no. 68 of the research project Errors in Surveys, Institute of Statistics, University of Stockholm.*

ERIKSSON, S. (1973). Randomized response. *Ph. D. thesis, Department of Statistics, University of Göteborg.*

GREENBERG, B. G. - KUEBLER, R. R., JR. - ABERNATHY, J. R. - HORWITZ, D. G. (1971). Application of the randomized response technique in obtaining quantitative data. *Journ. Amer. Statist. Assoc.*, **66,** 243—250.

POOLE, W. K. (1974). Estimation of the distribution function of a continuous type random variable through randomized response. *Journ Amer. Statist. Assoc.*, 69, 1002—1005.

WARNER, S. L. (1965). Randomized response: a survey technique for eliminating evasive answer bias. *Journ. Amer. Statist. Assoc.*, **60,** 63—69.

(1), (2) BROWN UNIVERSITY, PROVIDENCE, RHODE ISLAND, U.S.A.

Received October 1975

HEURISTICS FOR SURVEY SAMPLING DESIGN: A CASE STUDY

by

TORE DALENIUS

0. Introduction

It is a characteristic element in the development of survey sampling methods and theory in the last few decades that it has been increasingly oriented towards the construction of designs which – as judged from the viewpoint of the applications – may be called *realistic*. Consequently, today's survey statisticians are often in the position of using an analytic approach to the problem of constructing an efficient survey sampling design.

But sometimes it happens that the design problem is not easily – if at all – amenable to an analytic approach. This does not necessarily mean that the problem cannot be tackled in a way which yields satisfactory results. In this paper we will show – by means of a case study – how heuristics may be adhered to in this situation. More specifically, we will illustrate how an electronic computer may play an instrumental role in coping with the design problem.*

I. THE DESIGN PROBLEM

1. The origin of the problem

The case study referred to above has been developed in conjunction with some work that we have carried out on a research project "Semi-automatic analysis of large data bases" carried out by Professor Ulf Grenander, Division of Applied Mathematics, Brown University; for details, the reader is referred to Grenander (1972a).

---

* This illustration supports the assertion made in Dalenius (1971) that "the potential role of the electronic computer in the realm of survey sampling is a virgin field for exploration". Two additional examples are

(1) computer-supported instruction; cf. Dalenius (1973) for a discussion; and

(2) construction of complex and/or advanced schemes for sample selection or for estimation.

A large-scale computer system – IBM 360/67 – is available at Brown University. Certain changes have been made in the original operating system. In order to provide a basis for an evaluation of these changes, observations have been recorded concerning a variety of technical aspects: frequency of page transaction, and user-initiated I/0 operations and interrupts are examples in kind.

The observations thus collected are available in a data bank which comprises an unusually large volume of data: there are virtually millions of data. It is characteristic of this data bank that the underlying mechanism that has generated the data *is only vaguely known.**

Consequently, the size and structure of the data bank act together to make it unfeasible to resort to traditional means of analyzing the data. In their place, various heuristic methods are to be adhered to.

One of the problems for the solution of which heuristic methods are being developed concerns the identification of "regimes". In order to clarify the notion of a regime, we consider one of the many data vectors in the data bank:

$$x = (x_1, x_2, \ldots, x_i, \ldots, x_N)$$

where N is a very large number. Each observation x_i is of the following form:

$$x_i = m_i + \varepsilon_i$$

where m_i is a systematic component and ε_i reflects the "noise". It is a characteristic of this x-vector that sequences of consecutive elements have the same systematic component; moreover, $E\varepsilon_i = 0$ and $E\varepsilon_j^2 = \sigma^2$. In all, there are R such components or regimes.

Thus, the geometrical representation of the x-vector looks as follows. Fig. 1 on p. 51 illustrates four regimes:

Regime No 1: $x_1 \ldots x_4$	with length $L_1 = 4$
Regime No 2: x_5, x_6	with length $L_2 = 2$
Regime No 3: $x_7 \ldots x_9$	with length $L_3 = 3$
Regime No 4: $x_{10} \ldots x_{14}$	with length $L_4 = 5$

One of the topics of the research project mentioned above concerns methods for the identification of the regimes on the basis of sample observations.

* We will not elaborate upon the formal structure of the a priori information.

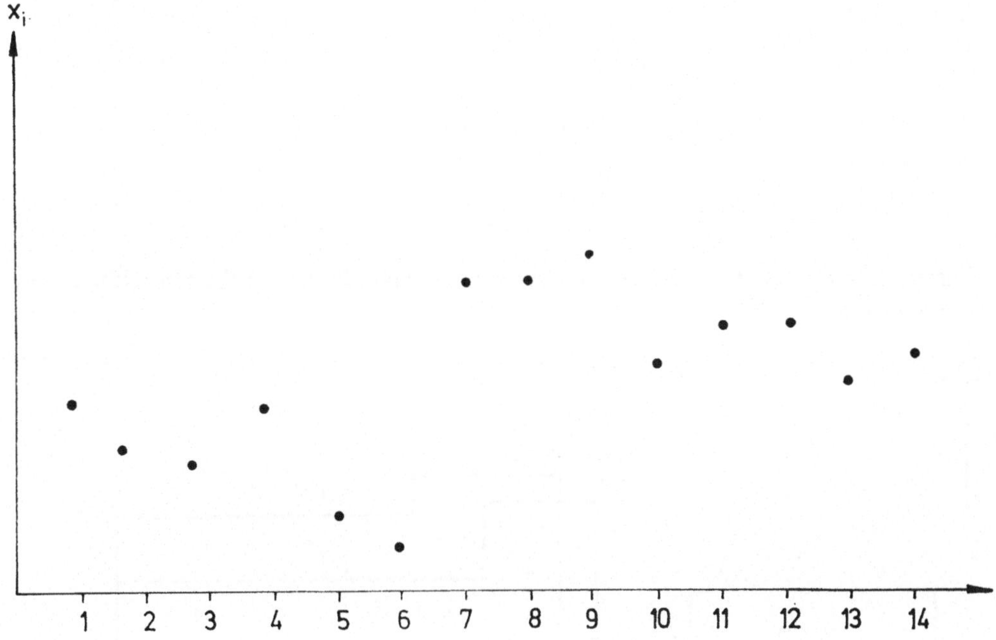

Fig. 1.

2. A reformulation of the problem

For the purpose of this paper, we have found it useful to reformulate the problem outlined in section 1 as follows. Thus, the m-values associated with the x-vector are represented by a histogram as shown in figure 2, which refers to the same vector as that in figure 1:

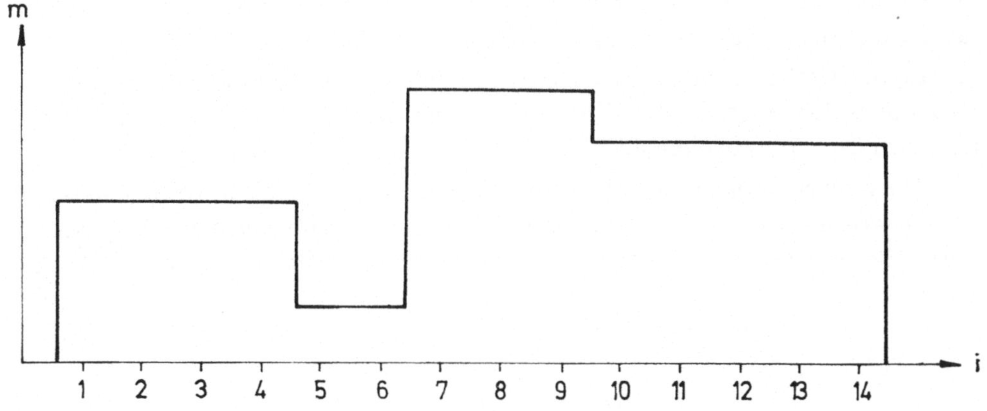

Fig. 2.

In the terminology of Grenander (1972b), this histogram may be looked upon as a pure image *I*; in traditional survey sampling terminology, the histogram is a population.

The design problem may now be formulated as follows. A sample of x-values is selected from the x-vector, yielding observations

$$y_1, y_2, \ldots, y_j, \ldots, y_n .$$

These observations are to be used to derive an estimate I^* of I, in the sense illustrated in figure 3.

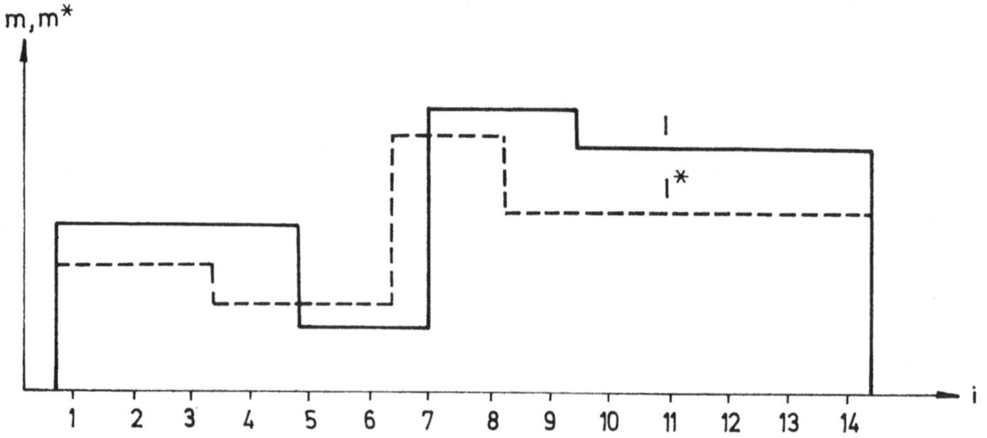

Fig. 3.

The design problem focuses on finding a "good" scheme for selecting the sample observations, and then a "good" estimator I^*.

As formulated, the problem is indeed not similar to design problems as found in the literature on survey sampling, possibly with the exception of the problem discussed in the context of "mapping surveys" in Mahalanobis (1944).

Consequently, the literature on survey sampling provides rather limited explicit guidance to the solution of the design problem! By the same token, an analytic approach may not prove feasible.

The problem at hand may, however, be tackled by way of a heuristic approach, using a large-scale computer as the basis for generating data which are then used to reflect the performance of some "trial design", as we shall show in part II of this paper.

II. A HEURISTIC APPROACH

3. Strategic consideration

At first sight, the use of a heuristic approach may seem to involve no more and no less than an analysis of a sequence of schemes $S_1, S_2, \ldots, S_k \ldots$. A moment's reflection shows, however, that the analysis of a sequence of schemes is by itself far from sufficient; something more is called for, if we are to hope for useful results. We will distinguish two vital aspects.

i) Each scheme S_k must be designed and analyzed in the light of the outcome of the preceding schemes: the heuristic program must be adaptive. In order to emphasize this point, we will denote the sequence of schemes by:

$$S_1, S_2(S_1), S_3(S_2), \ldots, S_k(S_{k-1}) \ldots .$$

ii) It is of decisive importance to choose a "good" starting scheme S_1. In sections 4 and 5 we will illustrate the operational significance of these two aspects.

4. The choice of the starting scheme

In view of the fact that little is known in advance of the population to be observed, it seems reasonable to adhere to a starting scheme which allocates observations evently over the population. The scheme actually chosen – vide infra – is rather unsophisticated; consequently, it may not perform well if applied to the kind of data base under consideration. The scheme may, however, serve to clarify certain concepts and ideas which are basic to the choice of a more advanced scheme to be presented in section 5.

4.1. Sampling

We select a sample of x-values using "mid-center systematic sampling". Let q^* be an approximation to the average regime length $q = N/R$. We select for the first observation (the "start") the x-value that has index

$$s = \tfrac{1}{2}q^*$$

and every q^*th value thereafter. Thus, we select the x-values having indices

$$s, s + q^*, s + 2q^*, \ldots .$$

In this way we get a sample which we denote by

$$y_1, y_2, \ldots, y_j, \ldots, y_n .$$

4.2. *Estimation*

We estimate I by I^*, where I^* is computed as follows. We use y_j to estimate the m-value associated with the x-value thus observed; we denote this estimate by m_j^*. We assign the same estimate m_j^* to the $q^*/2$ x-values preceding and succeeding y_j.

This technique of estimation is illustrated in figure 4 for three consecutive observations:

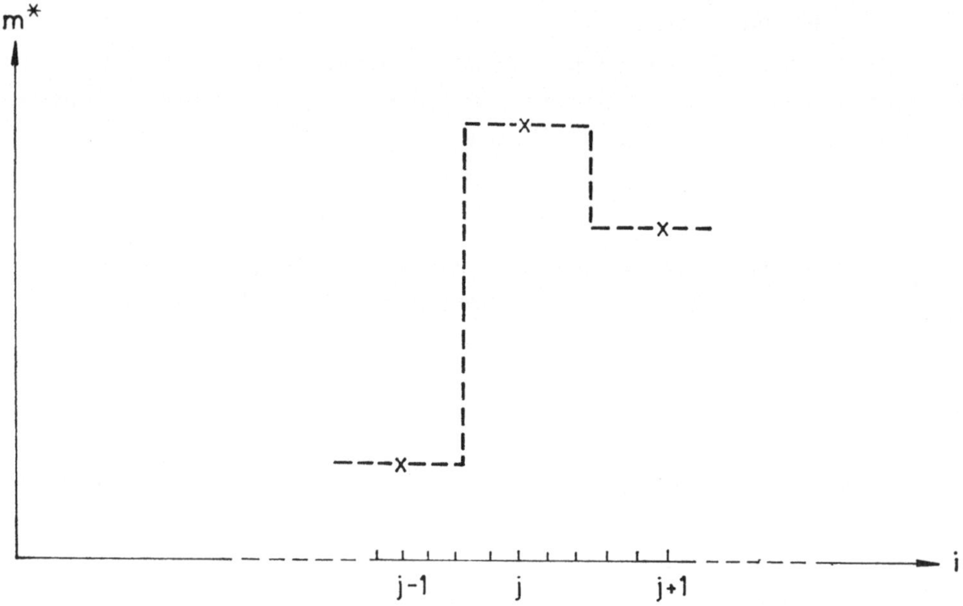

Fig. 4.

Using this technique, we arrive at an estimate I^* of I, which in a concrete case may look as in figure 3 above.

4.3. *Measure of the "goodness" of* I^*

As I^* is based on a sample of x-values, we must take into account that I^* is only an approximation of I. This assertion calls for the construction of some measure of how well I^* approximates I. Such a measure may be constructed as follows.

For each index i, $i = 1, 2, \ldots, N$, there is a true value m_i and an estimate m_i^*. It seems natural to base a measure of the "goodness" of I on the difference $m_i^* - m_i$.

One such measure is:

$$g_1(I^*) = \frac{1}{N} \sum_{i=1}^{N} |m_i^* - m_i| \, .$$

An alternative measure is

$$g_2(I^*) = \frac{1}{N} \sum_{i=1}^{N} (m_i^* - m_i)^2 .$$

The treatment of the analogous problem in statistical sampling theory suggests that $g_2(I^*)$ may have some worthwhile advantages over $g_1(I^*)$. In what follows we will consequently use $g_2(I^*)$ to measure how well I^* approximates I.

As defined above, $g_2(I^*)$ does not explicitly reflect the impact of the sample size. Generally speaking, we would expect the difference $m_i^* - m_i$ to decrease with increasing sample size. Another disadvantage of $g_2(I^*)$ is that it does not necessarily equal zero when $n = N$.

4.4. *An evaluation of the scheme*

It is characteristic of the heuristic approach that we analyze – in a more or less formal way – realizations of a certain scheme in order to detect possibilities for improvements.

In this case, we had access to a few computer runs. By and large, these runs looked like figure 4. As revealed by that figure, the "jump" $|y_{j-1} - y_j|$ is much larger than the "jump" $|y_j - y_{j+1}|$. Under certain reasonable assumptions about the process that has generated the population, – cf the footnote in section 1 – it seems plausible to expect that (on the average) $g_2(I^*)$ gets a larger contribution from the interval (y_{j-1}, y_j) than from the interval (y_j, y_{j+1}). This suggests that it may pay off to make some additional observations in intervals with large jumps! This suggestion takes us to the next scheme, i.e. to $S_2(S_1)$.

5. A sequential scheme for regime identification

We use in this case a two-step procedure for sample selection.

In the primary step, we select a sample of x-values by means of the scheme discussed in section 4. This yields the sample:

$$y_1, y_2, \ldots, y_j, \ldots, y_n$$

in the order selected.

Having access to this sample, we compute the successive differences

$$d_1 = |y_2 - y_1| ,$$
$$d_2 = |y_3 - y_2|$$

etc. These d-values serve as indicators of differences between the corresponding m-values.

It is convenient to order these values by decreasing size:

$$d_{(1)} > d_{(2)} > \ldots > d_{(p)} > d_{(p+1)} \ldots .$$

In the secondary step, a sample of p additional x-values is selected: one x-value is selected in the midst of each one of the p "intervals" corresponding to the p largest d-values. We denote these values by

$$z_1, z_2, \ldots, z_p .$$

The estimation makes use of the same technique as that discussed in section 4. The computation is illustrated in figure 5 for the case $n = 3$ and $p = 1$.

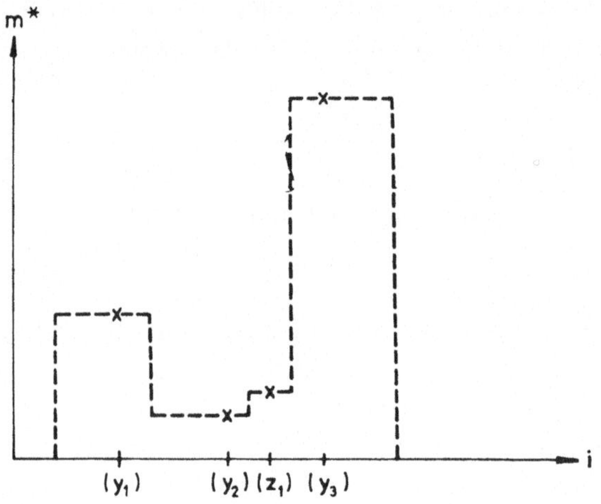

Fig. 5.

In order to get some insight into the efficiency of the scheme, a sequence of computer runs has been made. In figure 6 a typical run is illustrated.

Thus, for this run

$$\sqrt{[q_2(I^*)]} = 0.48$$

which is 2.5 times σ, the level of the "noise".

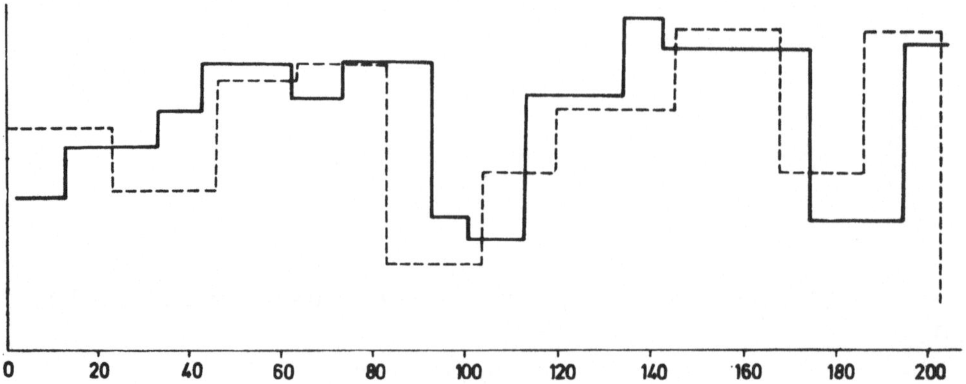

Fig. 6.

An analysis of figure 6 indicates that the choice of design parameters produces too many regimes. One possible way of tackling this problem is to choose a larger q^*-value than that used here.

Alternatively, and no doubt better, we may supplement the scheme by some specific device which reduces the number of estimated regimes. Such a device may be constructed as follows.

Consider again the differences

$$d_{(1)} > d_{(2)} > \dots > d_{(n-1)} .$$

Those regimes for which $d < \alpha$, where for example $\alpha = \sqrt{(2)}\,\sigma$, are merged.

We will not elaborate on the successive improvements* which may be arrived at on the basis of additional "trials and errors".

III. FINAL REMARKS

6. An essential question

The use of a heuristic approach to survey sampling design in part II concerns clearly a special case. With this fact in mind, it is natural to ask for some kind of general conclusions.

7. Two general conclusions

The heuristic approach to survey sampling design is, of course, by no means new. One example is provided by the 1935/36 population sample census in Sweden, S.O.S. (1938). The reader is also reminded of the use of "model sampling experiments" as discussed by Mahalanobis (1961). Additional examples can easily be given.

As a first general conclusion, we state that access to a large-scale computer should serve to increase significantly the power of the heuristic approach. Such an approach will be no less needed in the future than today and in the past; we are indeed likely to face design problems which are not amenable to an analytic approach.

As a second general conclusion, we state that much work is necessary in order to develop a comprehensive "set of rules", that is a strategy, for the efficient use of computer-based heuristics in survey sampling design.

---

* One of the improvements called for concerns the measure of how well I^* approximates I.

References

DALENIUS, T. (1971). Survey sampling in a computerized environment. *Review of the International Statistical Institute*, **39**, 3, 373—397.

DALENIUS, T. (1973). Computer-supported instruction in survey sampling. *Comp. Prob.*, **12**. *Center for Computer and Information Sciences, and Division of Applied Mathematics*, Brown University, Providence, R.I.

GRENANDER, U. (1972a). Semi-automatic analysis of large data bases, Working Paper No. 1. *Center for Computer and Information Sciences, and Division of Applied Mathematics*, Brown University, Providence. R.I.

GRENANDER, U. (1972b). Lecture notes on compumetrics. *Center for Computer and Information Sciences, and Division of Applied Mathematics*, Brown University, Providence, R.I.

MAHALANOBIS, P. C. (1944). On large-scale sample surveys. *Phil. Trans*: *Roy. Soc. London* B, **231**, 584, 329—451.

MAHALANOBIS, P. C. (1961). Experiments in statistical sampling in the Indian Statistical Institute. Calcutta.

S.O.S. (1938). Särskilda folkräkningen 1935/36, IV. Stockholm.

BROWN UNIVERSITY, PROVIDENCE, RHODE ISLAND, U.S.A.

Received November 1975

ASYMPTOTIC NORMALITY OF THE CONTINUOUS ROBBINS-MONRO STOCHASTIC APPROXIMATION PROCEDURE

by

VÁCLAV DUPAČ

0. Introduction

Asymptotic normality for the continuous Robbins-Monro stochastic approximation procedure was established by Nevel'son and Has'minskii (1972) under fairly general assumptions; see also Has'minskii (1972). An earlier result for a rather special continuous stochastic approximation procedure is due to Holevo (1967). The constants $a(t)$ appearing in the Robbins-Monro procedure (see formula (1) below) have been chosen as a/t in all quoted papers; the purpose of the present paper is to extend the result on asymptotic normality to the case $a(t) = a/t^\alpha$ with $\frac{1}{2} < \alpha < 1$, which is of importance, e.g., for dynamical stochastic approximation.

1. Notation and overall assumptions

We shall consider the continuous RM procedure for finding the root $\mathbf{x}_0$ of the system of equations $R(\mathbf{x}) = \mathbf{0}$, when the values of $R(\mathbf{x})$ are observed (or realized) with white noise type errors. More precisely, let $R(\mathbf{x})$ be a mapping from E_l into E_l such that the equation $R(\mathbf{x}) = \mathbf{0}$ has the unique root $\mathbf{x}_0$. For some $t_0 > 0$, let $\sigma_r(t, \mathbf{x})$, $1 \leqq r \leqq k$ be continuous mappings from $[t_0, +\infty) \times E_l$ into E_l; suppose that for each bounded region $D \subset [t_0, +\infty) \times E_l$ there exists a constant K_D such that

$$\|R(\mathbf{x}) - R(\mathbf{y})\| = \sum_{r=1}^{k} \|\sigma_r(t, \mathbf{x}) - \sigma_r(t, \mathbf{y})\| \leqq K_D \|\mathbf{x} - \mathbf{y}\|$$

everywhere in D.

Suppose there exists a function $V(\mathbf{x})$ with continuous second order derivatives such that

$$V(\mathbf{x}) > 0\,, \quad \left(R(\mathbf{x}), \frac{\partial}{\partial x} V(\mathbf{x})\right) < 0 \qquad \text{for} \quad \mathbf{x} \neq \mathbf{x}_0\,,$$

$$V(\mathbf{x}_0) = 0\,, \quad V(\mathbf{x}) \to +\infty \qquad \text{for} \quad \|\mathbf{x}\| \to +\infty\,,$$

$$\sum_{r=1}^{k}\left(\sigma_r(t,\mathbf{x}),\ \frac{\partial}{\partial x}\right)^2 V(\mathbf{x}) \leqq K(1+V(\mathbf{x})) \quad \text{for all} \quad \mathbf{x}\in E_l$$

$$\text{and some} \quad K > 0\,.$$

Let $\xi_r(t)$, $1 \leqq r \leqq k$, $0 \leqq t < +\infty$, be independent Wiener processes, i.e., continuous Gaussian processes with $E\,\xi_r(t) = 0$, $E\,\xi_r(s)\xi_r(t) = \min(s, t)$. Let $\mathscr{F}_t$ be the σ-field induced by processes $\{\xi_r(s), s \leqq t\}$, $1 \leqq r \leqq k$.

Let $a(t)$, $t_0 \leqq t < +\infty$, be a positive function such that

$$\int_{t_0}^{\infty} a(t)\,\mathrm{d}t = +\infty\,, \quad \int_{t_0}^{\infty} a^2(t)\,\mathrm{d}t < +\infty\,.$$

Consider the Itô stochastic differential equation

$$\text{(1)} \qquad \mathrm{d}\mathbf{X} = a(t)\,(R(\mathbf{X})\,\mathrm{d}t + \sum_{r=1}^{k}\sigma_r(t,\mathbf{X})\,\mathrm{d}\xi_r(t))\,, \quad t \geqq t_0\,, \quad \mathbf{X}(t_0) = \mathbf{x}\,,$$

describing the continuous RM procedure for finding the root $\mathbf{x}_0$.

It is known (Nevel'son-Has'minskii, Theor. 4.4.1) that under the assumptions listed above, for each $\mathbf{x}\in E_l$ there exists a unique (a.s.) continuous, $\mathscr{F}_t$-measurable solution $\mathbf{X}^{\mathbf{x}}(t)$ of the equation (1) with the initial condition $\mathbf{X}^{\mathbf{x}}(t_0) = \mathbf{x}$, which tends to $\mathbf{x}_0$ a.s. for $t \to +\infty$.

Note that typically the mappings $R(\mathbf{x})$, $\sigma_r(t,\mathbf{x})$, and the realizations of $\xi_r(t)$ are not known, whereas the function $a(t)$ is known and the quantity $R(\mathbf{X})\,\mathrm{d}t + \sum_{r=1}^{k}\sigma_r(t,\mathbf{X})\,\mathrm{d}\xi_r$ observable.

The symbol K with or without a subscript will always denote a positive constant; the same symbol may stand for different constants in different formulae.

The monograph by Nevel'son and Has'minskii will be referred to as NH.

2. The result

Throughout the paper, all the assumptions made in Section 1 are supposed to be fulfilled. We shall formulate further assumptions:

(i) $R(\mathbf{x}) = \mathbf{B}(\mathbf{x}-\mathbf{x}_0) + \delta(\mathbf{x})$, where $\delta(\mathbf{x}) = o(\|\mathbf{x}-\mathbf{x}_0\|)$ when $\mathbf{x}\to\mathbf{x}_0$, and $\mathbf{B}$ is a constant $l \times l$ matrix, all eigenvalues of which have negative real parts.

(ii) $\lim_{\substack{t\to\infty \\ \mathbf{x}\to\mathbf{x}_0}} \sigma_r(t,\mathbf{x}) = \sigma_{r,0}$, $1 \leqq r \leqq k$.

(iii) $a(t) = a/t^{\alpha}$, $a > 0$, $\frac{1}{2} < \alpha < 1$.

Theorem. *Under assumptions* (i), (ii), (iii), *the asymptotic distribution of* $t^{\alpha/2}(\mathbf{X}^x(t) - \mathbf{x}_0)$ *for* $t \to +\infty$ *is normal with zero mean and variance matrix*

$$\mathbf{S} = a \int_0^\infty e^{\mathbf{B}v} \mathbf{S}_0 e^{\mathbf{B}^T v} \, dv \,, \tag{2}$$

where

$$\mathbf{S}_0 = \sum_{r=1}^{k} \sigma_{r,0} \sigma_{r,0}^T \,.$$

Corollary. *If in addition,* $\mathbf{B}$ *is negative definite, and* $\mathbf{P}$ *is an orthogonal matrix such that* $-\mathbf{P}^T\mathbf{B}\mathbf{P} = \mathbf{\Lambda}$ *is diagonal, then the asymptotic variance matrix can be written as* $\mathbf{S} = a\mathbf{P}\mathbf{M}\mathbf{P}^T$, *where*

$$\mathbf{M}^{(ij)} = (\mathbf{P}^T\mathbf{S}_0\mathbf{P})^{(ij)} / (\mathbf{\Lambda}^{(ii)} + \mathbf{\Lambda}^{(jj)}) \,. \tag{3}$$

Remark. Our Theorem is a pendant to the Theor. 6.5.1 in NH, which treats the case $a(t) = a/t$ and, under the same assumptions strengthened only in that the real parts of all eigenvalues of $\mathbf{B}$ should be less than $-(2a)^{-1}$, asserts the asymptotic normality of $t^{1/2}(\mathbf{X}^x(t) - \mathbf{x}_0)$ with zero mean and variance matrix

$$\mathbf{S} = a \int_0^\infty e^{(\mathbf{B} + (2a)^{-1} I)v} \, \mathbf{S}_0 \, e^{(\mathbf{B}^T + (2a)^{-1} I)v} \, dv \,;$$

in the case of negative definite $\mathbf{B}$, the matrix $\mathbf{S}$ can be again written as $\mathbf{S} = a\mathbf{P}\mathbf{M}\mathbf{P}^T$, where

$$\mathbf{M}^{(ij)} = (\mathbf{P}^T\mathbf{S}_0\mathbf{P})^{(ij)} / (\mathbf{\Lambda}^{(ii)} + \mathbf{\Lambda}^{(jj)} - a^{-1}) \,.$$

3. The proof

Lemma 1. *Let* a, b, c, c_1, T *be positive numbers,* $a < 1$, $bT^{a-1} < \frac{1}{2}c$. *Let* $x(t)$ *be a differentiable function satisfying*

$$dx/dt \leqq -ct^{-a}x + c_1 t^{-a-b} \quad for \quad t \geqq T \,.$$

Then we have

$$x(t) \leqq qt^{-b} \quad for \quad t \geqq T \,,$$

with $q = \max(2c_1c^{-1}, 2\,x(T)\,T^b)$.

Proof. Denote $x^*(t) = x(t) - qt^{-b}$. We have

$$dx^*/dt \leqq -ct^{-a}x + c_1 t^{-a-b} + bqt^{-b-1}$$
$$= -ct^{-a}x^* + (-cq + c_1 + bqt^{a-1})\, t^{-a-b} \,.$$

The expression in brackets is negative, because $c_1 \leqq \frac{1}{2}cq$ and $bqt^{a-1} \leqq bqT^{a-1} < \frac{1}{2}cq$, according to the assumptions made; hence

$$\mathrm{d}x^*/\mathrm{d}t < -ct^{-a}x^*\,, \quad t \geqq T\,. \tag{4}$$

From the definition of q it further follows that $-qT^{-b} < x(T)$, hence $x^*(T) < 0$.

Now if there exists a $t > T$ such that $x^*(t) > 0$, then for some $T < t^* < t$ we must have both $x^*(t^*) = 0$ and $(\mathrm{d}x^*/\mathrm{d}t)_{t=t^*} \geqq 0$. This, however, contradicts the inequality (4). Hence, $x^*(t) \leqq 0$, $t \geqq T$. □

We shall formulate further assumptions, that will appear in some auxiliary results:

(iv) There is a positive definite matrix $\mathbf{C}$ and a $\lambda > 0$ such that
$$(\mathbf{C}R(\mathbf{x}), \mathbf{x} - \mathbf{x}_0) \leqq -\lambda(\mathbf{C}(\mathbf{x} - \mathbf{x}_0), \mathbf{x} - \mathbf{x}_0)\,, \quad \forall \mathbf{x} \in E_l\,.$$

(v) $\|R(\mathbf{x})\| + \sum_{r=1}^{k} \|\sigma_r(t, \mathbf{x})\| \leqq K(1 + \|\mathbf{x}\|)$, $\forall \mathbf{x} \in E_l$, $t \geqq t_0$.

(vi) ζ is a l-dimensional, $\mathscr{F}_{t_0}$-measurable random vector, $E\|\zeta\|^2 < +\infty$.

Lemma 2. *Under assumptions* (iii), (iv), (v), *there exist* $K_1 > 0$, $T_1 > 0$ (*independent of* $\mathbf{x}$) *such that*

$$E\|\mathbf{X}^{\mathbf{x}}(t) - \mathbf{x}_0\|^2 \leqq K_1(1 + \|\mathbf{x}\|^2)\, t^{-\alpha}\,, \quad \forall t \geqq T_1\,.$$

Proof. We may assume $\mathbf{x}_0 = \mathbf{0}$ without loss of generality. Denote as

$$L(= \partial/\partial t + at^{-\alpha}(R(\mathbf{x}), \partial/\partial x) + \tfrac{1}{2}a^2t^{-2\alpha}\sum_{r=1}^{k}(\sigma_r(t, \mathbf{x}), \partial/\partial x)^2)$$

the differential generating operator for $\mathbf{X}^{\mathbf{x}}(t)$. Put $V(\mathbf{x}) = (\mathbf{C}\mathbf{x}, \mathbf{x})$. Then

$$\begin{aligned} LV &= 2at^{-\alpha}(\mathbf{C}R(\mathbf{x}), \mathbf{x}) + a^2t^{-2\alpha}\sum_{r=1}^{k}(\mathbf{C}\sigma_r, \sigma_r) \\ &\leqq -2a\lambda t^{-\alpha}V(\mathbf{x}) + K_2t^{-2\alpha}(1 + V(\mathbf{x})) \\ &\leqq -a\lambda t^{-\alpha}V(\mathbf{x}) + K_2t^{-2\alpha}\,, \quad \text{for} \quad t \geqq T, \end{aligned} \tag{5}$$

with T satisfying $K_2T^{-\alpha} < a\lambda$.

Hence (see NH, formula 3.5.5),

$$(\mathrm{d}/\mathrm{d}t)\, EV(\mathbf{X}^{\mathbf{x}}(t)) = ELV(\mathbf{X}^{\mathbf{x}}(t)) \leqq -a\lambda t^{-\alpha}EV(\mathbf{X}^{\mathbf{x}}(t)) + K_2t^{-2\alpha}\,, \quad t \geqq T\,.$$

Making use of Lemma 1, we get

$$EV(\mathbf{X}^{\mathbf{x}}(t)) \leqq \max\left(2K_2(a\lambda)^{-1}\,,\ 2EV(\mathbf{X}^{\mathbf{x}}(T)\,T^{\alpha}\right).\, t^{-\alpha}$$

(for t larger than some T_1).

Hence,

(6) $$E\|\mathbf{X}^{\mathbf{x}}(t)\|^2 \leqq K_3(1 + E\|\mathbf{X}^{\mathbf{x}}(T)\|^2)\, t^{-\alpha}.$$

However, our assumptions ensure (NH, Theor. 3.7.1) that

$$E\|\mathbf{X}^{\mathbf{x}}(T)\|^2 \leqq K_4(1 + \|\mathbf{x}\|^2)$$

with a K_4 depending on T but not on $\mathbf{x}$, which together with (6) entails the asesrtion of the Lemma. □

As an immediate corollary we get the following

Lemma 3. *Let* (iii), (iv), (v), (vi) *be fulfilled. Denote* $\mathbf{X}^{\zeta}(t)$ *the solution of* (1) *with the initial condition* $\mathbf{X}^{\zeta}(t_0) = \zeta$. *Then we have*

$$E\|\mathbf{X}^{\zeta}(t) - \mathbf{x}_0\|^2 \leqq K_5 t^{-\alpha},\ t \geqq T_1 .$$

Lemma 4. *Under assumptions of Lemma* 3, *we have*

$$t^{\gamma}\|\mathbf{X}^{\zeta}(t) - \mathbf{x}_0\|^2 \to 0 \quad \text{a.s.},$$

for every $\gamma < 2\alpha - 1$.

Especially, $\mathbf{X}^{\zeta}(t) \to \mathbf{x}_0$ a.s., *and to each* $\varepsilon > 0$, $\varepsilon_1 > 0$ *there is a* $T > 0$ *such that*

$$P(\sup_{u \geqq T} \|\mathbf{X}^{\zeta}(u) - \mathbf{x}_0\| < \varepsilon_1) > 1 - \varepsilon .$$

Proof. Again, put $\mathbf{x}_0 = \mathbf{0}$, $V(\mathbf{x}) = (\mathbf{C}\mathbf{x}, \mathbf{x})$. Let $0 < \gamma < 2\alpha - 1$, $0 < \varepsilon < 2\alpha - -1 - \gamma$; put $V_1(t, \mathbf{x}) = t^{\gamma} V(\mathbf{x}) + t^{-\varepsilon}$. Using (5), we get for $t \geqq T_1$ (say),

$$LV_1(t, \mathbf{x}) = t^{\gamma} LV(\mathbf{x}) + \gamma t^{\gamma - 1} V(\mathbf{x}) - \varepsilon t^{-\varepsilon - 1}$$
$$\leqq -K_1 t^{\gamma - \alpha} V(\mathbf{x}) + K_2 t^{\gamma - 2\alpha} - \varepsilon t^{-\varepsilon - 1} \leqq 0 ,$$

as $-\varepsilon - 1 > \gamma - 2\alpha$. This means (NH, Corollary 3.8.1) that $\{V_1(t, \mathbf{X}^{\zeta}(t)),\ t \geqq T_1\}$ is a positive supermartingale, which implies the existence of finite $\lim_{t\to\infty} V_1(t, \mathbf{X}^{\zeta}(t))$ a.s., i.e., of finite $\lim_{t\to\infty} t^{\gamma} V(\mathbf{X}^{\zeta}(t))$ a.s. As the latter exists for *each* $0 < \gamma < 2\alpha - 1$, it must be zero; hence also $t^{\gamma}\|\mathbf{X}^{\zeta}(t)\|^2$ tends to zero a.s. □

In the sequel the following norm of a matrix will be used:

$$\|\mathbf{A}\| = (\sum_i \sum_j a_{ij}^2)^{1/2} .$$

Lemma 5. *Let* $a > 0$, $\frac{1}{2} < \alpha < 1$, *let all eigenvalues of* $\mathbf{B}$ *have negative real parts. Then there exist* $K > 0$, $\lambda_1 > 0$, $t_1 > 0$ *such that*

$$\left(\frac{t}{u}\right)^{\alpha/2} \|\exp\{a(1-\alpha)^{-1}\mathbf{B}(t^{1-\alpha} - u^{1-\alpha})\}\| \leqq K \exp\{-\lambda_1(t^{1-\alpha} - u^{1-\alpha})\}$$

for all $(0<)\ t_0 \leqq u \leqq t,\ t \geqq t_1$.

Proof. The inequality $\|e^{\mathbf{B}\tau}\| \leqq Ke^{-\lambda\tau}$, $\tau \geqq 0$, follows from the properties of the matrix norm. Further, the inequality $(t/u)^{\alpha/2} \leqq \exp\{\varepsilon(t^{1-\alpha} - u^{1-\alpha})\}$ holds true for $t_0 \leqq u \leqq t$, an arbitrary $\varepsilon > 0$ and t sufficiently large, as it is equivalent to the inequality

$$u^{\alpha/2} \exp\{-\varepsilon u^{1-\alpha}\} \geqq t^{\alpha/2} \exp\{-\varepsilon t^{1-\alpha}\},$$

the validity of which in the same domain easily follows from the shape of the function $u^{\alpha/2} \exp\{-\varepsilon u^{1-\alpha}\}$. □

Lemma 6. *Let* (i), (iii), (iv), (v), (vi) *be fulfilled. Then for* $t \to \infty$,

$$\int_{t_0}^{t} J(t, u)\, du \to 0 \quad \textit{in probability},$$

where

$$J(t, u) = \left(\frac{t}{u}\right)^{\alpha/2} \exp\left\{\frac{a}{1-\alpha} \mathbf{B}(t^{1-\alpha} - u^{1-\alpha})\right\} u^{-\alpha/2}\, \delta(\mathbf{X}^{\zeta}(u)).$$

Proof. Put $\mathbf{x}_0 = \mathbf{0}$. For each $\varepsilon > 0$, there is (according to (i)) a ε_1 such that

$$\|\mathbf{x}\| < \varepsilon_1 \Rightarrow \|\delta(\mathbf{x})\| \leqq \varepsilon^2 \|\mathbf{x}\|, \tag{7}$$

and a $T > 0$ (according to Lemma 4) such that

$$P(\sup_{u \geqq T} \|\mathbf{X}^{\zeta}(u)\| < \varepsilon_1) > 1 - \varepsilon. \tag{8}$$

Now, from Lemma 5 it follows

$$\lim_{t \to \infty} \int_{t_0}^{T} J(t, u)\, du = 0 \quad \text{a.s.}, \tag{9}$$

whereas

$$P\left(\left\|\int_T^t J(t, u)\, du\right\| > \varepsilon\right) = \varepsilon + P\left(\left\|\int_T^t J(t, u)\, du\right\| > \varepsilon,\ \sup_{u \geqq T} \|\mathbf{X}^{\zeta}(u)\| < \varepsilon_1\right)$$

$$\leqq \varepsilon + P\left(\int_T^t (t/u)^{\alpha/2} \left\|\exp\{a(1-\alpha)^{-1}\mathbf{B}(t^{1-\alpha} - u^{1-\alpha})\}\right\|\right.$$

$$\left.\cdot \|\delta(\mathbf{X}^{\zeta}(u)\| \cdot u^{-\alpha/2}\, du > \varepsilon,\ \sup_{u \geqq T} \|\mathbf{X}^{\zeta}(u)\| < \varepsilon_1\right)$$

$$\leqq \varepsilon + P\left(\int_T^t (t/u)^{\alpha/2} \left\|\exp\{a(1-\alpha)^{-1}\, \mathbf{B}(t^{1-\alpha} - u^{1-\alpha})\}\right\|\right.$$

$$\left.\cdot \|\mathbf{X}^{\zeta}(u)\|\, u^{-\alpha/2}\, du > \varepsilon^{-1}\right)$$

$$\leqq \varepsilon + \varepsilon \int_T^t (t/u)^{\alpha/2} \|\exp\{a(1-\alpha)^{-1} \mathbf{B}(t^{1-\alpha} - u^{1-\alpha})\}\| \,.\, E\|\mathbf{X}^{\zeta}(u)\| \, u^{-\alpha/2} \,\mathrm{d}u$$

$$\leqq \varepsilon + K\varepsilon \int_T^t \exp\{-\lambda_1(t^{1-\alpha} - u^{1-\alpha})\} \, E\|\mathbf{X}^{\zeta}(u)\| \, u^{-\alpha/2} \,\mathrm{d}u$$

$$\leqq \varepsilon + K_1\varepsilon \int_T^t \exp\{-\lambda_1(t^{1-\alpha} - u^{1-\alpha})\} \, u^{-\alpha} \,\mathrm{d}u \,,$$

where we used, in their turn, (8), (7), Chebyshev inequality, Lemma 5 and Lemma 3, and assumed that T had been chosen sufficiently large. Substituting $t^{1-\alpha} - u^{1-\alpha} = v$, we find that the integral

$$\int_T^t \exp\{-\lambda_1(t^{1-\alpha} - u^{1-\alpha})\} \, u^{-\alpha} \,\mathrm{d}u = (1-\alpha)^{-1} \int_0^{t^{1-\alpha} - T^{1-\alpha}} \mathrm{e}^{-\lambda_1 v} \,\mathrm{d}v$$

remains bounded for $t \to \infty$, hence finally

$$(10) \qquad P\left(\left|\int_T^t J(t, u) \,\mathrm{d}u\right| > \varepsilon\right) \leqq \varepsilon + K_2\varepsilon \,, \quad t \geqq T \,.$$

Combining (9) and (10), we obtain the desired assertion. □

Lemma 7. *Let σ, q, λ be positive numbers,* $\mathbf{F}(v)$ *a continuous matrix function such that*

$$\|\mathbf{F}(v)\| \leqq K\mathrm{e}^{-\lambda v} \,, \quad v \geqq 0 \,.$$

Then

$$\lim_{\tau \to \infty} \int_0^{\tau - \sigma} (1 - v\tau^{-1})^{-q} \, \mathbf{F}(v) \,\mathrm{d}v = \int_0^{\infty} \mathbf{F}(v) \,\mathrm{d}v \,.$$

Proof. For each $0 < \varrho < 1$,

$$\int_0^{\tau-\sigma} (1 - v\tau^{-1})^{-q} \, \mathbf{F}(v) \,\mathrm{d}v = \int_0^{\tau-\sigma} \mathbf{F}(v) \,\mathrm{d}v$$

$$+ \left(\int_0^{\varrho(\tau-\sigma)} + \int_{\varrho(\tau-\sigma)}^{\tau-\sigma}\right) ((1 - v\tau^{-1})^{-q} - 1) \, \mathbf{F}(v) \,\mathrm{d}v \,.$$

The norm of the integral $\int_0^{\varrho(\tau-\sigma)}$ on the right hand side is less than $K_1((1-\varrho)^{-q} - 1)$, as $0 \leqq (1 - v\tau^{-1})^{-q} - 1 \leqq (1-\varrho)^{-q} - 1$ for $0 \leqq v \leqq \varrho(\tau - \sigma)$; the norm of the integral $\int_{\varrho(\tau-\sigma)}^{\tau-\sigma}$ is less than $K_2\tau^{q+1}\mathrm{e}^{-\lambda\varrho\tau}$, as $0 \leqq (1 - v\tau^{-1})^{-q} - 1 \leqq (\tau/\sigma)^q - 1$ and $\|\mathbf{F}(v)\| \leqq K\mathrm{e}^{-\lambda\varrho(\tau-\sigma)}$ for $\varrho(\tau - \sigma) \leqq v \leqq \tau - \sigma$. Passing to the limit for $\tau \to \infty$ and then for $\varrho \to 0$, we get the assertion. □

Now, we proceed to the *proof of the Theorem*:

Put $\mathbf{x}_0 = \mathbf{0}$. From (i) it follows, that there exist $\lambda > 0$, $\varepsilon > 0$ and a positive definite matrix $\mathbf{C}$ such that

$$(\mathbf{C}R(\mathbf{x}), \mathbf{x}) \leqq -\lambda(\mathbf{C}\mathbf{x}, \mathbf{x}) \quad \text{for all} \quad \|\mathbf{x}\| < \varepsilon \,;$$

we may suppose that $\lambda < \lambda_1$ of Lemma 5.

Put

$$\hat{R}(\mathbf{x}) = \begin{cases} R(\mathbf{x}) & \text{for } \|\mathbf{x}\| < \varepsilon\,, \\ R(\varepsilon\mathbf{x}/\|\mathbf{x}\|)\,.\,\|\mathbf{x}\|/\varepsilon & \text{for } \|\mathbf{x}\| \geqq \varepsilon\,; \end{cases}$$

$$\hat{\sigma}_r(t, \mathbf{x}) = \begin{cases} \sigma_r(t, \mathbf{x}) & \text{for } \|\mathbf{x}\| < \varepsilon\,, \\ \sigma_r(t, \varepsilon\mathbf{x}/\|\mathbf{x}\|) & \text{for } \|\mathbf{x}\| \geqq \varepsilon\,, \end{cases} \qquad 1 \leqq r \leqq k\,.$$

Together with (1) (where $a(t) = a/t^{\alpha}$) consider the auxiliary stochastic differential equation

$$\mathrm{d}\hat{\mathbf{X}} = at^{-\alpha}\Big(\hat{R}(\hat{\mathbf{X}})\,\mathrm{d}t + \sum_{r=1}^{k} \hat{\sigma}_r(t, \hat{\mathbf{X}})\,\mathrm{d}\xi_r\Big)\,, \quad t \geqq t_0\,, \tag{11}$$

with initial condition

$$\hat{\mathbf{X}}(t_0) = \zeta\,, \quad \zeta \text{ satisfying (vi)}\,. \tag{12}$$

Notice that the equation (11), (12) satisfies all assumptions (i) through (vi). Its solution is

$$\hat{\mathbf{X}}(t) = t^{\alpha/2} \exp\{a(1-\alpha)^{-1}\,\mathbf{B}(t^{1-\alpha} - t_0^{1-\alpha})\}\,\zeta$$
$$+ a\int_{t_0}^{t}\left(\frac{t}{u}\right)^{\alpha/2} \exp\{a(1-\alpha)^{-1}\,\mathbf{B}(t^{1-\alpha} - u^{1-\alpha})\}\,u^{-\alpha/2}\big(\hat{\delta}(\hat{\mathbf{X}}^{\zeta}(u))\,\mathrm{d}u$$
$$+ \sum_{r=1}^{k} \hat{\sigma}_r(u, \hat{\mathbf{X}}^{\zeta}(u))\,\mathrm{d}\xi_r(u)\big)\,.$$

The process $\hat{\mathbf{Y}}(t) = t^{\alpha/2}\,\hat{\mathbf{X}}^{\zeta}(t)$ is then described by the following equation

$$\mathrm{d}\hat{\mathbf{Y}} = \big(\tfrac{1}{2}\alpha\mathbf{I}t^{-1} + a\mathbf{B}t^{-\alpha}\big)\,\hat{\mathbf{Y}}\,\mathrm{d}t$$
$$+ at^{-\alpha/2}\Big(\hat{\delta}(\hat{\mathbf{X}})\,\mathrm{d}t + \sum_{r=1}^{k} \hat{\sigma}_r(t, \hat{\mathbf{X}})\,\mathrm{d}\xi_r\Big)\,, \quad t \geqq t_0\,,$$
$$\hat{\mathbf{Y}}(t_0) = t_0^{\alpha/2}\zeta\,,$$

where we put $\hat{\delta}(\mathbf{x}) = \hat{R}(\mathbf{x}) - \mathbf{B}\mathbf{x}$.

Evidently, the first term on the right hand side tends to 0 a.s. for $t \to \infty$. Further,

$$\int_{t_0}^{t} \left(\frac{t}{u}\right)^{\alpha/2} \exp\{a(1-\alpha)^{-1}\,\mathbf{B}(t^{1-\alpha}-u^{1-\alpha})\}\, u^{-\alpha/2}\, \delta(\hat{\mathbf{X}}^{\zeta}(u))\, du \to 0$$

in probability, according to Lemma 6.

Let us investigate the process

$$\boldsymbol{\eta}(t) = \int_{t_0}^{t} \left(\frac{t}{u}\right)^{\alpha/2} \exp\{a(1-\alpha)^{-1}\,\mathbf{B}(t^{1-\alpha}-u^{1-\alpha})\}\, u^{-\alpha/2} \sum_{r=1}^{k} (\hat{\sigma}_r(u, \hat{\mathbf{X}}^{\zeta}(u)) - \sigma_{r,0})\, d\xi_r(u)\,.$$

Denote $\boldsymbol{\psi}(u) = \sum_{r=1}^{k} (\hat{\sigma}_r(u, \hat{\mathbf{X}}^{\zeta}(u)) - \sigma_{r,0})$; we have

$$E(\boldsymbol{\eta}(t)\,\boldsymbol{\eta}^T(t)) = \int_{t_0}^{t} \left(\frac{t}{u}\right)^{\alpha} \exp\{a(1-\alpha)^{-1}\,\mathbf{B}(t^{1-\alpha}-u^{1-\alpha})\}\, E(\boldsymbol{\psi}(u)\,\boldsymbol{\psi}^T(u)) \,.\, \exp\{a(1-\alpha)^{-1}\,\mathbf{B}^T(t^{1-\alpha}-u^{1-\alpha})\}\, u^{-\alpha}\, du\,.$$

Hence, using Lemma 5,

$$\|E(\boldsymbol{\eta}(t)\,\boldsymbol{\eta}^T(t))\| \leqq K^2 \int_{t_0}^{t} \exp\{-2\lambda(t^{1-\alpha}-u^{1-\alpha})\}\, E\|\boldsymbol{\psi}(u)\,\boldsymbol{\psi}^T(u)\|\, u^{-\alpha}\, du\,. \tag{13}$$

The fact that $\hat{\mathbf{X}}^{\zeta}(t) \to 0$ a.s. (Lemma 3) together with (ii) imply $\boldsymbol{\psi}(t) \to 0$ a.s. Moreover, from the definition of $\hat{\sigma}_r$ and from the assumptions made about σ_r in Section 1, it follows that $\|\boldsymbol{\psi}(t)\,\boldsymbol{\psi}^T(t)\|$ is bounded by a constant; hence (Lebesgue Theorem)

$$E\|\boldsymbol{\psi}(t)\,\boldsymbol{\psi}^T(t)\| \to 0 \quad \text{for} \quad t \to \infty\,. \tag{14}$$

The substitution $t^{1-\alpha} - u^{1-\alpha} = v$ changes the integral in (13) into

$$\int_{0}^{t^{1-\alpha}-t_0^{1-\alpha}} e^{-2\lambda v}\, E\|\boldsymbol{\psi}((t^{1-\alpha}-v)^{1/(1-\alpha)})\,\boldsymbol{\psi}^T((t^{1-\alpha}-v)^{1/(1-\alpha)})\|\, dv\,.$$

From (14) it easily follows that this integral tends to 0 for $t \to \infty$. Hence, $\|E(\boldsymbol{\eta}(t)\,\boldsymbol{\eta}^T(t))\| \to 0$, and consequently, $\boldsymbol{\eta}(t) \to 0$ in probability for $t \to \infty$.

The distribution of $\hat{\mathbf{Y}}(t)$ is thus asymptotically equivalent to the distribution of

$$\boldsymbol{\eta}_1(t) = a \int_{t_0}^{t} \left(\frac{t}{u}\right)^{\alpha/2} \exp\{a(1-\alpha)^{-1}\,\mathbf{B}(t^{1-\alpha}-u^{1-\alpha})\}\, u^{-\alpha/2} \sum_{r=1}^{k} \sigma_{r,0}\, d\xi_r(u)\,.$$

Obviously, $\boldsymbol{\eta}_1(t)$ is a Gaussian process with zero mean and the variance matrix

$$E(\boldsymbol{\eta}_1(t)\,\boldsymbol{\eta}_1^T(t))$$

$$= a^2 \int_{t_0}^{t} \left(\frac{t}{u}\right)^{\alpha} \exp\left\{a(1-\alpha)^{-1}\,\mathbf{B}(t^{1-\alpha}-u^{1-\alpha})\right\}$$

$$.\,\mathbf{S}_0\,.\,\exp\left\{a(1-\alpha)^{-1}\,\mathbf{B}^T(t^{1-\alpha}-u^{1-\alpha})\right\}\,u^{-\alpha}\,du$$

$$= \frac{a^2}{1-\alpha}\int_0^{t^{1-\alpha}-t_0^{1-\alpha}} \left(1-\frac{v}{t^{1-\alpha}}\right)^{-\alpha/(1-\alpha)} e^{a(1-\alpha)^{-1}\mathbf{B}v}\,\mathbf{S}_0\,e^{a(1-\alpha)^{-1}\mathbf{B}^Tv}\,dv$$

$$\to \frac{a^2}{1-\alpha}\int_0^{\infty} e^{a(1-\alpha)^{-1}\mathbf{B}v}\,\mathbf{S}_0\,e^{a(1-\alpha)^{-1}\mathbf{B}^Tv}\,dv = a\int_0^{\infty} e^{\mathbf{B}w}\,\mathbf{S}_0\,e^{\mathbf{B}^Tw}\,dw = \mathbf{S}\,.$$

We have used, in their turn, the substitution $t^{1-\alpha} - u^{1-\alpha} = v$, Lemma 7 and the substitution $av/(1-\alpha) = w$.

The distribution of $\hat{\mathbf{Y}}(t)$ is thus asymptotically normal with parameters $\mathbf{0}$ and $\mathbf{S}$.

The rest of the proof is exactly the same as in NH, Theor. 6.5.1. Let $\mathbf{X}^{\mathbf{x}}(t)$ be the solution of (1), satisfying $\mathbf{X}^{\mathbf{x}}(t_0) = \mathbf{x}$. Choose an $\varepsilon_1 > 0$; there is a $T > 0$ such that

$$P(\sup_{t \geqq T} \|\mathbf{X}^{\mathbf{x}}(t)\| < \varepsilon) > 1 - \varepsilon_1 \quad (\text{as } \mathbf{X}^{\mathbf{x}}(t) \to \mathbf{0} \text{ a.s.})\,.$$

Consider the equation (1) for $t \geqq T$ only; its solution $\mathbf{X}(t)$ satisfying the initial condition $\mathbf{X}(T) = \mathbf{X}^{\mathbf{x}}(T)$ coincides with $\mathbf{X}^{\mathbf{x}}(t)$ for all $t \geqq T$. Denote as Ω_0 the set $[\sup_{t \geqq T} \|\mathbf{X}^{\mathbf{x}}(t)\| < \varepsilon]$ and as Ω_1 the set $[\|\mathbf{X}^{\mathbf{x}}(T)\| < \varepsilon]$. Consider the equation (11) for $t \geqq T$ only; its solution $\mathbf{X}(t)$ satisfying the initial condition $\hat{\mathbf{X}}(T) = \mathbf{X}^{\mathbf{x}}(T)\,\chi_{\Omega_1}$ (χ_{Ω_1} being the indicator of Ω_1) coincides with $\mathbf{X}^{\mathbf{x}}(t)$ for all $t \geqq T$ on the set Ω_0. Hence $P(\sup_{t \geqq T} \|\mathbf{X}^{\mathbf{x}}(t) - \hat{\mathbf{X}}(t)\| > 0) < \varepsilon_1$.

Moreover, for every Borel set A,

$$\limsup_{t \to \infty} P(t^{\alpha/2}\,\mathbf{X}^{\mathbf{x}}(t) \in A)$$

$$\leqq \limsup_{t \to \infty} P(t^{\alpha/2}\,\mathbf{X}^{\mathbf{x}}(t) \in A,\ \sup_{t \geqq T} \|\mathbf{X}^{\mathbf{x}}(t) - \mathbf{X}(t)\| = 0)$$

$$+ \limsup_{t \to \infty} P(t^{\alpha/2}\,\mathbf{X}^{\mathbf{x}}(t) \in A,\ \sup_{t \geqq T} \|\mathbf{X}^{\mathbf{x}}(t) - \mathbf{X}(t)\| > 0)$$

$$\leqq \limsup_{t \to \infty} P(t^{\alpha/2}\,\mathbf{X}(t) \in A) + \varepsilon_1\,,$$

and, similarly,

$$\liminf_{t\to\infty} P(t^{\alpha/2}\, \mathbf{X}^x(t) \in A)$$

$$\geqq \liminf_{t\to\infty} P(t^{\alpha/2}\, \mathbf{X}^x(t) \in A,\ \sup_{t\geqq T} \|\mathbf{X}^x(t) - \mathbf{X}(t)\| = 0)$$

$$\geqq \liminf_{t\to\infty} P(t^{\alpha/2}\, \mathbf{X}(t) \in A) - P(\sup_{t\geqq T} \|\mathbf{X}^x(t) - \mathbf{X}(t)\| > 0)$$

$$\geqq \liminf_{t\to\infty} P(t^{\alpha/2}\, \mathbf{X}(t) \in A) - \varepsilon_1.$$

As the distribution of $\hat{\mathbf{Y}}(t) = t^{\alpha/2}\, \hat{\mathbf{X}}(t)$ is asymptotically normal with parameters $\mathbf{0}$ and $\mathbf{S}$, and as $\varepsilon_1 > 0$ is arbitrary, the asymptotic normality (with parameters $\mathbf{0}$ and $\mathbf{S}$) of $\mathbf{Y}(t) = t^{\alpha/2}\, \mathbf{X}^x(t)$ follows. □

References

[1] Nevel'son, M. B. - Has'minskii, R. Z. (1972). "Stochastic Approximation and Recursive Estimation". (In Russian). Nauka, Moscow.

[2] Has'minskii, R. Z. (1972). On the behaviour of stochastic approximation processes for large values of time-parameter. (In Russian.) *Problemy peredači informacii*, **8**, 81—91.

[3] Holevo, A. S. (1967). Estimation of drift parameters of a diffusion process by stochastic approximation method. (In Russian.) *Studies in the Theory of Selfadjusting Systems.* Computing Center Acad. Sci. U.S.S.R., Moscow.

CHARLES UNIVERSITY, PRAGUE, CZECHOSLOVAKIA

Received December 1975

A NOTE ON REJECTIVE SAMPLING

by

JITKA DUPAČOVÁ

1. Preliminaries

Consider a *population* S consisting of N units; identify them with numbers $1, 2, \ldots, N$. A *sample* is defined as a subset of S. Any probability distribution $P(s)$ on the space of all samples $s \subset S$ is called a sampling design (or *sampling*, briefly). Let Ω be the set of all samples of fixed size n; a sampling with $P(\Omega) = 1$ is called *sampling of size n.*

Poisson sampling with parameter $\boldsymbol{p} = (p_1, \ldots, p_N)^T$, $0 \leqq p_i \leqq 1$, $1 \leqq i \leqq N$, is defined by

$$P(s) = \prod_{i \in s} p_i \prod_{i \notin s} (1 - p_i), \quad s \subset S;$$

the corresponding conditional distribution

$$P(s \mid \Omega) = \begin{cases} \dfrac{P(s)}{P(\Omega)}, & \text{if } s \in \Omega, \\ 0, & \text{otherwise}, \end{cases} \tag{1}$$

defines the *rejective sampling of size n.*

The correspondence between the primary Poisson sampling and the rejective sampling defined by (1) is not one-to-one. Any other parameter value $\boldsymbol{p}^* = (p_1^*, \ldots, p_N^*)^T$, $0 \leqq p_i^* \leqq 1$, $1 \leqq i \leqq N$, provides the same rejective sampling (1), if

$$\frac{p_i}{1 - p_i} = \lambda \frac{p_i^*}{1 - p_i^*}, \quad 1 \leqq i \leqq N \tag{2}$$

for some $\lambda > 0$. In other words, there are infinitely many *representations* (1) of the same rejective sampling (except the case of a parameter consisting of zeros and ones only). The representation (1) is called *canonical*, if $\sum_{i=1}^{N} p_i = n$; obviously, the canonical representation is unique.

For any sampling, let π_i $[\pi_{ij}]$ denote the probability of including the unit i [the units i and j] in the sample, i.e., $\pi_i = \sum_{s \ni i} P(s)$, $\pi_{ij} = \sum_{s \ni i,j} P(s)$. Call π_i, π_{ij} inclusion probabilities. For (unconditional) Poisson sampling we have $\pi_i = p_i$, $\pi_{ij} = p_i p_j$. For any sampling of size n, the inclusion probabilities satisfy

$$\sum_{i=1}^{N} \pi_i = n\,, \tag{3}$$

$$\sum_{j \neq i} (\pi_i \pi_j - \pi_{ij}) = \pi_i(1 - \pi_i)\,, \quad 1 \leqq i \leqq N\,. \tag{4}$$

For rejective sampling, the inequality

$$\pi_i \pi_j - \pi_{ij} \geqq 0\,, \quad 1 \leqq i \neq j \leqq N\,,$$

holds true; in case of $0 < p_i < 1$, $1 \leqq i \leqq N$, the inequality is strict

$$\pi_i \pi_j - \pi_{ij} > 0\,, \quad 1 \leqq i \neq j \leqq N\,, \tag{5}$$

(Lanke (1972)). If $\boldsymbol{p}^0 = (p_1^0, \ldots, p_N^0)^T$ is the parameter value of the canonical representation of a rejective sampling, then the inequalities

$$\min_{1 \leqq i \leqq N} \pi_i < \min_{1 \leqq i \leqq N} p_i^0 \leqq \max_{1 \leqq i \leqq N} p_i^0 < \max_{1 \leqq i \leqq N} \pi_i \tag{6}$$

hold true except the cases when some p_i^0 equals zero or one or when all values p_i^0 are equal.

The following question was raised by Hájek in his seminar on sampling techniques in 1973:

Can the inclusion probabilities π_j, $1 \leqq j \leqq N$, of rejective sampling be arbitrarily prescribed?

The question is answered affirmatively in the present paper, the precise statement being given in Section 2.

2. The result

Theorem. *Let $n < N$, let $\pi_1^*, \ldots, \pi_N^*$ be arbitrary numbers satisfying $0 < \pi_j^* < 1$, $1 \leqq j \leqq N$, $\sum_{j=1}^{N} \pi_j^* = n$. Then there exist numbers p_j, $1 \leqq j \leqq N$, such that the conditional Poisson sampling (1) with parameter $\boldsymbol{p} = (p_1, \ldots, p_N)^T$ represents rejective sampling of size n with inclusion probabilities π_j^*, $1 \leqq j \leqq N$.*

The proof will make use of the following

Lemma. *For the conditional Poisson sampling conditioned by an arbitrary set of samples $\tilde{\Omega}$ (and hence, also for rejective sampling), we have*

$$\frac{\partial \pi_i}{\partial p_i} = \frac{\pi_i(1-\pi_i)}{p_i(1-p_i)}, \quad 1 \leqq i \leqq N, \tag{7}$$

$$\frac{\partial \pi_j}{\partial p_i} = \frac{\pi_{ij} - \pi_i \pi_j}{p_i(1-p_i)}, \quad 1 \leqq i, j \leqq N, \quad i \neq j. \tag{8}$$

Proof:

$$\frac{\partial P(s)}{\partial p_i} = \frac{P(s)}{p_i}, \quad i \in s,$$

$$= -\frac{P(s)}{1-p_i}, \quad i \notin s.$$

$$\frac{\partial P(\tilde{\Omega})}{\partial p_i} = \sum_{s \in \tilde{\Omega}} \frac{\partial P(s)}{\partial p_i} = \sum_{\substack{s \in \tilde{\Omega} \\ s \ni i}} \frac{P(s)}{p_i} - \sum_{\substack{s \in \tilde{\Omega} \\ s \not\ni i}} \frac{P(s)}{1-p_i}$$

$$= P(\tilde{\Omega}) \left[\frac{\pi_i}{p_i} - \frac{1-\pi_i}{1-p_i} \right].$$

According to (1), we have

$$\frac{\partial P(s \mid \tilde{\Omega})}{\partial p_i} = \frac{\dfrac{\partial P(s)}{\partial p_i} P(\tilde{\Omega}) - P(s) \dfrac{\partial P(\tilde{\Omega})}{\partial p_i}}{[P(\tilde{\Omega})]^2}$$

$$= \frac{P(s)}{P(\tilde{\Omega})} \left[\frac{1}{p_i} - \frac{\pi_i - p_i}{p_i(1-p_i)} \right] = P(s \mid \tilde{\Omega}) \frac{1-\pi_i}{p_i(1-p_i)} \quad \text{for} \quad i \in s,$$

$$\frac{\partial P(s \mid \tilde{\Omega})}{\partial p_i} = P(s \mid \tilde{\Omega}) \left[-\frac{1}{1-p_i} - \frac{\pi_i - p_i}{p_i(1-p_i)} \right]$$

$$= P(s \mid \tilde{\Omega}) \frac{-\pi_i}{p_i(1-p_i)} \quad \text{for} \quad i \notin s.$$

Finally,

$$\frac{\partial \pi_i}{\partial p_i} = \sum_{s \ni i} \frac{\partial P(s \mid \tilde{\Omega})}{\partial p_i} = \frac{\pi_i(1-\pi_i)}{p_i(1-p_i)}$$

$$\frac{\partial \pi_j}{\partial p_i} = \sum_{s \ni j} \frac{\partial P(s \mid \tilde{\Omega})}{\partial p_i} = \sum_{s \ni j, i} P(s \mid \tilde{\Omega}) \frac{1-\pi_i}{p_i(1-p_i)}$$

$$+ \sum_{\substack{s \ni j \\ s \not\ni i}} P(s \mid \tilde{\Omega}) \frac{-\pi_i}{p_i(1 - p_i)}$$

$$= \pi_{ij} \frac{(1 - \pi_i)}{p_i(1 - p_i)} - \frac{(\pi_j - \pi_{ij}) \pi_i}{p_i(1 - p_i)} = \frac{\pi_{ij} - \pi_i \pi_j}{p_i(1 - p_i)}, \quad i \neq j .$$

Proof of the Theorem:

Let π_k^*, $1 \leqq k \leqq N$, be the prescribed inclusion probabilities for the rejective sampling of size n, i.e. $0 < \pi_k^* < 1$, $1 \leqq k \leqq N$, $\sum_{k=1}^N \pi_k^* = n$. Suppose that $\min_{1 \leqq i \leqq N} \pi_i^* < \max_{1 \leqq i \leqq N} \pi_i^*$ (otherwise any parameter $\boldsymbol{p}$ whose components are all equal gives the desired solution). Let $\pi_k(\boldsymbol{p})$, $1 \leqq k \leqq N$ denote the inclusion probabilities corresponding to a parameter $\boldsymbol{p} = (p_1, \ldots, p_N)^T$, $0 \leqq p_i \leqq 1$, $1 \leqq i \leqq N$. The problem whether $\boldsymbol{\pi}^*$ are realizable through some $\boldsymbol{p}$ reduces to the problem of

$$(9) \qquad \min \{ \sum_{k=1}^N |\pi_k(\boldsymbol{p}) - \pi_k^*| : 0 \leqq p_i \leqq 1,\ 1 \leqq i \leqq N \} .$$

If the minimal value of $S(\boldsymbol{p}) = \sum_{k=1}^N |\pi_k(\boldsymbol{p}) - \pi_k^*|$ under condition $0 \leqq p_i \leqq 1$, $1 \leqq i \leqq N$, is equal to 0 then $\{\pi_k^*\}_{1 \leqq k \leqq N}$ are realizable and, in addition, they are realizable through the canonical representation $\boldsymbol{p}^0$ that fulfils (6). We can therefore confine ourselves to the problem of minimizing

$$(10) \qquad S(\boldsymbol{p}) = \sum_{k=1}^N |\pi_k(\boldsymbol{p}) - \pi_k^*|$$

subject to constraints

$$(11) \qquad \min_{1 \leqq k \leqq N} \pi_k^* \leqq p_i \leqq \max_{1 \leqq k \leqq N} \pi_k^*, \quad 1 \leqq i \leqq N .$$

Suppose that

$$\min \{ S(\boldsymbol{p}) : \min_{1 \leqq k \leqq N} \pi_k^* \leqq p_i \leqq \max_{1 \leqq k \leqq N} \pi_k^*,\ 1 \leqq i \leqq N \} = S(\hat{\boldsymbol{p}}) \neq 0 ,$$

i.e. the minimal value is obtained for $\hat{\boldsymbol{p}}$ such that $\pi_k(\hat{\boldsymbol{p}}) \neq \pi_k^*$ at least for one index $k \in \{1, \ldots, N\}$.

The equality $\sum_{k=1}^N \pi_k^* = n = \sum_{k=1}^N \pi_k(\hat{\boldsymbol{p}})$ implies that two indices, $i, j \in \{1, \ldots, N\}$, can be given such that

$$(12) \qquad \pi_j(\hat{\boldsymbol{p}}) - \pi_j^* > 0 \quad \text{and} \quad \pi_i(\hat{\boldsymbol{p}}) - \pi_i^* < 0 .$$

Discussing distinct possibilities, we shall get the contradiction.

A) Suppose first that $\hat{p}$ is an interior point of the admissibility set (11). For $p(\delta) = \hat{p} + \delta e_i$, $\delta > 0$, where e_i denotes the i-th N-dimensional unit vector, we get using (7), (8),

$$\pi_i(\hat{p} + \delta e_i) = \pi_i(\hat{p}) + \Delta_i \pi_i(\hat{p})\,(1 - \pi_i(\hat{p})) + o(\delta)\,,$$

$$\pi_k(\hat{p} + \delta e_i) = \pi_k(\hat{p}) + \Delta_i(\pi_{ik}(\hat{p}) - \pi_i(\hat{p})\,\pi_k(\hat{p})) + o(\delta)\,,$$

$$1 \leqq k \leqq N\,, \quad k \neq i\,;$$

here,

$\Delta_i = \delta/[\hat{p}_i(1 - \hat{p}_i)]$, and for δ small enough, $p(\delta)$ evidently satisfies (11). The corresponding value of the objective function for $\hat{p} + \delta e_i$ is

$$\begin{aligned} S(\hat{p} + \delta e_i) &\approx \left|\pi_i(\hat{p}) - \pi_i^* + \Delta_i\, \pi_i(\hat{p})\,(1 - \pi_i(\hat{p}))\right| \\ &\quad + \sum_{k \neq i} \left|\pi_k(\hat{p}) - \pi_k^* + \Delta_i(\pi_{ik}(\hat{p}) - \pi_i(\hat{p})\,\pi_k(\hat{p}))\right| \\ &\leqq S(\hat{p}) + \sum_{k \neq i,j} \left|\Delta_i(\pi_{ik}(\hat{p}) - \pi_i(\hat{p})\,\pi_k(\hat{p}))\right| \\ &\quad - \Delta_i\, \pi_i(\hat{p})\,(1 - \pi_i(\hat{p})) + \Delta_i(\pi_{ij}(\hat{p}) - \pi_i(\hat{p})\,\pi_j(\hat{p})) \\ &= S(\hat{p}) - \Delta_i\{\pi_i(\hat{p})\,(1 - \pi_i(\hat{p})) + \sum_{k \neq i,j} (\pi_{ik}(\hat{p}) - \pi_i(\hat{p})\,\pi_k(\hat{p})) \\ &\quad - \pi_{ij}(\hat{p}) + \pi_i(\hat{p})\,\pi_j(\hat{p})\} \\ &= S(\hat{p}) - \Delta_i\{\pi_i(\hat{p})\,(1 - \pi_i(\hat{p})) + 2(\pi_i(\hat{p})\,\pi_j(\hat{p}) - \pi_{ij}(\hat{p})) \\ &\quad - \sum_{k \neq i} (\pi_i(\hat{p})\,\pi_k(\hat{p}) - \pi_{ik}(\hat{p}))\} \\ &= S(\hat{p}) - 2\,\Delta_i(\pi_i(\hat{p})\,\pi_j(\hat{p}) - \pi_{ij}(\hat{p})) < S(\hat{p}) \end{aligned}$$

because of (5).

The objective function $S(p)$ decreases in $\hat{p}$ in the admissible direction e_i what contradicts to the original assumption that

$$S(\hat{p}) = \min \{S(p) : \min_{1 \leqq k \leqq N} \pi_k^* \leqq p_i \leqq \max_{1 \leqq k \leqq N} \pi_k^*,\ 1 \leqq i \leqq N\}\,.$$

B) Suppose that (after suitable renumeration) the minimum of $S(p)$ is attained at the point $\hat{p}$ such that

$$\min_{1 \leqq k \leqq N} \pi_k^* = \hat{p}_1 \leqq \hat{p}_2 \leqq \ldots \leqq \hat{p}_N < \max_{1 \leqq k \leqq N} \pi_k^*\,.$$

According to (2), parameter value p^0 such that

$$\frac{p_k^0}{1 - p_k^0} = (1 + \varepsilon)\,\frac{\hat{p}_k}{1 - \hat{p}_k}\,, \quad 1 \leqq k \leqq N\,,$$

gives the same values of $P(s \mid \Omega)$ and inclusion probabilties $\pi_k(\hat{\boldsymbol{p}}) = \pi_k(\boldsymbol{p}^0)$, $1 \leqq k \leqq \leqq N$. Choosing $\varepsilon > 0$ small enough so that $p_N^0 < \max \pi_k^*$, we have an interior minimal point $\boldsymbol{p}^0$ with $S(\boldsymbol{p}^0) \neq 0$ again and the problem reduces to the case A.

C) Suppose that (after suitable renumeration) the minimum of $S(\boldsymbol{p})$ is attained at the point $\hat{\boldsymbol{p}}$ such that

$$\min_{1 \leqq k \leqq N} \pi_k^* < \hat{p}_1 \leqq \hat{p}_2 \leqq \ldots \leqq \hat{p}_N = \max_{1 \leqq k \leqq N} \pi_k^* .$$

Using the transformation

$$\frac{p_k^0}{1 - p_k^0} = (1 - \varepsilon) \frac{\hat{p}_k}{1 - \hat{p}_k}, \quad 1 \leqq k \leqq N,$$

with $\varepsilon > 0$ small enough (so that $p_1^0 > \min_{1 \leqq k \leqq N} \pi_k^*$) the problem reduces to the case A again.

D) Suppose that (after suitable renumeration) the minimum of $S(\boldsymbol{p})$ is attained at the point $\hat{\boldsymbol{p}}$ such that

$$\min_{1 \leqq k \leqq N} \pi_k^* = \hat{p}_1 \leqq \hat{p}_2 \leqq \ldots \leqq \hat{p}_N = \max_{1 \leqq k \leqq N} \pi_k^* .$$

Using a transformation of type (2) with a properly choosen value of λ we can pass to the canonical representation $\boldsymbol{p}^0$. Now we have

$$p_N^0 = \max_{1 \leqq k \leqq N} p_k^0 \geqq \max_{1 \leqq k \leqq N} \pi_k^* \quad \text{in case of} \quad \lambda \geqq 1,$$

or,

$$p_1^0 = \min_{1 \leqq k \leqq N} p_k^0 \leqq \min_{1 \leqq k \leqq N} \pi_k^* \quad \text{in case of} \quad \lambda \leqq 1 .$$

In view of (6),

$$\pi_N(\hat{\boldsymbol{p}}) = \max_{1 \leqq k \leqq N} \pi_k(\hat{\boldsymbol{p}}) = \max_{1 \leqq k \leqq N} \pi_k(\boldsymbol{p}^0) > \max_{1 \leqq k \leqq N} \pi_k^* \tag{13}$$

or

$$\pi_1(\hat{\boldsymbol{p}}) = \min_{1 \leqq k \leqq N} \pi_k(\hat{\boldsymbol{p}}) = \min_{1 \leqq k \leqq N} \pi_k(\boldsymbol{p}^0) < \min_{1 \leqq k \leqq N} \pi_k^* . \tag{14}$$

In the case (14) with $\hat{p}_1 < \hat{p}_2$, the objective function $S(\boldsymbol{p})$ decreases at the point $\hat{\boldsymbol{p}}$ in the admissible direction $\boldsymbol{e}_1$. Similarly, in the case (13) with $\hat{p}_N > \hat{p}_{N-1}$, the objective function $S(\boldsymbol{p})$ decreases at the point $\hat{\boldsymbol{p}}$ in the admissible direction $\boldsymbol{e}_N$ what contradicts to the optimality of $\hat{\boldsymbol{p}}$.

Suppose now that in case (14), we have

$$\min_{1 \leqq k \leqq N} \pi_k^* = \hat{p}_1 = \ldots = \hat{p}_m < \hat{p}_{m+1} \leqq \ldots \leqq \hat{p}_N = \max_{1 \leqq k \leqq N} \pi_k^* , \quad m > 1 ,$$

i.e.,

$$\pi_1(\hat{\boldsymbol{p}}) = \ldots = \pi_m(\hat{\boldsymbol{p}}) < \min_{1 \leqq k \leqq N} \pi_k^* . \tag{15}$$

Consider an admissible point $\boldsymbol{p}(\delta) = \hat{\boldsymbol{p}} + \delta \sum_{i=1}^{m} \boldsymbol{e}_i$ with $\delta > 0$ small enough. The corresponding inclusion probabilities satisfy

$$\pi_i(\delta) = \pi_i(\hat{\boldsymbol{p}} + \delta \sum_{i=1}^{m} \boldsymbol{e}_i) = \pi_i(\hat{\boldsymbol{p}}) + \Delta\, \pi_i(\hat{\boldsymbol{p}})\,(1 - \pi_i(\hat{\boldsymbol{p}}))$$

$$+ \Delta \sum_{\substack{k=1 \\ k \neq i}}^{m} (\pi_{ki}(\hat{\boldsymbol{p}}) - \pi_k(\hat{\boldsymbol{p}})\, \pi_i(\hat{\boldsymbol{p}})) + o(\delta) = \pi_1(\delta) , \quad 1 \leqq i \leqq m ,$$

because of (15); here $\Delta = \delta/(\hat{p}_1(1 - \hat{p}_1))$.

Similarly,

$$\pi_k(\delta) = \pi_k(\hat{\boldsymbol{p}}) + \Delta \sum_{i=1}^{m} (\pi_{ik}(\hat{\boldsymbol{p}}) - \pi_i(\hat{\boldsymbol{p}})\, \pi_k(\hat{\boldsymbol{p}})) + o(\delta)$$

$$= \pi_k(\hat{\boldsymbol{p}}) + m\Delta(\pi_{1k}(\hat{\boldsymbol{p}}) - \pi_1(\hat{\boldsymbol{p}})\, \pi_k(\hat{\boldsymbol{p}})) + o(\delta) , \quad m < k \leqq N .$$

The corresponding value of the objective function is

$$S(\hat{\boldsymbol{p}} + \delta \sum_{i=1}^{m} \boldsymbol{e}_i) = \sum_{i=1}^{m} \Big| \pi_i(\hat{\boldsymbol{p}}) - \pi_i^* + \Delta(\pi_1(\hat{\boldsymbol{p}})\,(1 - \pi_1(\hat{\boldsymbol{p}}))$$

$$+ \sum_{k=2}^{m} (\pi_{1k}(\hat{\boldsymbol{p}}) - \pi_1(\hat{\boldsymbol{p}})\, \pi_k(\hat{\boldsymbol{p}})))\Big| + \big|\pi_j(\hat{\boldsymbol{p}}) - \pi_j^* + m\Delta(\pi_{1j}(\hat{\boldsymbol{p}}) - \pi_1(\hat{\boldsymbol{p}})\, \pi_j(\hat{\boldsymbol{p}}))\big|$$

$$+ \sum_{\substack{k=m+1 \\ k \neq j}}^{N} \big|\pi_k(\hat{\boldsymbol{p}}) - \pi_k^* + m\Delta(\pi_{1k}(\hat{\boldsymbol{p}}) - \pi_1(\hat{\boldsymbol{p}})\, \pi_k(\hat{\boldsymbol{p}}))\big| + o(\delta)$$

with $\pi_j(\hat{\boldsymbol{p}}) - \pi_j^* > 0$ according to (12). For $\delta > 0$ small enough we have

$$S(\hat{\boldsymbol{p}} + \delta \sum_{i=1}^{m} \boldsymbol{e}_i) \leqq - \sum_{i=1}^{m} (\pi_i(\hat{\boldsymbol{p}}) - \pi_i^*) - m\Delta\{\pi_1(\hat{\boldsymbol{p}})\,(1 - \pi_1(\hat{\boldsymbol{p}}))$$

$$+ \sum_{i=2}^{m} (\pi_{1i}(\hat{\boldsymbol{p}}) - \pi_1(\hat{\boldsymbol{p}})\, \pi_i(\hat{\boldsymbol{p}}))\} + \pi_j(\hat{\boldsymbol{p}}) - \pi_j^*$$

$$+ m\Delta(\pi_{1j}(\hat{\boldsymbol{p}}) - \pi_1(\hat{\boldsymbol{p}})\, \pi_j(\hat{\boldsymbol{p}})) + \sum_{\substack{k=m+1 \\ k \neq j}}^{n} (|\pi_k(\hat{\boldsymbol{p}}) - \pi_k^*| - m\Delta(\pi_{1k}(\hat{\boldsymbol{p}})$$

$$- \pi_1(\hat{\boldsymbol{p}})\, \pi_k(\hat{\boldsymbol{p}}))) = S(\hat{\boldsymbol{p}}) - 2m\, \Delta(\pi_1(\hat{\boldsymbol{p}})\, \pi_j(\hat{\boldsymbol{p}}) - \pi_{1j}(\hat{\boldsymbol{p}})) < S(\hat{\boldsymbol{p}})$$

in contradiction with the optimality of $\hat{p}$. The remaining case (13) with

$$\hat{p}_{s+1} = \dots = \hat{p}_N = \max_{1 \leq k \leq N} \pi_k^*, \quad s + 1 < N$$

leads to contradiction in a similar way.

References

HÁJEK, J. (1964). Asymptotic theory of rejective sampling with varying probabilities from a finite population. *Ann. Math Statist.*, **35**, 1491—1523.

LANKE, J. (1972). On non-negative variance estimators in survey sampling. *The Manchester-Sheffield School of Probability and Statistics, Research Report* 106/JL 2.

CHARLES UNIVERSITY, PRAGUE, CZECHOSLOVAKIA

Received December 1975

ON TESTS OF FIT BASED ON GROUPED DATA

by

L. G. GVANCELADZE (1), AND D. M. CHIBISOV (2)

1. Introduction

Concerning the choice of the number of class intervals in Pearson's chi-square test there is a well-known result by Mann and Wald [9] stating that the optimal number k_n increases with n as $k_n \sim Cn^{2/5}$ where C depends on the test size α. The optimality of k_n is meant in a minimax sense, i.e. this k_n maximizes (asymptotically) the power against the least favorable alternative among those with the same distance from the hypothesis. The distance in [9] is the distance between the distribution functions (d.f.'s) in the sup metric. Another approach (see, e.g. [3]) might be to choose k_n maximizing the power against some particular alternatives. The aim of the present paper is to prove (Theorem 2) that in this case the optimal k_n does not tend to infinity with n but tends to a finite limit (in a sense to be specified since k is integer-valued). This result may be used to justify the determination of the optimal k_n from asymptotic considerations. In order to keep the power away from unity we adopt an approaching sequence of the form (2.1) as a model of a "particular alternative".

The distribution of the chi-square statistic converges under the alternatives (2.1) to the noncentral chi-square distribution with $k - 1$ degrees of freedom and the noncentrality parameter depending on $h(\cdot)$ (which is fixed in the present problem) and k. Theorem 2 is obtained by investigating the asymptotic power determined by this convergence, but there is a difficulty due to the fact that this convergence is not uniform in k. For this reason we need the assertion that for any sequence of integers $\{k(n)\}$ tending to infinity as $n \to \infty$ the power of the chi-square test with $k(n)$ class intervals tends to α. This follows from a general theorem (Theorem 1) stating that it is so for any sequence of tests symmetrically depending on frequencies.

The tests of this kind are based on $\mu_0, \mu_1, \mu_2, \ldots$ where μ_l is the number of class intervals containing exactly l observations, $l = 0, 1, \ldots$. To obtain a nontrivial limit of the power of such tests when $k \to \infty$ as $n \to \infty$ one should consider alternatives approaching H_0 slower than in (2.1) at a rate depending on the rate of the increase of k. There are results of this kind for k depending on n in some specific way (e.g. $k \sim n^\gamma$, $\gamma > 0$), see, e.g. [11]. The point of Theorem 1 giving only a degenerate asymptotic power is that no conditions on the growth of k are imposed.

Subsequently the following notational conventions will be used. Variables i and j run over the integers $1, \ldots, n$ and $1, \ldots, k$ respectively, unless otherwise stated; the limits of summations over i from 1 to n and over j from 1 to k are omitted. The passages to the limit are for $n \to \infty$ unless otherwise stated.

2. Formulation of results

Let $P_0, P_1, P_2, \ldots$ be probability measures on the real line having d.f.'s $F_n(x)$ and densities (w.r.t. the Lebesgue measure) $p_n(x)$, $n = 0, 1, 2, \ldots$ respectively.

Assumption (*A*). *There exist functions $h(x)$ and $r_n(x)$, $n = 1, 2, \ldots,$ such that*

$$\frac{p_n(x)}{p_0(x)} = 1 + n^{-1/2} h(x) + r_n(x) \tag{2.1}$$

on the set $C = \{x : p_0(x) > 0\}$ and satisfying the following conditions:

$$E_0 h = 0\,, \quad 0 < \lambda = E_0 h^2 < \infty \tag{2.2}$$

(*the notation like $E_0 h$ means $\int h(x)\, dF_0(x)$*); *for a sequence $X_1, X_2, \ldots$ of independent identically distributed r.v.'s with the common distribution P_0,*

$$\sum r_n(X_i) \to 0 \quad \textit{in probability}\,. \tag{2.3}$$

Let for $k = 2, 3, \ldots$ the numbers $a_{k1}, \ldots, a_{k,k-1}$ be such that $F_0(a_{kj}) = j/k$, $j = 1, \ldots, k-1$. Such numbers exist since F_0 is continuous but they may be not unique; then we choose them arbitrarily and fix. Let formally $a_{k0} = -\infty$, $a_{kk} = +\infty$ so that $F_0(a_{k0}) = 0$, $F_0(a_{kk}) = 1$. Put $\Delta_{kj} = \{x : a_{k,j-1} \leqq x < a_{kj}\}$. Then

$$P_0(\Delta_{kj}) = F_0(a_{kj}) - F_0(a_{k,j-1}) = 1/k\,. \tag{2.4}$$

Let for each $n = 1, 2, \ldots$ there be n i.i.d.r.v.'s $X_1, \ldots, X_n$; we test the hypothesis H_0 that their common distribution is P_0 and consider the alternative H_{1n} that this distribution is P_n. The joint distributions of $X^{(n)} = (X_1, \ldots, X_n)$ under H_0 and H_{1n} will be denoted by P_{0n} and P_{1n} and the corresponding expectations by E_{0n} and E_{1n} respectively.

Denote by ν_{nkj} the sample frequency of the interval Δ_{kj}, i.e. the number of X's among $X_1, \ldots, X_n$ belonging to Δ_{kj}. Let $\nu_{nk} = (\nu_{nk1}, \ldots, \nu_{nkk})$. Denote by M_{nk} the set of possible values of ν_{nk}, i.e. the set of vectors $x = (x_1, \ldots, x_k)$ with non-negative integer components such that $x_1 + \ldots + x_k = n$.

Theorem 1. *Let the assumption* (A) *be fulfilled. Let $\{k(n)\}$ be a sequence of integers, $k(n) \to \infty$, and $\{\Psi_n\}$ be any sequence of test functions defined on $M_{n,k(n)}$ such that*

$$\Psi_n(x_1, \ldots, x_{k(n)}) = \Psi_n(x_{j_1}, \ldots, x_{j_{k(n)}}),\ (x_1, \ldots, x_{k(n)}) \in M_{n,k(n)}\,,$$

for any permutation $j_1, \ldots, j_{k(n)}$ *of* $1, \ldots, k(n)$. *Then*

$$E_{1n}\,\Psi_n(v_{n,k(n)}) - E_{0n}\,\Psi_n(v_{n,k(n)}) \to 0\,. \tag{2.5}$$

Corollary 1. *Let for each n and k there be a test function* Ψ_{nk} *on* M_{nk}. *Then for any* $\varepsilon > 0$ *there exist* k_1 *and* n_1, *such that*

$$\sup\left[\left|E_{1n}\Psi_{nk} - E_{0n}\Psi_{nk}\right|;\ n > n_1,\ k > k_1\right] < \varepsilon\,.$$

This corollary is easily derived from Theorem 1 by contradiction.

Consider now the chi-square test statistic

$$\chi^2_{nk} = \frac{k}{n}\sum\left(v_{nkj} - \frac{n}{k}\right)^2, \quad k = 2, 3, \ldots\,. \tag{2.6}$$

Suppose that the test based on χ^2_{nk} is determined by constants C_{nk} and γ_{nk} such that H_0 is rejected with probability 1 if $\chi^2_{nk} > C_{nk}$ and with probability γ_{nk} if $\chi^2_{nk} = C_{nk}$. Denote by α_{nk} and β_{nk} the size and the power of this test, i.e.

$$\alpha_{nk} = P_{0n}\{\chi^2_{nk} > C_{nk}\} + \gamma_{nk}\,P_{0n}\{\chi^2_{nk} = C_{nk}\}\,, \tag{2.7}$$

$$\beta_{nk} = P_{1n}\{\chi^2_{nk} > C_{nk}\} + \gamma_{nk}\,P_{1n}\{\chi^2_{nk} = C_{nk}\}\,. \tag{2.8}$$

Theorem 2. *Let the assumption* (A) *be fulfilled. Let a number* α, $0 < \alpha < 1$, *be given and* C_{nk}, γ_{nk} *be such that* (i) $\alpha_{nk} \to \alpha$ *for each* $k = 2, 3, \ldots$ *and* (ii) $\sup[\alpha_{nk};\ k \geqq 2] \to \alpha$. *Then there exists an integer* $k^* \geqq 2$ *such that* $\sup[\beta_{nk};\ k \geqq 2] - \beta_{nk^*} \to 0$.

3. Preliminary results

First, some consequences of the assumption (A) will be stated as lemmas. The distributions P_{0n} and P_{1n} have the densities

$$p_{0n}(x^{(n)}) = \prod p_0(x_i)\,, \quad p_{1n}(x^{(n)}) = \prod p_n(x_i)\,, \quad x^{(n)} = (x_1, \ldots, x_n)\,.$$

Let

$$L_n(x^{(n)}) = \log\left[p_{1n}(x^{(n)})/p_{0n}(x^{(n)})\right] \ \text{if}\ \ p_{0n}(x^{(n)})\,p_{1n}(x^{(n)}) > 0\,, \tag{3.1}$$
$$0 \qquad \text{otherwise}\,.$$

(The value 0 in the second line of (3.1) is taken arbitrarily.)

Lemma 3.1. *Under the assumption* (A) *one has*

$$L_n(X^{(n)}) = n^{-1/2} \sum h(X_i) - \tfrac{1}{2}\lambda + \eta_n(X^{(n)}) \tag{3.2}$$

where $\eta_n(X^{(n)}) \to 0$ *in probability* P_{0n}.

For the *proof* see [2], Theorem 3.1.

It follows from (3.2) that the sequences $\{P_{0n}\}$ and $\{P_{1n}\}$ are contiguous ([4], Corollary VI.1.3).

Lemma 3.2. *Let* $g_1(x), \ldots, g_m(x)$ *be functions such that* $E_0 g_r = 0$, $E_0 g_r^2 < \infty$, $r = 1, \ldots, m$. *Define* $S_n = (S_{n1}, \ldots, S_{nm})$ *by* $S_{nr} = n^{-1/2} \sum g_r(X_i)$, $r = 1, \ldots, m$, *and* $B = (b_{rs})$ *as the covariance matrix* ($b_{rs} = E_0 g_r g_s$, $r, s = 1, \ldots, m$). *Then under the assumption* (A) S_n *is asymptotically normally distributed with parameters* $(0, B)$ *and* (a, B) *under* P_{0n} *and* P_{1n} *respectively, where* $a = (a_1, \ldots, a_m)$, $a_r = E_0 g_r h$, $r = 1, \ldots, m$.

This follows from the Central Limit Theorem and a multivariate version of the "Le Cam's third lemma" ([4], Lemma VI, 1.4).

It will be convenient for the subsequent proofs to modify the alternatives as follows. Let $h^+ = \max(h, 0)$, $h^- = \max(-h, 0)$, then $h = h^+ - h^-$. Take a sequence $\{d_n\}$ such that $0 < d_n < 1$, $d_n \to 0$ and $d_n n^{1/2} \to \infty$. Set

$$h_n^{\pm}(x) = \max\left(h^{\pm}(x), d_n n^{1/2}\right), \quad c_n^{\pm} = E_0 h_n^{\pm}, \tag{3.3}$$

$$h_n(x) = (c_n^- / c_n^+)\, h_n^+(x) - h_n^-(x) \quad \text{if} \quad c_n^- \leqq c_n^+, \tag{3.4}$$

$$h^+(x) - (c_n^+ / c_n^-)\, h_n^-(x) \quad \text{if} \quad c_n^- > c_n^+,$$

$$p_n'(x) = p_0(x)\left(1 + n^{-1/2} h_n(x)\right), \quad n = 1, 2, \ldots . \tag{3.5}$$

Then $p_n'(x)$ is a density function. Denote by P_n' and P_{1n}' the measures on the real line and on the sample space determined by $p_n'(x)$. It follows from (3.3)–(3.5) that

$$p_n'(x) = 0 \quad \text{if} \quad p_0(x) = 0, \tag{3.6}$$

hence $P_n'(P_{1n}')$ is absolutely continuous w.r.t. $P_0(P_{0n})$;

$$\sup n^{-1/2} |h_n(x)| \leqq d_n \to 0; \tag{3.7}$$

$$E_0 h_n = 0, \quad E_0(h - h_n)^2 \to 0. \tag{3.8}$$

The relation (3.8) follows from (3.3), (3.4), $d_n n^{1/2} \to \infty$ and $c_n^+ / c_n^- \to 1$; the last convergence holds because of $E_0 h = 0$ and $c_n^{\pm} \to E_0 h^+ = E_0 h^- > 0$. Moreover, we have

$$\|P_{1n}' - P_{1n}\| = \int |p_{1n}'(x^{(n)}) - p_{1n}(x^{(n)})|\, dx^{(n)} \to 0. \tag{3.9}$$

This follows from the fact that $p_n'(x)/p_0(x) = 1 + n^{-1/2}\,h(x) + r_n'(x)$ with $r_n' = n^{-1/2}(h_n - h)$ satisfying (2.3) because of (3.8). Thus p_n' also satisfies the assumption (A) with the same function h, and (3.2) holds with some η_n' for L_n' defined as in (3.1). Then $L_n'(X^{(n)}) - L_n(X^{(n)}) = \eta_n'(X^{(n)}) - \eta_n(X^{(n)}) \to 0$ in P_{0n} − probability, i.e. the likelihood ratio of P_{1n}' and P_{1n} converges to 1 in P_{0n} − probability. Together with the contiguity of $\{P_{1n}\}$ and $\{P_{0n}\}$ this implies (3.9).

It follows from (3.9) that, for any sequence of tests, their powers against P_{1n} and P_{1n}' differ by $o(1)$. Hence we may prove the Theorems 1 and 2 with $\{p_n(x)\}$ replaced by $\{p_n'(x)\}$. From now on we omit the prime assuming that (3.5−8) are fulfilled for $p_n(x)$ itself.

It will be convenient also to make the probability integral transformation, i.e. to introduce the r.v.'s $U_i = F_0(x_i)$. Then ν_{nkj} is the number of U's in the interval $\Delta_{kj} = [(j-1)/k, j/k)$. The r.v.'s U_i are uniformly distributed on $[0, 1]$ under H_0 and have the density $p_n(F_0^{-1}(x))/p_0(F_0^{-1}(x))$ under H_{1n}. Not to complicate the notation, we shall assume that P_0 itself is uniform on $[0,1]$. Then, together with the above convention, we have

$$p_0(x) \equiv 1\,, \quad p_n(x) = 1 + n^{-1/2}\,h_n(x)\,, \quad x \in [0, 1]\,, \tag{3.10}$$

and $p_0(x) = p_n(x) = 0$ outside $[0, 1]$ where $h_n(x)$ satisfies (3.7), (3.8). Now $E_0 h = \int h(u)\,du$. The variable u will always take values in $[0, 1]$; when integrating over u from 0 to 1, the limits of integration will be omitted.

Concluding this section we mention some sufficient conditions for the assumption (A) to be fulfilled. Consider a family of densities $\{p(x, \theta), \theta \in \Theta\}$, where Θ is an open set on the real line containing $\theta = 0$, define $p_0(x) = p(x, 0)$, $p_n(x) = p(x, tn^{-1/2})$, t being a fixed number. Then the condition that $[p(X_1, \theta)/p(X_1, 0)]^{1/2}$ is differentiable at $\theta = 0$ in quadratic mean ([10], p. 46, assumptions (A1′−2′)) is sufficient for (A) and so is the condition that the Fisher's information is finite and continuous at $\theta = 0$ ([5], p. 189, assumptions (A1−3)). Note that in [5] and [10] these are proved to be sufficient for (3.2) but, as shown in [2], (2.3) is necessary for (3.2) provided (2.1) and (2.2) hold.

4. Proof of Theorem 1

To simplify the notation we shall usually suppress the dependence of $k(n)$ on n and of a_{kj}, ν_{nkj} etc. on k and n. By the arguments of Section 3 we assume that (3.10), (3.7), (3.8) hold. Let

$$\delta_n(j) = kn^{-1/2} \int_{\Delta_j} h_n(u)\,du\,, \quad p_{1j} = F_n(j/k) - F_n((j-1)/k)\,. \tag{4.1}$$

By (3.10) and (4.1), the probabilities of Δ_j under H_0 and H_{1n} are

$$p_{0j} = 1/k\,, \quad p_{1j} = (1 + \delta_n(j))/k \tag{4.2}$$

respectively.

Let $Q_l(A) = Q_{nkl}(A) = P_{ln}\{\nu_{nk} \in A\}$ for $A \subset M_{nk}$, $l = 0, 1$. Then for $x = (x_1, \ldots, x_k) \in M_{nk}$

$$Q_l(\{x\}) = \frac{n!}{x_1! \ldots x_k!} \prod p_{lj}^{x_j}\,, \quad l = 0, 1\,. \tag{4.3}$$

Let $r = (r_1, \ldots, r_k)$ be a permutation of $(1, \ldots, k)$ and R_k the set of all $k!$ such permutations. For $x \in M_{nk}$ and $r \in R_k$, let $rx = (x_{r_1}, \ldots, x_{r_k})$. Denote by $g(x)$, $x \in M_{nk}$, the number of $r \in R_k$ such that $rx = x$ (e.g. if $x_1, \ldots, x_k$ are different then $g(x) = 1$, if $x_1 = x_2$ and $x_2, \ldots, x_k$ are different then $g(x) = 2$ etc.). Let

$$q_l(x) = Q_l(\{x\})/g(x)\,, \quad x \in M_{nk}\,, \quad l = 0, 1\,, \tag{4.4}$$

$$M_0 = M_{0nk} = \{x \in M_{nk} : x_1 \leqq x_2 \leqq \ldots \leqq x_k\}\,. \tag{4.5}$$

Then for any permutation invariant function $\Psi(x)$, $x \in M_{nk}$,

$$\begin{aligned} E_{1n}\,\Psi(\nu) &= \sum_{M_{nk}} \Psi(x)\,g(x)\,q_1(x) = \sum_{M_0} \Psi(x) \sum_{r \in R_k} q_1(rx) \\ &= k! \sum_{M_0} \Psi(x)\,q_0(x) \left[(k!)^{-1} \sum_{r \in R_k} \frac{q_1(rx)}{q_0(rx)}\right], \end{aligned} \tag{4.6}$$

$$E_{0n}\,\Psi(\nu) = k! \sum_{M_0} \Psi(x)\,q_0(x)\,. \tag{4.7}$$

Let

$$U_n(x) = (k!)^{-1} \sum_{r \in R_k} \frac{q_1(rx)}{q_0(rx)}\,, \quad x \in M_{n,k(n)}\,. \tag{4.8}$$

Then if Ψ is a test function then

$$|E_{1n}\,\Psi(\nu) - E_{0n}\,\Psi(\nu)| \leqq k! \sum_{M_0} |U_n(x) - 1|\,q_0(x) = E_{0n}|U_n(\nu) - 1|\,. \tag{4.9}$$

Thus we have to prove that

$$E_{0n}|U_n(\nu) - 1| \to 0\,. \tag{4.10}$$

Denote by $S_n(x)$, $x \in M_{nk}$, the random variable $\log\,[q_1(Rx)/q_0(Rx)]$ where R is a random permutation taking its values on R_k with equal probabilities $(k!)^{-1}$. Then

$$U_n(x) = E[\exp\,(S_n(x))]\,. \tag{4.11}$$

Lemma 3. *There exist subsets* $B_n \subset M_{nk}$ *such that* $Q_0(B_n) \to 1$ *and for any sequence* $x_{(n)} \in B_n$ *the distribution of* $S_n(x_{(n)})$ *is asymptotically normal* $N(-\lambda/2, \lambda)$.

The *proof* will be given in Section 5.

Note that if a random variable Z is distributed normally $N(-\lambda/2, \lambda)$ then $Ee^Z = 1$. Together with Lemma 3 and (4.8) this implies that for any sequence $x_{(n)} \in B_n$

$$\liminf U_n(x_{(n)}) \geqq 1$$

([8], Theorem 11.4A(i)). Hence $\limsup \varepsilon_n \leqq 0$ where $\varepsilon_n = \max\left[1 - U_n(x);\, x \in B_n\right]$. Since $E_{0n}\, U_n(v) = 1$ we have

$$E_{0n}\left|U_n(v) - 1\right| = 2E_{0n}(1 - U_n(v))^+ \leqq 2[\varepsilon_n\, Q_0(B_n) + Q_0(M_{nk} \setminus B_n)]\,.$$

The last expression is non-negative and its lim sup is $\leqq 0$ hence it tends to 0. Thus (4.10) is proved.*)

5. Proof of Lemma 3

Let

$$A_n = \left\{x \in M_{nk} : \left|\frac{k}{n}\sum\left(x_j - \frac{n}{k}\right)^2 - (k-1)\right| \leqq k^{3/4}\right\}. \tag{5.1}$$

Since $E_{0n}\chi^2_{nk} = k - 1$, $\mathrm{var}_{0n}\chi^2_{nk} = 2(k-1)(1 - n^{-1})$ (see, e.g., [7], (30.47) and Exercise 30.5), we have by Chebyshev inequality that

$$1 - Q_0(A_n) \leqq 2(k-1)(1 - n^{-1})/k^{3/2} \to 0\,. \tag{5.2}$$

Now let $\varrho_n = [(3n + k)/kn]^{1/8}$ and

$$C_n = \left\{x \in M_{nk} : \max_j \left|x_j - \frac{n}{k}\right| \leqq \varrho_n n^{1/2}\right\}. \tag{5.3}$$

Using the facts that v_j has a binomial distribution with parameters $(n, 1/k)$ and that the 4th central moment, μ_4, of this distribution is ([6], (3.13))

$$\mu_4 = 3n^2p^2q^2 + npq(1 - 6pq)\,, \quad p = 1 - q = 1/k\,, \tag{5.4}$$

* (This note belongs to Chibisov.) The method of the above proof is similar to that used in [1]. Now we give a correction to the proof in [1]. Namely, in that proof a likelihood ratio g_n, is expressed as $E[\exp(T_n)]$ (cf. (4.11)) and an assertion similar to Lemma 3 is proved for T_n. Then it is stated that this implies $g_n(z^{(n)}) \to 1$ for $z^{(n)} \in B^{(n)}$ (a counterpart of B_n in Lemma 3) by properties of moment generating functions. The last argument is wrong. However, the proof can be completed using the same arguments as in the last paragraph of the present proof.

we obtain

$$(5.5)\qquad 1 - Q_0(C_n) \leqq kP_{0n}\left\{\left|v_1 - \frac{n}{k}\right| > \varrho_n n^{1/2}\right\} \leqq k\,\frac{\mu_4}{\varrho_n^4 n^2} \to 0\,.$$

Set $B_n = A_n \cap C_n$. Then (5.2) and (5.5) imply $Q_0(B_n) \to 1$. Let $\{x_{(n)}\}$ be a sequence such that $x_{(n)} \in B_n$. The subscript (n) of $x_{(n)}$ will be usually suppressed. It follows from (5.1) and (5.3) that such sequence has the following properties $(x = x_{(n)} = (x_1, \ldots, x_k))$:

$$(5.6)\qquad \frac{1}{n}\sum\left(x_j - \frac{n}{k}\right)^2 \to 1\,,$$

$$(5.7)\qquad \max_j\left(x_j - \frac{n}{k}\right)^2 \Big/ \sum\left(x_j - \frac{n}{k}\right)^2 \to 0\,.$$

Now we establish some properties of $\delta_n(j)$ defined by (4.1). First, (3.7) implies that

$$(5.8)\qquad \max_j|\delta_n(j)| \leqq d_n \to 0\,.$$

For a function $f(x)$, $x \in [0, 1]$, denote by $f^{(k)}(x)$ the step function taking the value $k\int_{\Delta_j} f(u)\,du$ for $x \in \Delta_j = [(j-1)/k, j/k]$. Then if $E_0 f^2 < \infty$ one has $E_0(f^{(k)})^2 \leqq E_0 f^2$ by Cauchy-Schwarz inequality and $E_0(f - f^{(k)})^2 \to 0$ as $k \to \infty$ (see, e.g., [4], Lemma V.1.6b). Note that

$$(5.9)\qquad h_n^{(k)}(x) = \sqrt{(n)}\,\delta_n(1 + [xk])\,,\quad x \in [0,1]\,.$$

Now (3.8) implies that

$$(5.10)\qquad E_0(h_n^{(k)} - h^{(k)})^2 = E_0[(h_n - h)^{(k)}]^2 \leqq E_0(h_n - h)^2 \to 0$$

hence

$$(5.11)\qquad E_0(h_n^{(k)} - h)^2 = \int[\sqrt{(n)}\,\delta_n(1 + [uk]) - h(u)]^2\,du \to 0\,,$$

$$(5.12)\qquad \lambda_n = \frac{n}{k}\sum\delta_n^2(j) = E_0(h_n^{(k)})^2 \to \lambda\,.$$

It follows from (4.2–4) that

$$(5.13)\qquad S_n(x) = \sum_{j=1}^{k} x_{R_j}\log(1 + \delta_n(j))\,.$$

Expanding the logarithm as

$$\log(1 + \delta) = \delta - \tfrac{1}{2}\delta^2 + \frac{1}{3}\frac{\delta^3}{(1 + \theta\delta)^3}\,,\quad 0 \leqq \theta \leqq 1\,,$$

we obtain $S_n(x) = S_1 - \frac{1}{2}S_2 + \frac{1}{3}S_3$ with

$$S_1 = \sum x_{R_j}\,\delta_n(j)\,, \quad S_2 = \sum x_{R_j}\,\delta_n^2(j)\,, \quad S_3 = \sum x_{R_j}\frac{\delta_n^3(j)}{(1+\theta_j\,\delta_n(j))^3}\,. \tag{5.14}$$

We shall show that (i) S_1 is asymptotically normal $N(0, \lambda)$, (ii) $S_2 \to \lambda$ and (iii) $S_3 \to 0$ in probability. Thus the lemma will be proved.

Note that S_1 has the same distribution as

$$S_1' = \sum x_j\,\delta_n(R_j) = \sum (n^{-1/2}x_j)\,(\sqrt{(n)}\,\delta_n(R_j))\,.$$

This is a simple linear rank statistic and its asymptotic normality follows from Hájek's theorem ([4], Theorem V.1. 6a). The conditions of this theorem on the regression constants $n^{-1/2}x_j$ and the scores $n^{1/2}\,\delta_n(j)$ are fulfilled by (5.7) and (5.11). Recall that generally, for a statistic $S = \sum c_j\,a(R_j)$,

$$ES = k\bar{c}\bar{a} = \frac{1}{k}\sum c_j \sum a(j)\,, \tag{5.15}$$

$$\operatorname{var} S = \frac{1}{k-1}\sum (c_j - \bar{c}) \sum (a(j) - \bar{a})^2 \tag{5.16}$$

([4], Theorem II.3.1b). In our case

$$\frac{1}{k}\sum (n^{1/2}\,\delta_n(j)) = \int h_n(u)\,du = 0\,,$$

hence $ES_1 = 0$, and (5.6), (5.12) imply that $\operatorname{var} S_1 \to \lambda$. Thus (i) is proved.

For S_2 we have from (5.12) and (5.15) that

$$ES_2 = \frac{1}{k}\,n\sum \delta_n^2(j) \to \lambda\,. \tag{5.17}$$

Next, (5.6), (5.8), (5.12), (5.16) and the inequality

$$\sum\left(\delta_n^2(j) - \frac{\lambda_n}{n}\right)^2 = \sum \delta_n^4(j) - k\left(\frac{\lambda_n}{n}\right)^2 \leqq d_n^2 \sum \delta_n^2(j) = d_n^2\,\frac{k\lambda_n}{n}$$

imply

$$\operatorname{var} S_2 = \frac{1}{k-1}\sum\left(x_j - \frac{n}{k}\right)^2 \sum\left(\delta_n^2(j) - \frac{\lambda_n}{n}\right)^2 \to 0\,. \tag{5.18}$$

Thus (ii) follows from (5.17), (5.18).

Finally, we have $|S_3| \leqq d_n(1 - d_n)^{-3}\,S_2$ which proves (iii). The proof of the lemma is completed.

6. Proof of Theorem 2

Denote by χ^2_{k-1} and $\chi^2_{k-1}(\lambda)$ random variables having the chi-square distributions with $k-1$ degrees of freedom, central and non-central with non-centrality parameter λ respectively. It is well-known ([7], 30.27) that for a fixed k, χ^2_{nk} is distributed asymptotically as χ^2_{k-1} and $\chi^2_{k-1}(\lambda^{(k)})$ under H_0 and H_{1n} respectively, where (see (4.2), (5.9), (5.10))

$$\lambda^{(k)} = \lim_{n\to\infty} n \sum \frac{(p_{1j}-p_{0j})^2}{p_{0j}} = \lim_{n\to\infty} \frac{n}{k} \sum \delta_n^2(j) = E_0(h^{(k)})^2 .$$

Note that the asymptotic distribution of χ^2_{nk} might be derived from Lemma 2. Namely, one should take $g_j(x) = 1_{\Delta_j}(x) - (1/k)$, $x \in [0,1]$ where $1_\Delta(x)$ is the indicator function of the set Δ. Then Lemma 2 implies the asymptotic normality of the vector $\{v_j - (n/k)\}$ under H_0 and H_{1n} with the asymptotic means 0 and $\{E_0 1_{\Delta_j} h\}$ and equal asymptotic covariance matrices, from which the above convergence is derived.

It follows now that C_{nk} tends to C_k defined by $P\{\chi^2_{k-1} > C_k\} = \alpha$ and

$$\beta_{nk} \to \beta_k = P\{\chi^2_{k-1}(\lambda^{(k)}) > C_k\} \tag{6.1}$$

for each k. It is well known that $\beta_k > \alpha$ provided $\lambda^{(k)} > 0$. Take an arbitrary k_0 such that $\lambda^{(k_0)} > 0$ (it exists since $\lambda^{(k)} \to \lambda > 0$ as $k \to \infty$) and a number β, $\alpha < \beta < \beta_{k_0}$. It follows from Corollary 1 and the assumption (ii) that there exist n_1, k_1 such that

$$\sup\,[\beta_{nk};\ n > n_1,\ k > k_1] \leqq \beta . \tag{6.2}$$

A passage to the limit for each k gives $\sup\,[\beta_k;\ k \geqq k_1] \leqq \beta$. Since $\beta < \beta_{k_0}$ this implies that the max β_k is attained at some $k = k^*$ with $2 \leqq k^* \leqq k_1$. Now $\beta_{nk^*} \to \beta_{k^*}$ and $\max\,[\beta_{nk};\ 2 \leqq k \leqq k_1] \to \beta_{k^*}$ by (6.1), and, for sufficiently large n, $\sup\,[\beta_{nk};\ k \geqq 2] = \max\,[\beta_{nk};\ 2 \leqq k \leqq k_1]$ by (6.2). Thus the assertion of the theorem follows.

References

[1] Chibisov, D. M. (1961). On tests of fit based on spacings. *Teor. Verojatnost. i Primenen.*, **6**, 3, 354—358.

[2] Chibisov, D. M. (1969). Transition to the limiting process for deriving asymptocially optimal tests. *Sankhyā* A, **31**, 3, 241—258.

[3] Dahya, R. C. - Gurland, J. (1973). How many classes in the Pearson chi-square test? *Journ. Amer. Statist. Assoc.*, **68**, 343, 707—712.

[4] Hájek, J. - Šidák, Z. (1967). "Theory of rank tests". Prague.

[5] Hájek, J. (1972). Local asymptotic minimax and admissibility in estimation. *Proc. 6th Berkeley Symposium Math. Statist. and Prob.*, vol. 1, 175—194.

[6] KENDALL, M. G. - STUART, A. (1958). "*The advanced theory of statistics*". Vol. 1. Distribution theory, 2nd ed.
[7] KENDALL, M. G. - STUART, A. (1973). "The advanced theory of statistics". Vol. 2. Inference and relationship, 3rd ed.
[8] LOÈVE, M. (1960). "Probability theory". Van Nostrand.
[9] MANN, H. B. - WALD, A. (1942). On the choice of the number of class intervals in the application of the chi-square test. *Ann. Math. Statist.*, **13**, 306—317.
[10] ROUSSAS, G. (1972). "Contiguity of probability measures: Some applications in statistics". Cambridge.
[11] VIKTOROVA, I. I. - ČISTJAKOV, V. P. (1966). Some generalizations of the empty boxes test. *Teor. Verojatnost. i Primenen.*, **11**, 2, 306—313.

(1) INSTITUTE OF ECONOMICS AND LAW, TBILISI, U.S.S.R.
(2) STEKLOV MATHEMATICAL INSTITUTE, MOSCOW, U.S.S.R.

Received December 1975

LOWER BOUND FOR THE RISKS OF NONPARAMETRIC ESTIMATES OF THE MODE

by

R. Z. HASMINSKII

0. Introduction

In his paper of fundamental importance [1], Hájek obtained asymptotically minimax lower bound for the risks of the parametric estimates for a wide class of loss functions. Levit [2] showed, that these ideas can be successfully employed also for obtaining asymptotically precise lower bounds for the risks of nonparametric estimates for a wide class of functionals. However, the nonparametric analogue of Fisher information evaluated accordingly [2], equals to zero in many interesting cases. It is so, e.g., for nonparametric estimates of a density at some point, of its derivatives, for estimates of the mode and so on.

Minimax lower bounds for these "degenerated" (from the point of view of [2]) problems were obtained firstly by Čentsov [3] and Farrell [4]. The problem of estimation of the density was considered there under various assumptions on the distribution. In the present paper, we obtain analogous bounds in the case of estimation of the mode. Our main goal is to show that Hájek's ideas and results can be fruitfully utilized for problem of that sort.

1. A lemma

In this paper, the main idea of obtaining lower bounds consists in looking for a parametric family in the nonparametric set of densities, that is (in some sense) most difficult to estimate (cf. [2]). Thus we reduce our problem to the problem of parametric estimation. The latter is then solved by the help of the following modification of Theorem 4.1 from [1].

Denote by $W = \{w\}$ the class of loss functions, which are defined on R^1 and have properties: $w(0) = 0$, $w(-x) = w(x)$, $w(x)$ increases for $x > 0$.

Lemma 1.1. *Let* $(\mathscr{X}, \mathscr{A})$ *be a measurable space, let* $f_n(x, \Theta)$, $x \in \mathscr{X}$, $|\Theta| < 1$, $n = 1, 2, 3, \ldots$ *be densities with respect to σ-finite measures* ν_n, *defined on* $\mathscr{A}$. *Let* X_{jn}, $j = \overline{1, n}$ *be a triangular array of observations, and assume for every n the observa-*

tions $X_{1n}, \ldots, X_{nn}$ to be independent and have common density $f_n(x, \Theta)$. Assume that this array satisfies conditions of local asymptotical normality (LAN) at $\Theta = 0$ for $n \to \infty$ with a constant (not depending of n) normalizing factor $c_0 > 0$, i.e., that the equality

$$\sum_{j=1}^{n} \ln \frac{f_n(X_{jn}, c_0 u)}{f_n(X_{jn}, 0)} = u\,\Delta_n - \tfrac{1}{2}u^2 + \delta_n(u) \tag{1.1}$$

holds true for $|u| < c_0^{-1}$ with

$$\Delta_n \to \mathcal{N}(0, 1), \quad \delta_n(u) \to 0 \tag{1.2}$$

in distribution $P_0^{(n)}$.*

Then the inequality

$$\varliminf_{n\to\infty} \sup_{|\Theta| < CC_0^{-1}} E_\Theta^{(n)} w[(T_n - \Theta)\,c_0] \geqq (2\pi)^{-1/2} \int_{|y|<b} w(y) \exp(-y^2/2)\,dy\,(1 - bc^{-1}) \tag{1.3}$$

holds true for any estimator T_n of parameter Θ, based on the sample $X_{jn}, j = \overline{1, n}$, any $w \in W$ and any constants c and b, satisfying the inequalities $0 < b < c \leqq c_0$.

The proof of this lemma follows the main lines of the proof of Theorem 4.1 from [1]. First of all, the inequality

$$\sup_{|\Theta| < C_0^{-1}C} E_\Theta^{(n)} w[(T_n - \Theta)\,c_0] \geqq (2c)^{-1} \int_{-C}^{C} E_{C_0^{-1}h}\, w(c_0 T_n - h)\,dh$$

is evident. Further, the measure $P_{C_0^{-1}h}^{(n)}(\cdot)$ is close to the measure

$$Q_h^{(n)}(A) = \int_A \exp(\hat{\Delta}_n u - u^2/2)\,dP_0^{(n)}$$

(here $\hat{\Delta}_n$ is the corresponding truncation of the random variable Δ_n, so that $\hat{\Delta}_n - \Delta_n \to 0$ in probability $P_0^{(n)}$) in the sense that

$$\operatorname{var} \left| P_{C_0^{-1}h}^{(n)}(\cdot) - Q_h^{(n)}(\cdot) \right| \to 0 \quad (n \to \infty) \tag{1.4}$$

if LAN condition is satisfied (see [5], [6]).

* Here and in the sequel, $P_\Theta^{(n)}(\cdot)$ denotes the measure generated by the sample $X_{jn}, j = \overline{1, n}$, with the parameter value Θ; the symbol $E_\Theta^{(n)}(\cdot)$ is used to denote the expectation with respect to this measure.

Denote $w_a = \min(w, a)$. Then for any $a > 0$ the inequalities

$$\frac{1}{2c}\int_{-c}^{c} E^{(n)}_{c_0^{-1}h}\, w_a(c_0 T_n - h)\,dh = \frac{1}{2c}\int_{-c}^{c} E_0^{(n)}\{w_a(c_0 T_n - h)\exp[\hat{\Delta}_n h - h^2/2]\,dh + o(1)$$

$$\geqq \frac{1}{2c}\,E_0^{(n)}\left\{\exp(\hat{\Delta}_n^2/2)\int_{-c+\hat{\Delta}_n}^{c+\hat{\Delta}_n} w_a(c_0 T_n - \hat{\Delta}_n - y)\exp(-y^2/2)\,dy\right\} + o(1)$$

$$\geqq \frac{1}{2c}\,E_0^{(n)}\left\{\chi(|\hat{\Delta}_n| < c - b)\exp(\hat{\Delta}_n^2/2)\int_{-c+\hat{\Delta}_n}^{c+\hat{\Delta}_n} w_a(c_0 T_n - \hat{\Delta}_n - y)\exp(-y^2/2)\,dy\right\} +$$

$$+ o(1) \geqq \frac{1}{2c}\,E_0^{(n)}\left\{\chi(|\hat{\Delta}_n| < c - b)\exp(\hat{\Delta}_n^2/2)\int_{-b}^{b} w_a(c_0 T_n - \hat{\Delta}_n - y)\right.$$

$$\left. .\exp(-y^2/2)\,dy\right\} + o(1)$$

hold when $n \to \infty$. The integral on the right side of the latter inequality can be bounded from below by the help of Anderson's lemma [7] (bearing in mind that $w_a(x) \in W$ for any $a > 0$):

$$\int_{-b}^{b} w_a(c_0 T_n - \hat{\Delta}_n - y)\exp(-y^2/2)\,dy \geqq \int_{-b}^{b} w_a(y)\exp(-y^2/2)\,dy = \Psi(a, b)\,.$$

Therefore

$$\sup_{|\Theta| < c_0^{-1}c} E_\Theta^{(n)} w[(T_n - \Theta)\,c_0] \geqq \Psi(a, b)\,\frac{1}{2c}\,E_0^{(n)}\{\chi(|\hat{\Delta}_n| < c - b)\exp(\hat{\Delta}_n^2/2)\} + o(1)$$

as $n \to \infty$. The equation (1.3) follows from the latter inequality by the help of Fatou lemma, if one takes into account, that $\hat{\Delta}_n$ is asymptotically normal with parameters 0, 1 and the constant a is arbitrary.

2. A lower bound for the risks

In papers [8], [9], [10] and others, nonparametric estimates for the mode μ_f of a distribution with the density f were proposed, based on a sample $X_1, \ldots, X_n$ from a population with that density. Particularly, when it is known, that f is twice differentiable in a neighbourhood of the mode and $f''(\mu_f) < 0$, estimators μ_n, for which

$$0 < \lim_{n\to\infty} P_f\{n^{1/5}|\mu_n - \mu_f| < K\} < 1 \tag{2.1}$$

were constructed in [10]. (Here and in the sequel, $P_f(\cdot)$, $E_f(\cdot)$ denote probability and expectation in the case, when the population density is f.) The question arises, whether it is possible to obtain estimates μ_n, which converge to μ_f faster, i.e., estimates, which satisfy for some (or any) $K_0 < \infty$ equality

$$\overline{\lim_{n\to\infty}} P_f\{n^{1/5}|\mu_n - \mu_f| < K_0\} = 1 .$$

Certainly it is easy to find estimators, with convergence rate $n^{-1/2}$ for particular parametric families. However, as we shall see later, it is impossible to find any improvement of that kind for the class of all functions twice differentiable at the point μ_f.

The density $f(x)$, $x \in R^1$, is called, as usual, unimodal, if there is a unique point μ_f (mode) such that $f(\mu_f) = \sup f(x)$. Let $\mathscr{F}$ be the family of unimodal densities $f(x)$, twice differentiable in a neighbourhood of the mode, the function $f''(x)$ being bounded in this neighbourhood and $f''(\mu_f) < 0$. Let us define the ε-neighbourhood of the density $f \in \mathscr{F}$ as the set of densities $\varphi(x)$, coinciding with f for $|x - \mu_f| > \varepsilon$ and satisfying the inequality

$$\sup_{|x-\mu_f|\leq\varepsilon} (|\varphi(x) - f(x)| + |\varphi'(x) - f'(x)|) < \varepsilon .$$

Denote this set by $U_\varepsilon(f)$. Denote also by $\mathscr{M}_n$ the class of all estimates of μ_f, based on $X_1, \ldots, X_n$ i.e., which are measurable functions of $X_1, \ldots, X_n$.

Theorem 2.1. *The inequality*

$$\underline{\lim_{n\to\infty}} \inf_{\mu_n\in\mathscr{M}_n} \sup_{f\in\mathscr{F}\cap U_\varepsilon(f_0)} E_f w[(\mu_n - \mu_f)\, n^{1/5}] > 0 \tag{2.2}$$

holds true for any density $f_0 \in \mathscr{F}$, for every sufficiently small $\varepsilon > 0$ and an arbitrary loss function $w \in W$.

Corollary. *Letting $w(x) = \chi(|x| < k)$ ($\chi(A)$ is indicator of the set A) one obtains from (2.2) for any $k > 0$ the inequality*

$$\overline{\lim_{n\to\infty}} \sup_{\mu_n\in\mathscr{M}_n} \inf_{f\in\mathscr{F}\cap U_\varepsilon(f_0)} P_f\{|\mu_n - \mu_f|\, n^{1/5} < k\} < 1 .$$

Comparing this inequality with (2.1) we establish that the estimates of μ_f in the paper [10] are the best minimax estimates with respect to the rate of convergence.

3. The proof of Theorem 2.1

Assume that $f_0 \in \mathscr{F}$. Denote $f_0''(\mu_{f_0}) = -a < 0$ and assume that the function $f_0''(x)$ is bounded for $|x - \mu_{f_0}| < \varepsilon_0$. Let us consider an auxiliary function $g(x)$, $x \in R^1$ with the following properties:

$$g(x) = x \quad \text{if} \quad |x| < 1/a , \tag{3.1}$$

(3.2) $$|g''(x)| < a/2\,, \quad x \in R^1\,,$$

(3.3) $$g(-x) = -g(x)$$

and

(3.4) $$g(x) \equiv 0$$

for some $k > 0$.

Let $f_n(x, \Theta)$, $|\Theta| < 1$ be the parametrical family of functions defined by the equation

(3.5) $$f_n(x, \Theta) = f_0(x) + \Theta n^{-2/5} g((x - \mu_{f_0})\, n^{1/5})\,.$$

Let us investigate the properties of this family, putting $\mu_{f_0} = 0$ without a loss of generality.

Lemma 3.1. *The function $f_n(x, \Theta)$ is a unimodal density for $n \geqq n_0$, with bounded first and second order derivatives for $|x - \mu_{f_n}| < \varepsilon_0/2$. The equation*

(3.6) $$\mu_{f_n} = \Theta a^{-1} n^{-1/5} + o(n^{-1/5}) \quad (n \to \infty)$$

holds true for the mode of this density. Finally, the inclusion $f_n(x, \Theta) \in U_\varepsilon(f_0)$ is true for any $\varepsilon > 0$ and $n \geqq n_0(\varepsilon)$.

Proof. It is clear from (3.3)-(3.5), that f_n is a density for $n \geqq n_0$. The equations (3.2), (3.5) imply the boundedness of the second derivative $f_0''(x, \Theta)$ for $|x| < \varepsilon_0$. Let us notice further, that the equations

$$f_n(x, \Theta) = f_0(x) \quad \text{for} \quad |x| > Kn^{-1/5}\,,$$
$$f_n(0, 0) = f_0(0)$$

imply the statement, that all modes of f_n belong to interval $|x| < Kn^{-1/5}$. The following equation is true in this interval

(3.7) $$\begin{aligned} f_n'(x, \Theta) &= f_0'(x) + \Theta n^{-1/5}\, g'(xn^{1/5}) \\ &= -ax + \Theta n^{-1/5}\, g'(xn^{1/5}) + o(n^{-1/5})\,. \end{aligned}$$

The equation $f_n'(x, \Theta) = 0$ has a single root (see (3.1)) for $|x| < a^{-1}n^{-1/5}$. This root satisfies (3.6). There are no other roots in the interval $|x| < Kn^{-1/5}$ in view of the condition (3.2). The inclusion $f_n(x, \Theta) \in U_\varepsilon(f_0)$ is a consequence of (3.5), (3.7). The lemma is proved.

Lemma 3.2. *The family $f_n(x, \Theta)$ satisfies the conditions* LAN (1.1), (1.2) *for* $\Theta = 0$; *the normalizing factor c_0 is determined by*

$$c_0 = (\max f_0(x))^{1/2} \left(\int g^2(x)\,dx \right)^{-1/2}.$$

Proof. Introduce the notation

$$\Psi_n^2 = n \int \frac{\left[\frac{\partial}{\partial\Theta} f_n(x, \Theta)\right]^2}{f_n(x, \Theta)}\,dx = n^{1/5} \int \frac{g^2(xn^{1/5})}{f_n(x, \Theta)}\,dx\,.$$

The equation $\Psi_n^2 = c_0^{-1} + o(1)$ $(n \to \infty)$ is evident. The lemma will be proved, if the following conditions of the Theorem 1.2 from [6] are verified: 1) the equation

$$\lim_{n\to\infty} \int_{A_n(\varepsilon)} \left|\frac{\partial \ln f_n(x, \Theta)}{\partial\Theta}\right|^2 f_n(x, \Theta)\,dx = 0$$

holds true for every $\varepsilon > 0$, where

$$A_n(\varepsilon) = \left\{ x : \left|\frac{\partial \ln f_n(x, \Theta)}{\partial\Theta}\right| > \varepsilon c_0^{-1/2} \right\}.$$

2)

$$\lim_{n\to\infty} \sup_{|\Theta|<1} n \int \left[\frac{\partial}{\partial\Theta} f_n^{1/2}(x, \Theta) - \frac{\partial}{\partial\Theta} f_n^{1/2}(x, 0)\right]^2 dx = 0\,.$$

The fulfilment of the condiction 1) follows from the fact, that the set $A_n(\varepsilon)$ is empty for $n \geqq n_0(\varepsilon)$. Further

$$n \int \left[\frac{\partial}{\partial\Theta} f_n^{1/2}(x, \Theta) - \frac{\partial}{\partial\Theta} f_n^{1/2}(x, 0)\right]^2 dx$$

$$\leqq cn^{1/5} \int g^2(xn^{1/5}) \left[(f_0(x) + \Theta n^{-2/5} g(xn^{1/5}))^{1/2} - f_0^{1/2}(x)\right]^2 dx \to 0$$

as $h \to \infty$. The lemma is proved.

Now one can complete the proof of Theorem 2.1. Lemma 3.1 implies the inequality

$$\sup_{f\in\mathcal{F}\cap U_\varepsilon(f_0)} E_f w[(\mu_n - \mu_f)\, n^{1/5}] \geqq \sup_{|\Theta|<1} E_{f_n(x,\Theta)} w[(\mu_n - \mu_f)\, n^{1/5}] \tag{3.8}$$

for $n \geqq n_0$, $\mu_n \in \mathcal{M}_n$.

Taking an arbitrary estimate μ_n of the mode of the distribution in the parametric family $f_n(x, \Theta)$, we can construct an estimate Θ_n of the parameter Θ according to the formula $\Theta_n = a\, n^{1/5}\mu_n$. The equations (3.6), (3.8) imply for any estimate μ_n and any $c \leqq c_0$ the inequality

$$\sup_{f\in\mathscr{F}\cap U_\varepsilon(f_0)} E_f w[(\mu_n - \mu_f)\, n^{1/5}] \geqq \sup_{|\Theta|<c_0^{-1}c} E_\Theta^{(n)} w[a^{-1}(\Theta_n - \Theta) + o(1)]\,, \tag{3.9}$$

Let us introduce now a new loss function $\tilde{w}(x)$, defined by

$$\tilde{w}(x) = \begin{cases} w(x/(2ac_0))\,, & \text{if } |x| > 1\,, \\ 0\,, & \text{if } |x| \leqq 1\,. \end{cases}$$

It is obvious, that for $n \geqq n_0$ the inequality

$$E_\Theta^{(n)} w[a^{-1}(\Theta_n - \Theta) + o(1)] \geqq E_\Theta^{(n)} \tilde{w}((\Theta_n - \Theta)\, c_0) \tag{3.10}$$

is true. Lemma 1.1 can now be used to obtain a lower bound to the least upper bound over $|\Theta| < c_0^{-1}c$ of the right side of inequality (3.10) (Lemma 3.2 guarantees the fulfilment of the conditions of Lemma 1.1). Choosing, for example, $c = c_0$, $b = c_0/2$ and recalling, that the estimate μ_n is arbitrary, one obtains

$$\varliminf_{n\to\infty} \inf_{\mu_n\in\mathscr{M}_n} \sup_{f\in\mathscr{F}\cap U_\varepsilon(f_0)} E_f w[(\mu_n - \mu_f)\, n^{1/5}] \geqq \tfrac{1}{2}(2\pi)^{-1/2} \int_{|x|<c_0/2} \tilde{w}(y) \exp(-y^2/2)\, \mathrm{d}y\,.$$

The theorem is proved.

References

[1] HÁJEK, J. (1972). Local asymptotic minimax and admissibility in estimation. *Proc. 6th Berkeley Symp. Math. Stat. and Probab.*, Univ. Calif. Press, 175—194.

[2] LEVIT, B. (1973). On optimality of some statistical estimates. *Proc. Prague Symp. on Asymptotic Statistics*, vol. 2, 215—238.

[3] ЧЕНЦОВ, Н. Н. (1972). „Статистические решаюшие правила и оптимальные выводы" Наука, § 25.

[4] Farrell, R. H. (1972). On best obtainable asymptotic rates of convergence in estimation of a density function at a point. *Ann. Math. Statist.*, **43**, 1, 170—180.

[5] LE CAM, L. (1960). Locally asymptotically normal families of distributions. *Univ. Calif. Publ. Statist.*, **3**, 37—98.

[6] HÁJEK, J. (1970). A characterization of limiting distributions of regular estimates. *Z. Wahrscheinlichkeitshteorie verw. Geb.* **14**, 323—330.

[7] ANDERSON, J. T. (1966). The integral of a symmetrical unimodal function. *Proc. Amer. Math. Soc.*, **6**, 1, 170—176.

[8] CHERNOFF, H. (1964). Estimation of the mode. *Ann. Inst. Statist. Math.*, 16, 31—41.

[9] GRENANDER, U. (1965). Some direct estimates of the mode. *Ann. Math. Statist.*, **36**, 131—138.

[10] VENTER, J. (1967). On estimation of the mode. *Ann. Math. Statist.*, **38**, 1446—1455.

INSTITUTE OF PROBLEMS OF INFORMATION TRANSMISSION, MOSCOW, U.S.S.R.

Received December 1975

THE RATE OF CONVERGENCE OF SIMPLE LINEAR RANK STATISTICS UNDER ALTERNATIVES

by

MARIE HUŠKOVÁ

1. Introduction

Let $X_1, \ldots, X_N$ be independent random variables with continuous distribution functions $F_1, \ldots, F_N$. Let R_i be the rank of X_i, $1 \leqq i \leqq N$. The rate of the convergence of simple linear rank statistics

(1) $$S_c = \sum_{i=1}^{N} c_i\, a_N(R_i)$$

is investigated under very mild conditions imposed on the distribution of $X_1, \ldots, X_N$. Here $c_1, \ldots, c_N$ are regression constants, $a_N(1), \ldots, a_N(N)$ denote the scores given in either of the following way:

(2) $$a_N(i) = \varphi\left(\frac{i}{N+1}\right), \quad 1 \leqq i \leqq N,$$

(3) $$a_N(i) = E\,\varphi(U_N^{(i)}), \quad 1 \leqq i \leqq N,$$

where $U_N^{(i)}$ denotes the i-th order statistic in a sample of size N from the uniform distribution on $(0, 1)$. The score-generating function φ is assumed to satisfy:

(4) φ being a nonconstant function defined on $\langle 0, 1\rangle$ with absolutely continuous second derivative and square-integrable third derivative on $\langle 0, 1\rangle$.

Recently, several papers treating the rate of convergence of simple linear rank statistics under hypothesis or near alternatives have appeared (see Hušková (1977) for a review). As for general alternatives, Bergström and Puri (1977) showed under conditions similar to those considered here, that the rate is

$$\left(\sum_{v=1}^{N} |c_v|^3\right)^{1/3} (\operatorname{var} S_c)^{-2/3} N^{-1/3} \left[\max_{1 \leqq i \leqq N} |c_i| (\operatorname{var} S_c)^{-1/2}\right]^{2r/(2r+1)}$$

which differs from that of the "nearest" (in the quadratic mean) sum of independent random variables (which is $\sum_{v=1}^{N} |c_v|^3 (\text{var } S_c)^{-3/2}$). The purpose of this paper is to prove that the rate of convergence of S_c and of the "nearest" of independent random variables is the same.

To prove main result we use Hájek's projection methods (to find the "nearest" sum of independent random variables) and Bjerve's methods (to treat the difference of characteristic function). The making use of the second method is similar to that in Hušková (1977).

The main results of the present paper are the following:

Theorem A. *Let* $X_1, \ldots, X_N$ *be independent random variables with continuous distribution functions, assumption* (4) *be satisfied, the scores be given in either* (2) *or* (3). *Then there exists a constant* A_1 *(not depending on* N *and* $c_1, \ldots, c_N$*) such that*

$$\sup_x \left|P(S_c - ES_c < x(\text{var } S_c)^{1/2}) - \Phi(x)\right| \leqq A_1 \sum_{i=1}^{N} |c_i|^3 (\text{var } S_c)^{-3/2}$$

where Φ *is the normal distributions* (0, 1). *The assertion remains true if we replace* var S_c *by* var S_c^0 *given by* (10).

Corollary. *Let the assumptions of Theorem A be satisfied, let*

$$\sum_{i=1}^{N} c_i = 0\,, \quad N \geqq k_2 \left[\int_0^1 (\varphi(u) - \bar{\varphi})^2 \, du\right]^{-1}$$

and

$$\max_{\substack{1 \leqq i,j \leqq N \\ x}} \left|F_i(x) - F_j(x)\right| \leqq (4(2k_1 + k_2))^{-1} \int_0^1 (\varphi(u) - \bar{\varphi})^2 \, du\,,$$

where $\bar{\varphi} = \int_0^1 \varphi(u)\, du$ *and* k_1, k_2 *is given by* (9) *below. Then there exists a constant* A_2 *(not depending on* N *and* $c_1, \ldots, c_N$*) such that*

$$\sup_x \left|P(S_c - ES_c < x(\text{var } S_c)^{1/2}) - \Phi(x)\right| \leqq A_2 \sum_{i=1}^{N} |c_i|^3 \left(\sum_{i=1}^{N} c_i^2\right)^{3/2}.$$

Using the approach of Jurečková (1973), Lemma 1 and assuming that

(5) the score-generating function φ is nonconstant defined on (0, 1) with a bounded second derivative,

one can prove the following weaker assertion:

Theorem B. *Let* $X_1, \ldots, X_N$ *be independent random variables with continuous distribution functions, assumptions* (4) *and* (5) *be satisfied and the scores be given* *by either* (2) *or* (3). *Then there exists a constant* A_3 (*not depending on* N *and* $c_1, \ldots, c_N$) *such that*

$$\sup_x \left|P(S_c - ES_c < x(\operatorname{var} S_c)^{1/2}) - \Phi(x)\right| \leqq A_3 \sum_{i=1}^{N} |c_i|^{3+\delta} (\operatorname{var} S_c)^{-(3+\delta)/2}$$

where $\delta > 0$ *is arbitrary.*

2. Some lemmas

First, for statistics with certain properties (more general then that we need for our purposes), the upper bound for the $2k$-th moment will be derived:

Lemma 1. *Let* $\{Z_j\}_{j=1}^N$ *be a sequence of random variables of the form* $Z_j = c_j g_j(X_1, \ldots, X_j)$, *where* $\{X_i\}_{i=1}^N$ *are independent random variables, and such that*

$$E(Z_j \mid X_1, \ldots, X_{j-1}) = 0 . \tag{6}$$

Let $V_N = \sum_{j=1}^{N} Z_j$ *and* $m_{N,k} = \max_{1 \leqq j \leqq N} E(g_j(X_1, \ldots, X_j))^{2k}$, $k \geqq 1$. *Then for* $k \leqq N$

$$EV_N^{2k} \leqq k^k \Big(\sum_{j=1}^{N} c_j^2\Big)^k (4e)^k\, m_{N,k} , \tag{7}$$

where $c_1 \geqq c_2 \geqq \ldots \geqq c_N$.

Proof: First, we shall prove by induction on l for k fixed that

$$E\Big(\sum_{j=1}^{l} Z_j\Big)^{2k} \leqq \Big(\sum_{j=1}^{l} |c_j|\Big)^{2k} m_{l,k} , \quad l \leqq k . \tag{8}$$

For $l = 2$ we have

$$E\Big(\sum_{j=1}^{2} Z_j\Big)^{2k} = \sum_{j=0}^{2k} \binom{2k}{j} EZ_1^j Z_2^{2k-j} \leqq (|c_1| + |c_2|)^{2k} m_{2,k} .$$

Assume that (8) is true for all l', $2 \leqq l' \leqq l$, then

$$E\Big(\sum_{j=1}^{l+1} Z_j\Big)^{2k} = \sum_{v=0}^{2k} \binom{2k}{v} E\Big(\sum_{j=1}^{l} Z_j\Big)^{2k-v} Z_{l+1}^v$$

$$\leqq \sum_{v=0}^{2k} \binom{2k}{v} \Big(\sum_{j=1}^{l} |c_j|^{2k-v} |c_{l+1}|^v m_{l+1,k}\Big) = \Big[\sum_{j=1}^{l+1} |c_j|\Big]^{2k} m_{l+1,k} .$$

Thus (8) is true. In particular,

$$E\Big(\sum_{j=1}^{k} Z_j\Big)^{2k} \leqq \Big(\sum_{j=1}^{k} |c_j|\Big)^{2k} m_{N,k} \leqq \Big(k \sum_{i=1}^{k} c_i^2\Big)^k m_{N,k} .$$

Now we can follow the proof of Lemma 6.1 in Bickel (1974). Assuming that (7) is true for $N = l$ we prove it for $N = l + 1$

$$EV_{l+1}^{2k} = E(\sum_{j=1}^{l+1} Z_j)^{2k} = \sum_{v=0}^{2k} \binom{2k}{v} E(\sum_{j=1}^{l} Z_j)^{2k-v} Z_{l+1}^v$$

$$= E(\sum_{j=1}^{l} Z_j)^{2k} + \sum_{v=2}^{2k} \binom{2k}{v} (E(\sum_{j=1}^{l} Z_j)^{2k})^{(2k-v)/2k} (EZ_{l+1}^{2k})^{v/2k}$$

$$\leqq k^k(\sum_{j=1}^{l} c_j^2)^k (4e)^k \left[1 + \frac{c_{l+1}\, 4k^2}{4ek \sum_{i=1}^{l} c_j^2} \left(1 + \frac{|c_{l+1}|}{(4ek \sum_{i=1}^{l} c_i^2)^{1/2}}\right)^{2k-2}\right] m_{l+1,k}$$

$$\leqq k^k(\sum_{j=1}^{l} c_j^2)^k (4ek)^k [1 + c_{l+1}^2 k/(\sum_{v=1}^{l} c_v^2)]\, m_{l+1,k}$$

$$\leqq (4ek)^k (\sum_{i=1}^{l+1} c_i^2)^k m_{l+1,k} . \quad \text{Q.E.D.}$$

Denote

(9) $$k_i = \sup_{u\in(0,1)} |\varphi^{(i)}(u)| , \quad i = 0, 1, 2 .$$

(10) $$S_c^0 = \sum_{j=1}^{N} c_j \left[\varphi\left(\frac{E(R_j \mid X_j)}{N+1}\right) - E\varphi\left(\frac{E(R_j \mid X_j)}{N+1}\right) + (N+1)^{-1}\right.$$

$$\left. \cdot \sum_{\substack{i=1\\ j\neq i}}^{N} c_i \int (u(x - X_j) - F_j(x))\, \varphi'\left(\frac{E(R_i \mid X_i = x)}{N+1}\right) dF_i(x)\right],$$

where $u(x) = 1$, if $x \geqq 0$, $u(x) = 0$, if $x < 0$.

In this section we shall assume that

(11) $$\sum_{j=1}^{N} |c_i|^3 (\operatorname{var} S_c^0)^{-3/2} \leqq 1 .$$

Lemma 2. *Let $X_1, \ldots, X_N$ be independent random variables with continuous distribution functions. Then under assumptions (4) and (11)*

$$P(|\sum_{i=1}^{N} c_i(a_N(R_i) - \varphi(R_i/(N+1))| \geqq (\operatorname{var} S_c^0)^{-1} \sum_{j=1}^{N} |c_j|^3)$$

$$\leqq \sum_{j=1}^{N} |c_j|^3 (\operatorname{var} S_c^0)^{-3/2} .$$

where $a_N(i)$ is given by (3).

Proof: By some elementary inequalities and the Taylor expansion we have

$$E\{\sum_{i=1}^{N} c_i[a_N(R_i) - \varphi(R_i/(N+1)]\}^{2k} \leqq (\sum_{i=1}^{N} |c_i|)^{2k},$$

$$\max_{1 \leqq i \leqq N} [\sum_{j=1}^{N} \operatorname{var} U_N^{(j)} k_2 P(R_i = j)/2]^{2k} \leqq [k_2 (\operatorname{var} S_c^0)^{-1} \sum_{j=1}^{N} |c_j|^3]^{2k} (k_1^2 + k_2^2)^{2k}.$$

The proof can be easily conclude choosing

$$2k \geqq \log(2k_2(k_1^2 + k_2^2)) \log\{\sum_{j=1}^{N} |c_j|^3 (\operatorname{var} S_c^0)^{-3/2}\}^{-1}. \quad \text{Q.E.D.}$$

Using the Taylor expansion for φ the statistic S_c with scores given by (3) can be rewritten as follows:

$$(S_c - ES_c)(\operatorname{var} S_c^0)^{-1/2} = S_c^* + S_c^{**} + S_c^{***} - ES_c^{***}, \tag{12}$$

where

$$S_c^* = \sum_{i=1}^{N} c_i^* \{\varphi(E(R_i \mid X_i)/(N+1)) - E\varphi(E(R_i \mid X_i)/(N+1)) \tag{13}$$
$$+ (R_i - E(R_i \mid X_i))\, \varphi'(E(R_i \mid X_i)/(N+1))(N+1)^{-1}\},$$

$$S_c^{**} = \tfrac{1}{2} \sum_{i=1}^{N} c_i^* \{[(R_i - E(R_i \mid X_i))/(N+1)]^2 \varphi''(E(R_i \mid X_i)/(N+1)) \tag{14}$$
$$- E[(R_i - E(R_i \mid X_i))/(N+1)]^2 \varphi''(E(R_i \mid X_i)/(N+1))\},$$

$$S_c^{***} = \tfrac{1}{2} \sum_{i=1}^{N} c_i^* ((R_i - E(R_i \mid X_i))^2 (N+1)^{-2} \tag{15}$$
$$\cdot \left\{\int_0^1 2(1-\lambda)\, \varphi''\left(\frac{E(R_i \mid X_i)}{N+1} + \lambda \frac{R_i - E(R_i \mid X_i)}{N+1}\right) d\lambda - \varphi''\left(\frac{E(R_i \mid X_i)}{N+1}\right)\right\},$$

$$c_i^* = c_i (\operatorname{var} S_c^0)^{-1/2}, \tag{16}$$

where S_c^0 is given by (10).

By Lemma 4.1 Hájek (1968) the "nearest" sum of independent random variables (in the sense of quadratic mean) to S_c^* has the following form:

$$\hat{S}_c^* = \sum_{j=1}^{N} E(S_c^* \mid X_j) - (N-1)\, ES_c^* \tag{17}$$
$$= \sum_{j=1}^{N} \Big\{c_j^* [\varphi(E(R_j \mid X_j)/(N+1)) - E\varphi(E(R_j \mid X_j)/(N+1))]$$
$$+ (N+1)^{-1} \sum_{\substack{i=1 \\ j \neq i}}^{N} c_i^* \int (u(x - X_j) - F_j(x))\, \varphi'(E(R_i \mid X_i = x)/(N+1))\, dF_i(x)\Big\}.$$

Denote

(17a) $$S_c^* + S_c^{**} - \hat{S}_c^{**} - ES_c^{**} = \sum_{j=1}^{N} \sum_{\substack{i=1 \\ i \neq j}}^{N} c_j^* T_{ji} + \tfrac{1}{2} \sum_{j=1}^{N} c_j^* \sum_{\substack{i=1 \\ i \neq j}}^{N} \sum_{\substack{k=1 \\ k \neq j}}^{N} (P_{jik} - EP_{jik}),$$

where

(18) $$T_{ji} = (N+1)^{-1} (u(X_j - X_i) - F_i(X_j))\, \varphi'\left(\frac{E(R_j \mid X_j)}{N+1}\right)$$
$$- \int (u(x - X_i) - F_i(x))\, \varphi'\left(\frac{E(R_j \mid X_j)}{N+1}\right) \mathrm{d}F_j(x), \quad 1 \leqq i, \ j \leqq N,$$

(19)
$$P_{jik} = (N+1)^{-2} (u(X_j - X_i) - F_i(X_j))(u(X_j - X_k) - F_k(X_j))\, \varphi''\left(\frac{E(R_j \mid X_j)}{N+1}\right),$$
$$1 \leqq i, \ j \leqq N.$$

Notice that by definition of T_{ji} and P_{jik} we have

(20) $$E(T_{ji} \mid X_i) = E(T_{ji} \mid X_j) = 0, \quad j \neq i,$$

(21) $$E(P_{jik} \mid X_j) = E(P_{jik} \mid X_i) = E(P_{jik} \mid X_k) = 0, \quad j \neq i, \quad j \neq k, \quad i \neq j,$$

and by definition of c_j and (11)

(22) $$\sum_{j=1}^{N} |c_j^*|^3 \leqq 1, \quad \log\left(\sum_{j=1}^{N} |c_j^*|^3\right)^{-1} \leqq 2\left(\max_{1 \leqq j \leqq N} |c_j^*|\right)^{-1},$$
$$\left(N \sum_{j=1}^{N} |c_j^*|^3\right)^{-1} \leqq (2(k_0^2 + k_1^2))^{3/2}, \quad \sum_{j=1}^{N} c_j^{*2}(k_0^2 + k_1^2) \geqq 1/2,$$
$$N^{-1/2} \sum_{j=1}^{N} c_j^{*2} \leqq (2(k_0^2 + k_1^2))^{1/2} \sum_{j=1}^{N} |c_j^*|^3.$$

Now we shall state several lemmas we need to prove the main theorem. The first one concerns S_c^{***}:

Lemma 3. *Let $X_1, \ldots, X_N$ be independent random variables with continuous distribution functions, let assumptions (4) and (11) be satisfied and scores be given by (2). Then there exists a constant B_1 (not depending on N and $c_1, \ldots, c_N$) such that*

$$P\left(|S_c^{***} - ES_c^{***}| \geqq \left(\sum_{j=1}^{N} |c_j^*|^3\right)\right) \leqq \sum_{j=1}^{N} c_j^{*3} B_1.$$

Proof: The Lemma follows easily, similarly as Lemma 2.1 (Hušková (1977)), using elementary inequalities and Lemma 1 for $Z_j = u(X_i - X_j) - F_j(X_i)$ given X_i. Q.E.D.

Applying Lemma 1 several times we can assert:

Lemma 4. *Let assumptions of Lemma* 3 *be satisfied; then for positive integers k we have*

$$E(S_c^* + S_c^{**} - ES_c^{**} - \hat{S}_c^*)^{2k}$$

$$\leqq 3^{2k-1}(4e)^{2k}\left[2k_1^{2k} + (2k_2)^{2k}\right]\left[k^2N^{-1}\sum_{i=1}^{N} c_i^{*2}\right]^{2k}.$$

Proof: Obviously,

$$E(S_c^* + S_c^{**} - ES_c^{**} - \hat{S}_c^*)^{2k} \leqq 3^{2k-1}\{E(\sum_{j=1}^{N}\sum_{i<j} c_j^* T_{ji})^{2k}$$

$$+ E(\sum_{j=1}^{N}\sum_{i>j} c_j^* T_{ji})^{2k} + E(S_c^{**} - ES_c^{**})^{2k}\}.$$

Put $Z_j = c_j^* \sum_{\substack{j=1 \\ i<j}}^{N} T_{ji}$ then the assumptions of Lemma 1 are satisfied and we have

$$E(\sum_{j=1}^{N} c_j^* \sum_{i<j} T_{ji})^{2k} \leqq k^k(\sum_{j=1}^{N} c_j^{*2})^k (4e)^k \max_{1\leqq j\leqq N} E(\sum_{i<j} T_{ji})^{2k} \tag{23}$$

and, similarly.

$$E(\sum_{j=1}^{N} c_j^* \sum_{i>j} T_{ji})^{2k} \leqq k^k(\sum_{j=1}^{N} c_j^{*2})^k (4e)^k \max_{1\leqq j\leqq N} E(\sum_{i>j} T_{ji})^{2k}. \tag{24}$$

By definition, the sums $\sum_{i=j+1}^{N} T_{ji}$ and $\sum_{i=1}^{j-1} T_{ji}$, given X_j, are sums of independent random variables; thus we can again apply Lemma 1

$$E(\sum_{i=j+1}^{N} T_{ji})^{2k} \leqq k^k(N-j)^k (4e)^k k_1^{2k}. \tag{25}$$

As for $E(S_c^{**} - ES_c^{**})^{2k}$, by Lemma 1 and by elementary calculations we can write

$$E(S_c^{**} - ES_c^{**})^{2k} \leqq (2k_2)^{2k} (\sum_{j=1}^{N} |c_j^*|)^{2k}, \tag{26}$$

$$\max_{1\leqq i\leqq N} E(R_i - E(R_i \mid X_i))^{4k}(N+1)^{-4k} \leqq (8ek_2)^{2k} k^{2k}(\sum_{j=1}^{N} c_j^{*2})^k N^{-k}.$$

Our lemma can be concluded from (23–26). Q.E.D.

Lemma 5. *Under assumptions of Lemma 3 there exists a constant B_2 such that*

$$\int_{|t|<1/2\log(\sum_{j=1}^{N}|c_j^*|^3)^{-1}} \left|E(S_c^* - \hat{S}_c^* + S_c^{**} - ES_c^{**})\exp\{it\hat{S}_c^*\}\right| dt \leqq B_2 \sum_{j=1}^{N} |c_j^*|^3 .$$

Proof: For the characteristic function one can write

$$E\exp\{itE(S_c \mid X_j)\} = 1 - \frac{t^2}{2}\operatorname{var}[E(S_c \mid X_j)] + \frac{|t|^3}{3!} E \mid E(S_c \mid X_j)|^3 \eta_j ,$$

where $|\eta_j| \leqq 1$ and then by elementary considerations we get

$$\left|E\exp\{it\hat{S}_c^*\}\right| \leqq \exp\left\{-\frac{t^2}{4}\right\}$$

for $|t| \leqq \frac{3}{8}(\sum_{j=1}^{N}|c_j^*|^3(k_0^2 + k_1^2))^{-1}$. The rest of the proof is very close to that of Lemma 2.4 in Hušková (1977), where instead of $\sum_{j=1}^{N} c_j = 0$ we use (20) and (21); hence it is omitted. Q.E.D.

Lemma 6. *Under assumptions of Lemma 3 there exist constants B_3 and D_1 such that*

$$E(S_c^* - \hat{S}_c^* + S_c^{**} - ES_c^{**})^m \exp\{it\hat{S}_c^*\} \leqq (B_3|t|)^m$$
$$\cdot \exp\{-t^2/4 + 3t^2 m \max_{1\leqq j\leqq N} c_j^{*2}(k_0^2 + k_1^2)\} ,$$

for $m > 0$ integer, $|t| \leqq D_1(\sum_{j=1}^{N}|c_j^*|^3)^{-1}$.

Proof: Obviously,

$$(27)\qquad \left|E(S_c^* - \hat{S}_c^* + S_c^{**} - ES_c^{**})^m \exp\{it\hat{S}_c^*\}\right| \leqq \sum_{v=0}^{m}\binom{m}{v} \sum_{(A_1,\ldots,A_l)\in\mathfrak{A}}$$
$$\sum_{(i_1,\ldots,i_{3m-v})\in f(A_1,\ldots,A_l)} \left|Eh(i_1,\ldots,i_{3m-v})\right| ,$$

where $\mathfrak{A}$ denotes all decomposition of the set $(1, \ldots, 3m - v)$ such that for every $\alpha(=1, \ldots, m)$ the pairs $(\alpha, \alpha + m)$ and $(\alpha, \alpha + 2m)$ do not belong to the same subset of decomposition,

$$f(A_1, \ldots, A_l)$$
$$= \{(i_1, \ldots, i_{3m-v});\ \beta, \alpha \in A_j \Rightarrow i_\alpha = i_\beta,\ \alpha \in A_j,\ \beta \in A_k,\ j \neq k \Rightarrow i_\alpha \neq i_\beta\} ,$$

$$h(i_1, \ldots, i_{3m-v}) = \prod_{j=1}^{v} c_{i_j}^* T_{i_j i_{m+j}} \prod_{j=v+1}^{m} c_{i_j}^*(P(i_j, i_{j+m}, i_{j+2m})$$
$$- EP(i_j, i_{j+m}, i_{j+2m}))\exp\{it\hat{S}_c^*\} .$$

Denote the right side of (27) by $C(m)$. The proof of our lemma will be done by induction. By direct computations we get for $m = 1$

$$(28)\qquad E(S_c^* - \hat{S}_c^* + S_c^{**} - ES_c^{**}) \exp\{it\hat{S}_c^*\} \leqq |t|\,(k_0 + k_1)^3 + k_2 N^{-1}.$$

For $m + 1$ we can write

$$(29)\qquad E(S_c^* - \hat{S}_c^* + S_c^{**} - ES_c^{**})^{m+1} \exp\{it\hat{S}_c^*\} = \sum_{v=0}^{m} \binom{m}{v} \sum_{(A_1,\ldots,A_l)\in\mathfrak{U}}$$

$$\cdot \sum_{(i_1,\ldots,i_{3m-v})\in f(A_1,\ldots,A_l)} Eh(i_1,\ldots,i_{3m-v})$$

$$\cdot \{\sum_{i=1}^{4} \sum_{j\in B_i} \sum_{k=1}^{4} \sum_{\alpha\in B_k} [c_j^* T_{j\alpha} + \sum_{s=1}^{4} \sum_{\beta\in B_s} c_j^*(P_{j\alpha\beta} - EP_{j\alpha\beta})]\},$$

where $B_1 = \{j;\ 1 \leqq j \leqq N,\ j \neq i_1, \ldots, i_{3m-v}\}$, $B_2 = \{j;\ 1 \leqq j \leqq N$, there exist l, l^* such that $j = i_l = i_{l^*}\}$, $B_3 = \{j;\ 1 \leqq j \leqq N$, there exist just one i_l such that $j = i_l$ and the corresponding member contains $c_l^*\}$, $B_4 = \{j;\ 1 \leqq j \leqq N$, there exists just one i_l such that $j = i_l$, the corresponding member does not depend on $c_l^*\}$.

Using (20) one can easily obtain

$$|ET_{jk} \exp\{itE(S_c \mid X_j)\}| \leqq |t|\,(|c_j^*|\,k_0 + (N+1)^{-1} \sum_{j=1}^{N} |c_j^*|\,k_1)\,k_1, \quad j \neq k,$$

and thus

$$\sum_{(i_1,\ldots,i_{3m-v})\in f(A_1,\ldots,A_l)} |Eh(i_1,\ldots,i_{3m-v}) \sum_{\alpha\in B_1} \sum_{\beta\in B_3} T_{\alpha\beta} c_\alpha^*|$$

$$\leqq \sum_{(i_1,\ldots,i_{3m-v})\in f(A_1,\ldots,A_l)} Eh(i_1,\ldots,i_{3m-v})\,k_1 \exp\{t^2 \max_{1\leqq j\leqq N} c_j^{*2}(k_0^2 + k_1^2)\}$$

and

$$\sum_{(i_1,\ldots,i_{3m-v})\in f(A_1,\ldots,A_l)} |Eh(i_1,\ldots,i_{3m-v}) \sum_{\alpha\in B_2} \sum_{\beta\in B_4} c_\alpha^* T_{\alpha\beta}|$$

$$\leqq \sum_{(i_1,\ldots,i_{3m-v})\in f(A_1,\ldots,A_l)} Eh(i_1,\ldots,i_{3m-v})\,k_1((k_0 + k_1)\,|t|)^{-1}.$$

The other sums in (28) can be estimated in the same way and after some computations we get that there exists a constant B_3 such that

$$C(m+1) \leqq C(m)\,B_3|t| \exp\{3t^2 \max_{1\leqq j\leqq N} c_j^{*2}(k_0^2 + k_1^2)\}.$$

The assertion of our lemma can be concluded from the last inequality, (27) and (28). Q.E.D.

The proof of the Theorem A: Obviously,

$$\left(\frac{\operatorname{var} S_c}{\operatorname{var} S_c^0}\right)^{1/2} - 1 \leqq \left|\frac{\operatorname{var}(S_c - S_c^0)}{\operatorname{var} S_c^0}\right|^{1/2}$$

$$\leqq 2^{1/2} \sum_{i=1}^{N} |c_i^*|^3 (B_1 + 96e^2(k_1^2 + 2k_2^2))^{1/2} .$$

Thus it suffices to prove the assertion where var S_c is replaced by var S_c. The rest of the proof follows by Lemmas 2–6 in the same way as that Theorem A in Hušková (1975) if we assume that $\sum_{j=1}^{N} |c_j^*|^3 \leqq B_4$ with B_4 being a suitable constant not depending on N and $c_1, \ldots, c_N$. The validity of the assertion for $\sum_{j=1}^{N} |c_j^*|^3 \geqq B_4$ is obvious. Q.E.D.

References

BICKEL, P. J. (1974). Edgeworth expansion in nonparametric statistics. *Ann. Statist.*, **1**, 1–20.

BJERVE, S. (1973). Error bound and asymptotic expansions for linear combinations of order statistics. *Thesis.*

HUŠKOVÁ, M. (1977) The rate of convergence of simple linear rank statistics under hypothesis and alternatives. *Ann. Statist.*, **5**, 658–670.

HÁJEK, J. (1968). Asymptotic normality of simple linear rank statistic under alternatives. *Ann. Math. Statist.*, **39**, 325–346.

BERGSTRÖM, H. - PURI, M. L. (1977). Convergence and remainder terms in linear rank statistics. *Ann. Statist.* **5**, 671–680.

JUREČKOVÁ, J. (1973). Order of normal approximation for rank statistics distribution. *Report of Inst. Math. Stat.*, University of Copenhagen.

CHARLES UNIVERSITY, PRAGUE, CZECHOSLOVAKIA

Received December 1975

NUISSANCE MEDIANS IN RANK TESTING SCALE

by

JANA JUREČKOVÁ

1. Introduction

Consider two samples $X_1, \ldots, X_m$ and $X_{m+1}, \ldots, X_{m+n}$ and assume that their respective densities are of the form $\sigma_1^{-1} f(\sigma_1^{-1}(x - \mu))$ and $\sigma_2^{-1} f(\sigma_2^{-1}(x - \nu))$, $\sigma_1, \sigma_2 > 0$. If $\mu = \nu$, then proper rank tests for testing $\sigma_1 = \sigma_2$ against $\sigma_1 \neq \sigma_2$ (or $\sigma_1 > \sigma_2$, or $\sigma_1 < \sigma_2$) are based on the statistics

$$S_N = s_N(X_1, \ldots, X_N), \quad N = m + n \tag{1}$$

where

$$s_N(X_1, \ldots, X_N) = \sum_{i=1}^{m} a_N(R_{Ni}) \tag{2}$$

and

$$R_{Ni} = \sum_{j=1}^{N} u(X_i - X_j), \quad u(x) = 1 \quad \text{if} \quad x \geqq 0, \qquad = 0 \quad \text{otherwise} \tag{3}$$

is the rank of X_i among $X_1, \ldots, X_N$ and the scores $a_N(i)$ are generated by a function $\varphi(t)$, $0 < t < 1$ by either of the following two ways:

$$a_N(i) = \varphi\left(\frac{i}{N+1}\right), \quad i = 1, \ldots, N,$$

$$a_N(i) = E\,\varphi(U_N^{(i)}), \qquad i = 1, \ldots, N \tag{4}$$

where $U_N^{(i)}$ denotes the i-th order statistic in the sample of size N from the uniform (0, 1) distribution.

In this case, the statistics S_N have the asymptotically normal distribution under the null hypothesis as well as under the contiguous alternatives if some regularity conditions on φ and on the underlying distribution are satisfied. The proof of this result may be found in Hájek-Šidák [2] as well as a fairly complete survey of rank tests of scale.

If $\mu \neq \nu$ but we have some estimates $\hat{\mu}_N$ and $\hat{\nu}_N$ for μ and ν respectively, we may try to find the asymptotic distribution of the adjusted statistic

$$S_N^* = s_N(X_1 - \hat{\mu}_N, \ldots, X_m - \hat{\mu}_N, X_{m+1} - \hat{\nu}_N, \ldots, X_N - \hat{\nu}_N) . \tag{5}$$

Hájek in [4] proved that the asymptotic distribution of S_N^* will be the same as that of S_N under the null hypothesis as well as under the contiguous alternatives if

$$\int_0^1 \varphi(t)\, \varphi(t, f)\, \mathrm{d}t = 0 \tag{6}$$

where

$$\varphi(t, f) = -f'(F^{-1}(t))/f(F^{-1}(t)), \quad 0 < t < 1 , \tag{7}$$

$$F(x) = \int_{-\infty}^{x} f(y)\, \mathrm{d}y .$$

Such a situation occurs if $\varphi(t)$ is symmetric about $1/2$, i.e. $\varphi(t) = \varphi(1 - t), 0 < t < 1$ and if f is symmetric about 0.

In the present paper, we shall find the asymptotic distribution of the statistic (5) without the assumption (6), i.e. without assuming f being symmetric, if the estimates $\hat{\mu}_N$ and $\hat{\nu}_N$ could be approximated, in the sense of convergence in probability, by a sum of *iid* random variables in the following way:

$$m^{1/2}(\hat{\mu}_N - \mu) \sim m^{-1/2} \sum_{i=1}^{m} l(X_i - \mu) ,$$

$$n^{1/2}(\hat{\nu}_N - \nu) \sim n^{-1/2} \sum_{i=m+1}^{N} l(X_i - \nu)$$

where

$$\int l^2(x)\, \mathrm{d}F(x) - \left(\int l(x)\, \mathrm{d}F(x) \right)^2 < \infty .$$

This covers a large class of estimates, among others some of those to which Hájek's projection method applies (see Hájek [3]). The method being used here is thus applicable for such estimates as the mean, sample median (for justifying see J. K. Ghosh [1]), the trimmed mean and other linear order estimates with smooth weight function (for justifying see Stigler [7]). Hodges-Lehmann's estimate also belongs to this class. We must only take into account that, if we do not expect the symmetry of the basis distribution, we must utilize such versions of the estimates for which the assumption of symmetry is not a basic one. For instance, concerning the Hodges-Lehmann's estimate, we could recommend an estimate based on one-sample rank statistics considered by P. K. Sen and M. Ghosh [6] rather than an estimate based on the signed-rank statistics.

The method is straightforward and is based on the authors earlier results concerning the asymptotic linearity of rank tests statistics. Nevertheless, the method is applicable for rather general score-generating functions so that the theorems among others cover the following tests: *the Klotz test, the quartile test, the Mood test, the Capon test, the Ansari-Bradley test.*

As an illustration, the theorems are specialized to the case that $\hat{\mu}_N$ and $\hat{\nu}_N$ are the sample medians based on $X_1, \ldots, X_m$ and $X_{m+1}, \ldots, X_N$. Generally, the asymptotic distribution of S_N^* is no more the same as that of S_N; an increment in the asymptotic variance appears under the null hypothesis as well as under the contiguous scale alternatives and furthermore, under the contiguous scale alternative, there is an increment in the asymptotic expectation comparing with the asymptotic expectation of S_N.

2. Asymptotic distribution of S_N^* under the null hypothesis

The result will be proved under the following assumptions:

1° The random vector $X_{N1}, \ldots, X_{NN}$ has the density

(8) $$p_N(x_1, \ldots, x_N) = \prod_{i=1}^{m_N} f(x_i - \mu) \prod_{j=1}^{n_N} f(x_{m+j} - \nu)$$

$$(m_N + n_N = N)$$

where μ and ν are unknown real parameters and f is any density which is absolutely continuous and has finite Fisher's information,

(9) $$I(f) = \int [f'(x)/f(x)]^2 f(x)\,\mathrm{d}x < \infty\,.$$

We may assume, without any loss of generality, that median of f is 0 (we may put $f^*(x) = f(x + \xi)$ if $\operatorname{med} f = \xi$). Furthermore, suppose that $f(0) > 0$.

We say that two sequences V_N and W_N of random variables are asymptotically equivalent in probability and use the symbol $\sim$ for denoting this fact, if $V_N - W_N \to 0$ in probability for $N \to \infty$.

2° Let $\hat{\mu}_N$ and $\hat{\nu}_N$ be the respective estimates of μ and ν, $\hat{\mu}_N$ based on $X_1, \ldots, X_m$ and $\hat{\nu}_N$ based on $X_{m+1}, \ldots, X_N$ such that there exists the function $l(x)$, $x \in R^1$ satisfying

(10) $$m^{1/2}(\hat{\mu}_N - \mu) \sim m^{-1/2} \sum_{i=1}^{m} l(X_i - \mu)\,,$$

$$n^{1/2}(\hat{\nu}_N - \nu) \sim n^{-1/2} \sum_{i=m+1}^{N} l(X_i - \nu)$$

and

$$\varrho^2 = \int l^2(x)\, \mathrm{d}F(x) - \zeta^2 < \infty\,, \quad \zeta = \int l(x)\, \mathrm{d}F(x)\,.$$

3° The scores $a_N(i)$ are generated by a function $\varphi(t)$, $0 < t < 1$ either by (3) or by (4). We suppose that $\varphi(t)$ is a sum of a finite number of monotone functions square-integrable on $(0, 1)$.

Put

(11) $$\bar{\varphi} = \int_0^1 \varphi(t)\, \mathrm{d}t\,, \quad \alpha^2 = \int_0^1 (\varphi(t) - \bar{\varphi})^2\, \mathrm{d}t\,,$$
$$\beta = \int_0^1 \varphi(t)\, \varphi(t, f)\, \mathrm{d}t$$

where $\varphi(t, f)$ is given by (7).

4° Let

$$\min(m_N, n_N) \to \infty \quad \text{for} \quad N \to \infty\,.$$

Under the assumptions $1°-4°$, we have the following theorem (we shall suppress the indices N at m_N and n_N):

Theorem 1. *Under the assumptions* $1°-4°$

(12) $$\lim_{N\to\infty} P\{N^{-1/2}(\sigma_N^*)^{-1}\,(S_N^* - a_N^*) \leqq x\}$$
$$= (2\pi)^{-1/2} \int_{-\infty}^{x} \exp(-y^2/2)\, \mathrm{d}y \quad \textit{for} \quad x \in R^1$$

where

(13) $$a_N^* = mN^{-1} \sum_{i=1}^{N} a_N(i) \sim m\bar{\varphi}\,,$$

(14) $$\sigma_N^{*2} = mnN^{-2}\sigma^2$$

and

(15) $$\sigma^2 = \alpha^2 + \beta^2\varrho^2 - 2\beta \int \varphi(F(x))\, l(x)\, \mathrm{d}F(x) + 2\beta\bar{\varphi}\zeta\,.$$

Proof. Put $Y_i = X_i - \mu$ for $i = 1, \ldots, m$ and $Y_i = X_i - \nu$ for $i = m + 1, \ldots, N$. Let $S_N = s_N(Y_1, \ldots, Y_N)$. Theorem 2.1 of Hájek [4] (a corollary of the author's theorem on uniform asymptotic linearity of rank statistics (see [5])) implies that (see also Theorem 4.1 of Hájek [4])

(16) $$(\operatorname{var} S_N)^{-1/2}\,[S_N^* - S_N - mnN^{-1}(\mu - \hat{\mu}_N - \nu + \hat{\nu}_N)\,\beta] \to 0$$

in probability for $N \to \infty$. It means that $(\text{var } S_N)^{-1/2} S_N^*$ has the same asymptotic distribution as

$$(17) \qquad (\text{var } S_N)^{-1/2} \left[S_N + mnN^{-1}(\mu - \hat{\mu}_N - \nu + \hat{\nu}_N)\right].$$

We shall thus investigate the asymptotic distribution of (17) for $N \to \infty$.

It follows from Theorems V.1.5 and V.1.6a of Hájek-Šidák [2] that

$$(18) \qquad (\text{var } S_N)^{-1/2} S_N$$

$$\sim (mn)^{-1/2} N^{1/2} \alpha^{-1} \{nN^{-1} \sum_{i=1}^{m} \varphi(F(Y_i)) - mN^{-1} \sum_{i=m+1}^{N} \varphi(F(Y_i)) + mN^{-1} \sum_{i=1}^{N} a_N(i)\}$$

and that

$$mnN^{-1}\alpha^2(\text{var } S_N)^{-1} \to 1 \quad \text{for} \quad N \to \infty .$$

It follows from (10), (17) and (18) that

$$(19) \qquad N^{-1/2}[\lambda_N(1 - \lambda_N)]^{-1/2} \alpha^{-1} S_N^*$$

$$\sim N^{-1/2}[\lambda_N(1 - \lambda_N)]^{-1/2} \alpha^{-1} \{(1 - \lambda_N) \sum_{i=1}^{m} [\varphi(F(Y_i)) - \beta\, l(Y_i)]$$

$$- \lambda_N \sum_{i=m+1}^{N} [\varphi(F(Y_i)) - \beta\, l(Y_i)] + \lambda_N \sum_{i=1}^{N} a_N(i)\} .$$

Put

$$(20) \qquad Z_i \equiv (F(Y_i) - \beta\, l(Y_i)\,, \quad i = 1, \ldots, N .$$

Then $Z_1, \ldots, Z_N$ are iid random variables and

$$(21) \qquad EZ_1 = \bar{\varphi} - \int l(y)\, dF(y) = \xi ,$$

$$\text{var } Z_1 = \alpha^2 + \beta^2 \varrho^2 - 2\beta \int \varphi(F(y))\, l(y)\, dF(y) + 2\beta\bar{\varphi}\zeta = \tau^2 .$$

It follows from (19), (20) and (21) that

$$(22) \qquad N^{-1/2}[\lambda_N(1 - \lambda_N)]^{-1/2} \alpha^{-1}(S_N^* - \lambda_N \sum_{i=1}^{N} a_N(i))$$

is asymptotically equivalent to the linear combination of N independent copies of Z_1 and this, regarding the Theorem V.1.2 of [2] implies that (22) is asymptotically normal $\mathcal{N}(0, \alpha^{-2}\tau^2)$ which gives the desired result. □

3. Asymptotic distribution of S_N^* under contiguous scale alternatives

Suppose that the density of the random vector $X_{N1}, \ldots, X_{NN}$ is of the form

$$(23)\qquad q_N(x_1, \ldots, x_N)$$
$$= \prod_{i=1}^{m_N} f(x_i - \mu) \prod_{i=m_N+1}^{N} \exp(-N^{-1/2}\Delta) f[(x_i - \nu)\exp(-N^{-1/2}\Delta)]$$

where Δ is an unknown parameter, $\Delta > 0$. This means that the cdf of the second sample is $F((x - \nu)\exp(-N^{-1/2}\Delta))$ while that of the first sample is $F(x - \mu)$, $x \in R^1$. We shall investigate the asymptotic distribution of the adjusted rank statistic S_N^* given by (5), under the alternative q_N.

Let us denote

$$(24)\qquad \varphi_1(t) = \varphi_1(t, f) = -1 - F^{-1}(t)\,[f'(F^{-1}(t))/f(F^{-1}(t))]\,, \quad 0 < t < 1$$

and

$$(25)\qquad I_1(f) = \int_0^1 \varphi_1^2(t, f)\,\mathrm{d}t\,.$$

Theorem 2. *For $N = 1, 2, \ldots$, let q_N given by (23) be the density of the random vector $X_{N1}, \ldots, X_{NN}$. Let $I_1(f) < \infty$ and suppose that $\lim_{N\to\infty}(N^{-1}m_N) = \lambda$ exists and is strictly between 0 and 1. Then, under the assumptions 2°–4° of section 2,*

$$(26)\qquad Q_N\{N^{-1/2}(\sigma_N^*)^{-1}(S_N^* - a_N^{**}) \leqq x\} \to (2\pi)^{-1/2}\int_{-\infty}^{x} \exp(-y^2/2)\,\mathrm{d}y$$

holds for $N \to \infty$; Q_N is the probability distribution corresponding to q_N, σ_N^* is given by (14) or

$$(27)\qquad \sigma_N^{*2} = \lambda(1 - \lambda)\,\sigma^2$$

with σ^2 given by (15) and

$$(28)\qquad a_N^{**} = \lambda N\bar{\varphi} - \lambda\,(1 - \lambda)\,\Delta\left[\int_0^1 \varphi(t)\,\varphi_1(t, f)\,\mathrm{d}t - \beta\int l(y)\,\varphi_1(F(y))\,\mathrm{d}F(y)\right].$$

Proof. Similarly to section 2, we may suppose that $\operatorname{med} f = 0$. Denote again

$$Y_i = X_i - \mu\,, \quad i = 1, \ldots, m\,,$$
$$= X_i - \nu\,, \quad i = m + 1, \ldots, N$$

and introduce the densities

$$(29) \qquad p_N^*(y_1, \ldots, y_N) = \prod_{i=1}^{N} \exp(-nN^{-3/2}\Delta) f(y_i \exp(-nN^{-3/2}\Delta))$$

and

$$(30) \qquad q_N^*(y_1, \ldots, y_N) = \prod_{i=1}^{m} f(y_i) \prod_{i=m+1}^{N} \exp(-N^{-1/2}\Delta) f(y_i \exp(-N^{-1/2}\Delta)).$$

Let P_N^* and Q_N^* denote the respective probability distributions corresponding to p_N^* and q_N^*. Put $L_N^* = q_N^*/p_N^*$. It follows from Theorem VI.2.2 of [2] that

$$(31) \qquad \log L_N^* \sim T_N - (1/2)\, b^2$$

in P_N^* – probability, where

$$(32) \qquad b^2 = \lambda(1-\lambda)\, \Delta^2 I_1(f)$$

and

$$(33) \qquad T_N = -\sum_{i=1}^{m} nN^{-3/2}\Delta\varphi_1(U_i, f) + \sum_{i=m+1}^{N} mN^{-3/2}\Delta\varphi_1(U_i, f)$$

with $U_i = F_N^*(Y_i)$, $F_N^*(Y_i) = P_N^*(Y_i \leqq y)$.

Inspecting the forms of the left- and right-hand sides of (19) we see immediately that (19) holds also in P_N^*-probability, with F replaced by F_N^*. This together with (31) implies that the random vector

$$\left(N^{-1/2}\left(S_N^* - mN^{-1}\sum_{i=1}^{N} a_N(i)\right),\ \log L_N^*\right)$$

has under p_N^* the same asymptotic distribution as $(V_N, T_N - (b^2/2))$ where

$$(34) \qquad V_N = nN^{-3/2} \sum_{i=1}^{m} [\varphi(F_N^*(Y_i)) - \beta\, l(Y_i)]$$

$$- mN^{-3/2} \sum_{i=m+1}^{N} [\varphi(F_N^*(Y_i)) - \beta\, l(Y_i)]$$

and T_N is given by (33). Consequently, if we show that (V_N, T_N) is under P_N^* asymptotically jointly normal with expectations $\mu_1 = \mu_2 = 0$, variances $\sigma_1^2 = \sigma_N^{*2}$ and $\sigma_2^2 = b^2$ and covariance $\sigma_{12} = -\lambda(1-\lambda)\, \Delta[\int_0^1 \varphi(t)\, \varphi_1(t, f)\, dt - \beta \int l(y)\, \varphi_1(F(y)) \,.$ $.\, dF(y)]$ we can conclude that $(N^{-1/2}(S_N^* - a_N^*), \log L_N^*)$ are asymptotically jointly normal with the same parameters except for $\mu_2 = -b^2/2$ and the result will follow immediately from LeCam third lemma (for LeCam's lemmas, see [2]).

Actually, $EV_N = ET_N = 0$ directly from (33) and (34). Since Y_i's are *iid* random variables with *cdf* F_N^*, we get that

$$\operatorname{var} V_N = \sigma_N^{*2} \to \lambda(1-\lambda)\,\sigma^2\,,$$

$$\operatorname{var} T_N = mnN^{-2}\Delta^2 \int_0^1 \varphi_1^2(t,f)\,dt \to b^2\,,$$

$$\operatorname{cov}(V_N, T_N) = -mnN^{-2}\Delta\left[\int_0^1 \varphi(t)\,\varphi_1(t,f)\,dt - \beta\int l(y)\,\varphi_1(F(y))\,dF(y)\right]$$

$$\to -\lambda(1-\lambda)\,\Delta\left[\int_0^1 \varphi(t)\,\varphi_1(t,f)\,dt - \beta\int l(y)\,\varphi_1(F(y))\,dF(y)\right]$$

so that the limiting parameters have the required values. It remains to show for all real γ_1, γ_2 that $\gamma_1 V_N + \gamma_2 T_N$ is either asymptotically normal $(0, \varrho_{12})$ with $\varrho_{12} = \operatorname{var}(\gamma_1 V_N + \gamma_2 T_N)$ or that $\operatorname{var}(\gamma_1 V_N + \gamma_2 T_N) \to 0$. The rest of the proof will proceed in a way quite analogous to that used in the proof of Theorem VI.2.4 of [2] so that it may be omitted. □

4. Sample medians in the role of estimates

Let $\hat{\mu}_N$ and $\hat{\nu}_N$ be the sample medians based on $X_1, \ldots, X_m$ and $X_{m+1}, \ldots, X_N$, respectively. Then, it holds by J. K. Ghosh [1] that

$$m^{1/2}(\hat{\mu}_N - \mu) = [f(0)]^{-1}\, m^{-1/2} \sum_{i=1}^{m} [u(X_i - \mu) - (1/2)] + o_P(1)\,, \tag{35}$$

$$n^{1/2}(\hat{\nu}_N - \nu) = [f(0)]^{-1}\, n^{-1/2} \sum_{i=m+1}^{N} [u(X_i - \nu) - (1/2)] + o_P(1)$$

where $u(x) = 1$ if $x \geqq 0$ and $u(x) = 0$ otherwise; so that (10) holds with

$$l(y) = [f(0)]^{-1}\,[u(y) - (1/2)]$$

and

$$\zeta = 0\,,\quad \varrho^2 = [4f(0)]^{-1}\,.$$

The asymptotic variance (14) is then equal to $\sigma_N^{*2} = mnN^{-2}\sigma^2$ where

$$\sigma^2 = \alpha^2 + [4f(0)]^{-1}\,\beta^2 - 2\beta[f(0)]^{-1}\left[\int_{1/2}^{1} \varphi(t)\,dt - (\bar{\varphi}/2)\right] \tag{36}$$

while the asymptotic expectation (28) under the contiguous scale alternatives (23) turns out to be

$$a_N^{**} = \lambda N\bar{\varphi} - \lambda(1-\lambda)\,\Delta\int_0^1 \varphi(t)\,\varphi_1(t,f)\,dt\,. \tag{37}$$

References

[1] GHOSH, J. K. (1971). A new proof of the Bahadur representation of quantiles and an application. *Ann. Math. Statist.*, **42**, 1957—1961.

[2] HÁJEK, J. - ŠIDÁK, Z. (1967). "Theory of Rank Tests". Academia, Prague.

[3] HÁJEK, J. (1968). Asymptotic normality of simple linear rank statistics under alternatives. *Ann. Math. Statist.*, **39**, 325—46.

[4] HÁJEK, J. (1970). Miscellaneous problems of rank test theory. *Proc. First Intern. Symposium on Nonparametric Statistical Inference*, Cambridge Univ. Press, 3—19.

[5] JUREČKOVÁ, J. (1969). Asymptotic linearity of a rank statistic in regression parameter. *Ann. Math. Statist.*, **40**, 1889—1900.

[6] SEN, P. K. - GHOSH, M. (1971). On bounded length confidence intervals based on one-sample rank-order statistics. *Ann. Math. Statist.*, **42**, 189—203.

[7] STIGLER, S. M. (1974). Linear functions of order statistics with smooth weight function. *Ann. Statist.*, **2**, 676—93.

CHARLES UNIVERSITY, PRAGUE, CZECHOSLOVAKIA

Received September 1975

ON A THEOREM OF J. HÁJEK*

by

L. LE CAM

1. Introduction

The present paper is devoted to a generalization of Theorem 4.1 of J. Hájek's work [1]. The theorem in question consists of two different parts. In one of them Hájek gives an evaluation of the asymptotic minimax risk for sequences of experiments subject to his local asymptotic normality assumptions LAN.

The second part is essentially restricted to one-dimensional parameter sets. It says in effect that under the LAN assumptions, two sequences of estimates $\{T'_n\}$ and $\{T''_n\}$ which are both locally asymptotically minimax must also be such that $\sqrt{(n)}\,(T'_n - T''_n)$ tends to zero in probability. This result is reminiscent of a statement made by R. A. Fisher in his study [2] of maximum likelihood estimates. However, Fisher's statement was not correct. It differs from Hájek's in some essential respects.

In this paper we present an abstract version of Hájek's result, singling out the basic reasons for its validity. In this form the theorem becomes applicable to a variety of situations which are not necessarily related to the LAN assumptions.

Section 2 below recalls the necessary definitions. We have followed the general framework of [3] because it provides the maximum amount of flexibility with the least amount of technical difficulty. In this same section 2 we discuss two topologies on the space of decision functions. One of them, called "semistrong" corresponds very adequately to the convergences in probability mentioned in Hájek's theorem. Since it might not be familiar to all in this form we provide detailed explanations.

Section 3 gives two results which are directly related to the first part of Hájek's Theorem 4.1. One of them refers to convergences of the *distributions* of estimates. The other is an asymptotic minimax result already stated in [4]. The proof given there is inadequately written. Thus we provide a complete argument.

Section 4 investigates the relations between three different sets of assumptions. Two of them are concerned with special uniqueness properties of decision procedures. The third embodies "convergence in measure" assertions of the kind to be found

* This research was supported by National Science Foundation, Grant MPS-74-18967.

in Hájek's Theorem 4.1, part 2. The main impact of the section is that these three sets of assumptions are equivalent. This is the promised generalization of Hájek's result.

Section 5 elaborates on the LAN assumptions and shows how they can be combined usefully with the theorem of Section 4.

2. Notations and definitions

Unless otherwise stated, we shall follow the general notation of [3] which is briefly as follows.

The basic mathematical objects are abstract L-spaces in the sense of Kakutani and transitions between them. An L-space is a Banach lattice with a norm which satisfies the identity $\|\mu + \nu\| = \|\mu\| + \|\nu\|$ for all pairs (μ, ν) of positive elements. Consider two L-spaces, say L_1 and L_2. A *transition* from L_1 to L_2 is a positive linear map T from L_1 to L_2 such that $\|T\mu\| = \|\mu\|$ for every positive $\mu \in L_1$.

An *experiment* $\mathscr{E}$ indexed by a set Θ is a map $\theta \to P_\theta$ from Θ to some L-space L_1. The map is assumed to be such that $P_\theta \geqq 0$ and $\|P_\theta\| = 1$ for all θ. The band generated by the set $\{P_\theta; \theta \in \Theta\}$ in L_1 is called the L-space of $\mathscr{E}$. It is noted $L(\mathscr{E})$.

A *decision space* consists of a set Z and a uniform lattice Γ of bounded real functions defined on Z. A *loss function* W is a map from $\Theta \times Z$ to $(-\infty, +\infty]$ subject to the restriction that $\inf\{W_\theta(z); z \in Z\} > -\infty$ for all $\theta \in \Theta$. A loss function V is called *special* if for each $\theta \in \Theta$ the map $z \to V_\theta(z)$ is an element of the uniform lattice Γ.

A *decision procedure* for $\mathscr{E}$ and (Z, Γ) is a transition ϱ from $L = L(\mathscr{E})$ to the dual Γ' of the uniform lattice Γ. The transition ϱ can also be regarded as a positive normalized bilinear form on $\Gamma \times L$. Its value at $(\gamma, \lambda) \in \Gamma \times L$ is denoted $\gamma\varrho\lambda$. The symbol $\gamma\varrho$ denotes the image of γ in the dual M of L and $\varrho\lambda$ denotes the image of λ in Γ'.

The space $\mathscr{B}(\mathscr{E}, Z, \Gamma)$ is the space of all such transitions.

In the remainder of the present paper (except of course for the examples of Section 5) the space (Z, Γ) will be kept fixed. Thus we shall write $\mathscr{B}(\mathscr{E})$ instead of the more precise $\mathscr{B}(\mathscr{E}, Z, \Gamma)$.

For an experiment $\mathscr{E}$, a loss function W and a $\varrho \in \mathscr{B}(\mathscr{E})$ the *risk* of ϱ at θ is defined by the relation

$$R(\theta, \varrho) = W_\theta \varrho P_\theta = \sup_V V_\theta \varrho P_\theta$$

where the supremum is taken over all special loss functions V such that $V \leqq W$.

(Here $V_\theta \varrho P_\theta$ is well defined, but $W_\theta \varrho P_\theta$ is just a convenient symbol.)

A system $(\mathscr{E}, Z, \Gamma, W)$ defines a set $\mathscr{R}(\mathscr{E}, W)$ called the set of *available risk functions*. A function f from Θ to $(-\infty, +\infty]$ belongs to $\mathscr{R}(\mathscr{E}, W)$ if there is a $\varrho \in \mathscr{B}(\mathscr{E})$ such that $W_\theta \varrho P_\theta \leqq f(\theta)$ for all $\theta \in \Theta$.

An element f of $\mathscr{R}$ is called admissible if $W_\theta \varrho P_\theta \leqq f(\theta)$ for all θ implies $W_\theta \varrho P_\theta \equiv \equiv f(\theta)$.

An element ϱ of $\mathscr{B}(\mathscr{E})$ is called *pure* or nonrandomized if it is an extreme point of $\mathscr{B}(\mathscr{E})$. Equivalently ϱ is pure if the map $\gamma \to \gamma\varrho$ of Γ into M is multiplicative; that is, if $\gamma^2\varrho = (\gamma\varrho)^2$ for all $\gamma \in \Gamma$.

Let $\mathscr{E} = \{P_\theta;\ \theta \in \Theta\}$ and $\mathscr{F} = \{Q_\theta;\ \theta \in \Theta\}$ be two experiments indexed by Θ. Let S be a subset of Θ.

The *defficiency* $\delta_S(\mathscr{E}, \mathscr{F})$ of $\mathscr{E}$ relative to $\mathscr{F}$ on the set S is the number

$$\delta_S(\mathscr{E}, \mathscr{F}) = \inf_T \sup_{\theta \in S} \tfrac{1}{2}\|Q_\theta - TP_\theta\|$$

with an infimum taken over all transitions T from $L(\mathscr{E})$ to $L(\mathscr{F})$.

If $\mathscr{E}_S = \{P_\theta;\ \theta \in S\}$ one can show that $\delta_S(\mathscr{E}, \mathscr{F}) = \delta_S(\mathscr{E}_S, \mathscr{F}_S)$. The distance between $\mathscr{E}$ and $\mathscr{F}$ is the number $\Delta(\mathscr{E}, \mathscr{F}) = \max\{\delta(\mathscr{E}, \mathscr{F}), \delta(\mathscr{F}, \mathscr{E})\}$ in which δ without subscripts means that the deficiency is taken on Θ itself.

Two experiments $\mathscr{E}$ and $\mathscr{F}$ are called equivalent if $\Delta(\mathscr{E}, \mathscr{F}) = 0$.

For a fixed set Θ, let $\mathbf{E}(\Theta)$ be the set of equivalence classes of experiments indexed by Θ. This set can be metrized by Δ. However, one can also induce on $\mathbf{E}(\Theta)$ a weak topology by the pseudometrics $\Delta_S(\mathscr{E}, \mathscr{F}) = \max\{\delta_S(\mathscr{E}, \mathscr{F}), \delta_S(\mathscr{F}, \mathscr{E})\}$ where S runs through the family of finite subsets of Θ.

For its weak topology $\mathbf{E}(\Theta)$ is a compact Hausdorff space. (The proof of compactness given in [4] is marred by an apparently irreparable algebraic mistake. A different proof will be published elsewhere).

The results given in Sections 3 and 4 are properties of continuity or lower semicontinuity for the weak topology of $\mathbf{E}(\Theta)$.

Considering a fixed system $(\mathscr{E}, Z, \Gamma)$ one can topologize the space $\mathscr{B}(\mathscr{E})$ of decision procedures by the topology of pointwise convergence on $\Gamma \times L(\mathscr{E})$. This topology will also be called the weak topology of $\mathscr{B}(\mathscr{E})$. The definitions used here were selected to insure automatic fulfillment of two requirements:

(i) For its weak topology $\mathscr{B}(\mathscr{E})$ is a compact Hausdorff space;

(ii) For the same topology, the risk functions $\varrho \to W_\theta \varrho P_\theta$ are lower semicontinuous.

In Sections 3 and 4 we shall use two other topologies, or more exactly uniform structures on $\mathscr{B}(\mathscr{E})$. One of them is the structure of *convergence in distribution* defined as follows.

For a given $\theta \in \Theta$ and a given $\varrho \in \mathscr{B}(\mathscr{E})$ the image ϱP_θ is an element of the dual Γ' of Γ. It can be interpreted as the distribution of the estimate of $z \in Z$ when θ is true. The ordinary vague topology of Γ' is the weak topology $w(\Gamma', \Gamma)$ induced by Γ.

We shall say that decision procedures ϱ_ν converge in distribution to ϱ_0 if for every $\theta \in \Theta$ the images $\varrho_\nu P_\theta$ converge vaguely to $\varrho_0 P_\theta$.

Note that the topology of convergence in distribution is weaker than the weak topology of $\mathscr{B}(\mathscr{E})$. They coincide if and only if the identity $\varrho_1 P_\theta \equiv \varrho_2 P_\theta$, $\theta \in \Theta$ already implies $\varrho_1 = \varrho_2$.

Let A denote a strongly compact subset of Γ and let $B \subset \Theta$ denote a set such that $\{P_\theta;\ \theta \in B\}$ is strongly precompact in $L(\mathscr{E})$.

The structure of convergence in distribution may be defined by all the pseudometrics

$$\underline{d}_{A,B}(\varrho_1, \varrho_2) = \sup \{|\gamma\varrho_1 P - \gamma\varrho_2 P_\theta|;\ \gamma \in A,\ \theta \in B\}.$$

A much stronger structure, hereinafter called the semistrong structure of $\mathscr{B}(\mathscr{E})$ is the structure defined by all pseudometrics of the type

$$\bar{d}_{A,B}(\varrho_1, \varrho_2) = \sup_{\theta \in B} \{[\sup_{\gamma \in A} |\gamma\varrho_1 - \gamma\varrho_2|]\ P_\theta\}.$$

Here again A is compact and $\{P_\theta;\ \theta \in B\}$ precompact. The supremum $\sup_\gamma |\gamma\varrho_1 - \gamma\varrho_2|$ is taken in the complete lattice $M(\mathscr{E})$ dual of $L(\mathscr{E})$.

One could also take pseudometrics $\sup_{\theta \in B} \sup_{\gamma \in A} \langle |\gamma\varrho_1 - \gamma\varrho_2|, P_\theta \rangle$. These still define the same semistrong structure as above. We have taken the definition in the form $\bar{d}_{A,B}$ because it makes the relation with the usual convergence in measure more visible.

Suppose for instance that Z is a compact metric space with a distance noted d. Let Γ be the space $C(Z)$ of continuous functions on (Z, d). Take for A the set of Lipschitz functions γ such that $\|\gamma\| \leqq 1$ and $|\gamma(z_1) - \gamma(z_2)| \leqq d(z_1, z_2)$.

Suppose also that the P_θ are measures on a space $(\mathscr{X}, \mathscr{A})$. If ϱ_i, $i = 1, 2$, are decision procedures given by maps $x \to t_i(x)$ of $\mathscr{X}$ into Z then $\bar{d}_{A,B}(\varrho_1, \varrho_2)$ is simply the value

$$\sup_{\theta \in B} \int d[t_1(x), t_2(x)]\ P_\theta(dx).$$

In the foregoing expressions we have allowed arbitrary compacts $A \subset \Gamma$ and sets B for which $\{P_\theta : \theta \in B\}$ is precompact. Nothing would be changed by restricting both A and B to finite sets. The use of precompact sets makes results such as Theorem 3, Section 4 look stronger than if they were stated with finite sets. The difference, although noticeable, is not very great.

Consider two experiments $\mathscr{E} = \{P_\theta;\ \theta \in \Theta\}$ and $\mathscr{F} = \{Q_\theta;\ \theta \in \Theta\}$. Let $\mathscr{L}(\mathscr{E})$ be the linear subspace of $L(\mathscr{E})$ formed by finite linear combinations $\sum\{c_\theta P_\theta;\ \theta \in S\}$. Define $\mathscr{L}(\mathscr{F})$ similarly. The two experiments $\mathscr{E}$ and $\mathscr{F}$ are equivalent if and only if the correspondence $\sum c_\theta P_\theta \leftrightarrow \sum c_\theta Q_\theta$ is an isometry between $\mathscr{L}(\mathscr{E})$ and $\mathscr{L}(\mathscr{F})$.

It then extends in a unique way to an isometry of the vector sublattices $L_0(\mathscr{E})$ and $L_0(\mathscr{F})$ spanned by $\mathscr{L}(\mathscr{E})$ in $L(\mathscr{E})$ and $\mathscr{L}(\mathscr{F})$ in $L(\mathscr{F})$. The transitions T from $L(\mathscr{E})$ to $L(\mathscr{F})$ which achieve the identity $TP_\theta = Q_\theta$; $\theta \in \Theta$ are not necessarily uniquely determined. However, since T is well determined on $\mathscr{L}(\mathscr{E})$ and since it is an isometry, the corresponding adjoint map $\sigma \to \sigma T$ from $\mathscr{B}(\mathscr{F})$ to $\mathscr{B}(\mathscr{E})$ is an isomorphism of the structures of convergence in distribution. It is not usually an isomorphism of the semistrong structures.

Nor is it an isomorphism for the weak convergences on $\mathscr{B}(\mathscr{E})$ and $\mathscr{B}(\mathscr{F})$.

Finally let us note the following result, which indicates an important relation between the weak and semistrong topologies.

Lemma 1. *Let $\mathscr{E} = \{P_\theta;\ \theta \in \Theta\}$ be such that all the P_θ are nonatomic. Let $\varrho \in \mathscr{B}(\mathscr{E})$. In order that the filter of semistrong neighborhoods of ϱ coincide with the filter of weak neighborhoods of ϱ it is necessary and sufficient that ϱ be nonrandomized.*

The necessity is a well-known consequence of Liapunov's theorem. The sufficiency will appear as a consequence of Lemma 3, Section 4.

3. Convergence of minimax risks and distributions

In this section we consider an arbitrary index set Θ and an arbitrary loss function W. However, the first proposition immediately reduces arguments to finite subsets of Θ and special loss functions. Two consequences are derived. One concerns functions which cannot be risk functions. This may be regarded as a generalization of the first part of Hájek's Theorem 4.1. The second result is related to the convergence in distribution described in Section 2.

As already stated we assume that the pair (Z, Γ) which defines the decision space remains fixed throughout. The set Θ is also fixed. Furthermore, a certain loss function W plays a particular role.

With this in mind we shall introduce the following set of triplets.

Definition. The set $\mathscr{L}$ is the space of triplets (S, V, α) which consist of

(i) a finite set $S \subset \Theta$,

(ii) a special loss function $V \leqq W$,

(iii) a number $\alpha > 0$.

These triplets can be ordered by the relation $(S, V, \alpha) < (S', V', \alpha')$ if $S \subset S'$, $V \leqq V'$ and $\alpha' \leqq \alpha$. For this relation $\mathscr{L}$ is clearly a directed set.

Let $\mathscr{F} = \{Q_\theta;\ \theta \in \Theta\}$ be an experiment indexed by Θ and let f be any function from Θ to $(-\infty, +\infty]$. The set of decision procedures $\sigma \in \mathscr{B}(\mathscr{F})$ which satisfy the inequality $W_\theta\, \sigma\, Q_\theta \leqq f(\theta)$ for all $\theta \in \Theta$ will be noted $D(f, \mathscr{F})$. By $D(S, V, \alpha)$, or if necessary $D(S, V, \alpha, f, \mathscr{F})$ will be meant the set of procedures $\sigma \in \mathscr{B}(\mathscr{F})$ such that

$$V_\theta\, \sigma\, Q_\theta \leqq f(\theta) + \alpha$$

for all $\theta \in S$.

Proposition 1. *Let f, $\mathscr{F}$ and W be fixed. Let $U \subset \mathscr{B}(\mathscr{F})$ be a weak neighborhood of the set $D(f, \mathscr{F})$. Then, there exists a triplet $(S, V, \alpha) \in \mathscr{L}$ such that the set*

$$D(S, V, \alpha) = \{\sigma \in \mathscr{B}(\mathscr{F});\ V_\theta \sigma Q_\theta \leqq f(\theta) + \alpha,\ \theta \in S\}$$

is contained in U.

Proof. One can assume that U is open. Its complement $C = U$ in $\mathscr{B}(\mathscr{F})$ is then a compact set. The sets $C \cap D(S, V, \alpha)$ are compact and decreasingly directed. Thus, if their intersection D is empty one of the $C \cap D(S, V, \alpha)$ must be empty, implying $D(S, V, \alpha) \subset U$ as desired. However, if $\sigma \in D$ then $V_\theta \sigma Q_\theta \leqq f(\theta) + \alpha$ for all $\theta \in \Theta$, all $\alpha > 0$ and all special functions $V \leqq W$. This implies $W_\theta \sigma Q_\theta \leqq f(\theta)$ for all θ, since the risk $W_\theta \sigma Q_\theta$ is defined by a supremum. Therefore, $\sigma \in D$ implies $\sigma \in D(f, \mathscr{F}) \subset U$. This is impossible by construction. Hence the result.

The next statement is analogous to part 1 of Hájek's Theorem 4.1. To state it briefly, if $\mathscr{E} = \{P_\theta;\ \theta \in \Theta\}$ has a set of available risk functions $\mathscr{R}(\mathscr{E}, W)$, and if m is a probability measure with finite support on Θ, let

$$\chi(\mathscr{E}, W, m) = \inf_f \left\{ \int f \, dm;\quad f \in \mathscr{R}(\mathscr{E}, W) \right\}.$$

Theorem 1. *Let $\mathscr{F} = \{Q_\theta;\ \theta \in \Theta\}$ be an experiment indexed by Θ. Let f be a function from Θ to $(-\infty, +\infty]$. Assume that $f \notin \mathscr{R}(\mathscr{F}, W)$.*

Then there is a number $\varepsilon > 0$, a triplet $(S, V, \beta) \in \mathscr{L}$ and a probability measure m with support contained in S such that the inequality $\delta_S(\mathscr{F}, \mathscr{E}) < \varepsilon$ implies

$$\int (f + \beta) \, dm < \chi(\mathscr{E}, V, m).$$

Proof. According to Proposition 1 there is a triplet $(S, V, \alpha) \in \mathscr{L}$ such that the set $\{\sigma;\ \sigma \in \mathscr{B}(\mathscr{F}),\ V_\theta \sigma Q_\theta \leqq f(\theta) + \alpha,\ \text{all } \theta \in S\}$ is empty. This remains so if we replace f by a function g which is identical to f on the finite set S but is $+\infty$ on the complement of S. Thus, according to minimax theorems, there is a probability measure m carried by S such that $\int (f + \alpha) \, dm < \chi(\mathscr{F}, V, m)$;

Let $b = \sup \{|V_\theta(z)|;\ \theta \in S,\ z \in Z\}$. This is finite since V is a special loss function.

Take $\varepsilon = (1/4b)\,\alpha$ and let $\mathscr{E} = \{P_\theta;\ \theta \in \Theta\}$ be such that $\delta_S(\mathscr{F}, \mathscr{E}) \leqq \varepsilon$. Then, there is a transition T from $L(\mathscr{F})$ to $L(\mathscr{E})$ such that $\|P_\theta - TQ_\theta\| \leqq 2\varepsilon$ for all $\theta \in S$. Each element $\varrho \in \mathscr{B}(\mathscr{E})$ provides a $\sigma = \varrho T \in \mathscr{B}(\mathscr{F})$ such that

$$V_\theta \sigma Q_\theta \leqq V_\theta \varrho P_\theta + b \|P_\theta - TQ_\theta\| \leqq V_\theta \varrho P_\theta + \tfrac{1}{2}\alpha$$

for all $\theta \in S$. It follows that

$$\chi(\mathscr{F}, V, m) \leqq \chi(\mathscr{E}, V, m) + \tfrac{1}{2}\alpha.$$

Therefore,

$$\int (f + \alpha)\, dm < \chi(\mathscr{E}, V, m) + \tfrac{1}{2}\alpha .$$

This implies the desired result for $\beta = \frac{1}{2}\alpha$.

The minimax risk result of Hájek's Theorem 4.1 corresponds to the situation where one takes for f a function which is constant on the whole of Θ. We shall show later (Section 5) that Hájek's statement is obtainable from Theorem 1.

All the features of the present proof are already implicit in Hájek's arguments. The substitution for W of a special loss function V appears to complicate statements but it allows evaluations of bounds for variances of limiting distributions instead of limits of variances, which makes it worthwhile.

Our next result concerns distributions of estimates and their convergence. Here again we keep $\mathscr{F}$, Z, Γ, W and f fixed.

For simplicity of writing let us introduce another space $\mathscr{T}$ of triplets (A, S, ε).

Definition. The triplet (A, S, ε) belongs to $\mathscr{T}$ if

(i) A is a strongly precompact subset of Γ,

(ii) S is a subset of Θ such that $\{Q_\theta;\ \theta \in S\}$ is strongly precompact,

(iii) ε is a number $\varepsilon > 0$.

Theorem 2. *Let $\mathscr{F} = \{Q_\theta;\ \theta \in \Theta\}$, f and W be fixed. Let (A, S, ε) be a triplet $(A, S, \varepsilon) \in \mathscr{T}$.*

Then there is a number $\delta > 0$ and a triplet $(S_1, V, \alpha) \in \mathscr{L}$ with the following property:

Let $S_2 = S \cup S_1$ and let $\mathscr{E} = \{P_\theta;\ \theta \in \Theta\}$ be an experiment such that $\delta_{S_2}(\mathscr{F}, \mathscr{E}) \leqq \delta$. Let ϱ be any element of $\mathscr{B}(\mathscr{E})$ such that $V_\theta\, \varrho\, P_\theta \leqq f(\theta) + \alpha$ for all $\theta \in S_1$. Then there is a $\sigma \in D(f, \mathscr{F})$ such that

$$|\gamma\, \varrho\, P_\theta - \gamma\, \sigma\, Q_\theta| < \varepsilon$$

for all $\gamma \in A$ and all $\theta \in S$.

Proof. Let U be the set of $\tau \in \mathscr{B}(\mathscr{F})$ for which there is a $\sigma \in D(f, \mathscr{F})$ satisfying $|\gamma\, \tau\, Q_\theta - \gamma\, \sigma\, Q_\theta| < \varepsilon/2$ for all $\gamma \in A$ and $\theta \in S$. This is a weak neighborhood of $D(f, \mathscr{F})$ in $\beta(\mathscr{F})$. Thus, according to Proposition 1 there is a triplet (S_1, V, β) such that

$$D(S_1, V, \beta) = \{\sigma;\ V_\theta\, \sigma\, Q_\theta \leqq f(\theta) + \beta,\ \text{all } \theta \in S_1\}$$

is contained in U.

Let $a = \sup\{\|\gamma\|;\ \gamma \in A\}$ and let $b = \sup\{|V_\theta(z)|;\ \theta \in S_1,\ z \in Z\}$. Take $\alpha < \beta/2$ and select a $\delta > 0$ such that $4b\delta < \alpha$ and $4a\delta < \varepsilon$.

Let $\mathscr{E} = \{P_\theta; \theta \in \Theta\}$ be an experiment such that $\delta_{S_2}(\mathscr{F}, \mathscr{E}) \leqq \delta$ and let T be a transition from $L(\mathscr{F})$ to $L(\mathscr{E})$ such that $\|P_\theta - TQ_\theta\| \leqq 2\delta$ for all $\theta \in S_2$. If $\varrho \in \mathscr{B}(\mathscr{E})$ satisfies the inequality

$$V_\theta \, \varrho \, P_\theta \leqq f(\theta) + \alpha$$

for $\theta \in S_1$, its image $\tau = \varrho T$ satifies

$$V_\theta \, \tau \, Q_\theta \leqq f(\theta) + \alpha + 4b\delta < f(\theta) + \beta\,,$$

for all $\theta \in S_1$. Thus $\tau \in D(S_1, V, \beta) \subset U$.

By definition of U, this implies the existence of a $\sigma \in D(f, \mathscr{F})$ such that

$$|\gamma \, \varrho \, P_\theta - \gamma \, \sigma \, Q_\theta| \leqq |\gamma \, \varrho \, P_\theta - \gamma \, \varrho \, TQ_\theta| + |\gamma \, \varrho \, TQ_\theta - \gamma \, \sigma \, Q_\theta| \leqq 2\|\gamma\| \, \delta + \frac{\varepsilon}{2} < \varepsilon$$

for all $\gamma \in A$ and $\theta \in S$. This concludes the proof of the theorem.

Remark. In the preceding argument the set $D(f, \mathscr{F})$ may be empty. Then the same holds for U and a fortiori for $D(S_1, V, \alpha)$. In this case the theorem says that there are no $\varrho \in \mathscr{B}(\mathscr{E})$ satisfying the inequalities $V_\theta \, \varrho \, P_\theta \leqq f(\theta) + \alpha$, $\theta \in S_1$. This is analogous to Theorem 1.

The theorem is applicable also to the case where $D(f, \mathscr{F})$ consists of a single element σ. However, one can also consider a more general case as follows.

Definition. For a given system $(\mathscr{F}, Z, \Gamma, W)$ the function f is *unique in law* if any two σ_i, $i = 1, 2$, $\sigma_i \in \mathscr{B}(\mathscr{F})$ which satisfy $W_\theta \, \sigma_i \, Q_\theta \leqq f(\theta)$ for all $\theta \in \Theta$ must also be such that $\sigma_1 Q_\theta = \sigma_2 Q_\theta$ for all $\theta \in \Theta$.

If f is unique in law, Theorem 2 says that if $\mathscr{E}$ is close enough to $\mathscr{F}$ and if $\varrho \in \mathscr{B}(\mathscr{E})$ is such that $W_\theta \, \varrho \, P_\theta \leqq f(\theta) + \alpha$ on a suitable finite set, the distribution ϱP_θ must differ little from the distributions σQ_θ, at least for $\theta \in S$.

This appears to go in the general direction of Part 2 of Hájek's Theorem 4.1. One could be tempted to conjecture that an exact analogue of Hájek's result will hold if $D(f, \mathscr{F})$ contains only one element σ whose semistrong neighborhoods coincide with its weak neighborhoods.

As we shall see in Section 4, such a conjecture is correct in the nonatomic case, but not in general.

4. The second part of Hájek's theorem

As usual the system (Z, Γ, W) will be kept fixed. We shall consider pairs $(f, \mathscr{F})$ consisting of an experiment $\mathscr{F} = \{Q_\theta; \theta \in \Theta\}$ and a function f from Θ to $(-\infty, +\infty]$.

For such pairs we shall contemplate three properties, hereinafter called (A), (B) and (C). To state them recall that $D(f, \mathscr{F})$ is the set $D(f, \mathscr{F}) = \{\sigma \in \mathscr{B}(\mathscr{F});\ W_\theta \sigma Q_\theta \leqq f(\theta), \text{ all } \theta \in \Theta\}$ and that we have two systems of triplets $\mathscr{L}$ and $\mathscr{T}$. The triplets $\mathscr{L}$ are of the form (S, V, α) with S finite and V special, $V \leqq W$. The triplets of $\mathscr{T}$ are of the form (A, S, ε) with A precompact in Γ and $S \subset \Theta$ such that $\{Q_\theta;\ \theta \in \Theta\}$ is precompact.

Property (A). For every experiment $\mathscr{F}' = \{Q'_\theta;\ \theta \in \Theta\}$ which is equivalent to $\mathscr{F} = \{Q_\theta;\ \theta \in \Theta\}$ the set $D(f, \mathscr{F}')$ consists of one and only one element of $\mathscr{B}(\mathscr{F}')$.

Property (B). There is one and only one element $\sigma \in D(f, \mathscr{F})$ and it is nonrandomized.

Property (C). The pair $(f, \mathscr{F})$ enjoys property (C) if $D(f, \mathscr{F}) \neq \emptyset$ and if for every triplet $(A, S, \varepsilon) \in \mathscr{T}$ there are numbers $\delta > 0$ and triplets $(S_1, V, \alpha) \in \mathscr{L}$ satisfying the following requirements:

Let $S_2 = S \cup S_1$ and let $\mathscr{E} = \{P_\theta;\ \theta \in \Theta\}$ be such that $\delta_{S_2}(\mathscr{F}, \mathscr{E}) \leqq \delta$. Then two elements ϱ_i, $i = 1, 2$ of $\mathscr{B}(\mathscr{E})$ which both satisfy the inequalities

$$V_\theta\, \varrho_i\, P_\theta \leqq f(\theta) + \alpha$$

for $\theta \in S_1$ must also satisfy the relation

$$\Big[\sup_{\gamma \in A} |\gamma\varrho_1 - \gamma\varrho_2|\Big]\, P_\theta < \varepsilon$$

for all $\theta \in S$.

One first immediate remark is as follows.

Lemma 2. *Let $(f, \mathscr{F})$ satisfy* Property (C). *If $\delta(\mathscr{F}, \mathscr{E}) = 0$ then $\mathscr{E}$ enjoys* Property (C) *provided that $D(f, \mathscr{E})$ is not empty. The set $D(f, \mathscr{E})$ has then only one element. In particular $(f, \mathscr{F})$ satisfies* (A).

This follows from the wording of the relations themselves.

The next proposition shows that Property (A) implies property (B).

Proposition 2. *Assume that $(f, \mathscr{F})$ satisfies the uniqueness* Property (A). *Then the unique element σ of $D(f, \mathscr{F})$ is nonrandomized.*

Proof. Construct another experiment $\mathscr{E} = \{P_\theta;\ \theta \in \Theta\}$ as follows. Let M be the dual of $L(\mathscr{F})$. The tensor product $\Gamma \otimes M$ of Γ by M is obtained by taking pairs $(\gamma, u) \in \Gamma \times M$, forming finite linear combinations $\sum a_j \gamma_j \otimes u_j$ and identifying to zero the combinations which give value zero to all continuous bilinear forms on $\Gamma \times M$.

The procedure $\sigma \in D(f, \mathscr{F})$ induces a positive linear map Φ of $\Gamma \otimes M$ into M which extends linearly the map $\gamma \otimes u \to (\gamma \otimes u)\, \Phi = (\gamma\sigma)\, u$. It is easily seen that Φ

is well defined and that it preserves the constant elements of $\Gamma \otimes M$. Define $\mathscr{E} = \{P_\theta;\ \theta \in \Theta\}$ by

$$\langle \gamma \otimes u,\ P_\theta\rangle = \langle(\gamma\sigma)\,u,\ Q_\theta\rangle = (\gamma \otimes u)\,\Phi Q_\theta\,.$$

Since Φ operating on its right is a transition one has $\delta(\mathscr{F}, \mathscr{E}) = 0$. However, $\mathscr{E}$ is obviously at least as strong as $\mathscr{F}$ itself. Thus $\Delta(\mathscr{F}, \mathscr{E}) = 0$ and $D(f, \mathscr{E})$ contains one and only one element, say ϱ. Here it is easy to construct such a ϱ. In fact, let $(\Gamma \otimes M)'$ be the dual of $\Gamma \otimes M$. If $v \in (\Gamma \otimes M)'$ one can define its "marginal" Πv on Γ by $\langle \gamma, \Pi v\rangle = \langle \gamma \otimes 1, v\rangle$. This map Π is also a transition. It has the property that $\Pi P_\theta = \sigma Q_\theta$ for all $\theta \in \Theta$. The transpose of Π is the extension of the map $\gamma \to \gamma \otimes 1$ from Γ to $\Gamma \otimes M$. This is a multiplicative map. Hence Π is a nonrandomized procedure. Therefore, the unique $\varrho \in \mathscr{B}(\mathscr{E})$ which satisfies $W_\theta\,\varrho\,P_\theta \leqq f(\theta)$ is this nonrandomized Π.

Consider then an arbitrary transition T from $L(\mathscr{E})$ to $L(\mathscr{F})$ which achieves the identity $Q_\theta = TP_\theta$ for $\theta \in \Theta$. Take $\sigma \in D(f, \mathscr{F})$ and its image $\varrho = \sigma T$ in $\mathscr{B}(\mathscr{E})$ and note the relations

$$[\gamma^2\sigma - (\gamma\sigma)^2]\,T = \gamma^2\,\sigma\,T - (\gamma\sigma)^2\,T \leqq [\gamma^2\,\sigma\,T - (\gamma\sigma)^2\,T] + [(\gamma\sigma)^2\,T - (\gamma\,\sigma\,T)^2]$$
$$= \gamma^2\,\sigma\,T - (\gamma\,\sigma\,T)^2 = \gamma^2\varrho - (\gamma\varrho)^2\,.$$

From this, it follows that

$$[\gamma^2\sigma - (\gamma\sigma)^2]\,Q_\theta = [\gamma^2\sigma - (\gamma\sigma)^2]\,TP_\theta < [\gamma^2\varrho - (\gamma\varrho)^2]\,P_\theta = 0\,.$$

Thus σ is nonrandomized and the proposition is proved.

For nonrandomized procedures it is well known that weak and semistrong convergences coincide. The following lemma records this fact in a particular form.

Lemma 3. *Let σ be a nonrandomized element of $\mathscr{B}(\mathscr{F})$. Let (A, S, ε) be one of the triplets $(A, S, \varepsilon) \in \mathscr{T}$. Then the set of $\tau \in \mathscr{B}(\mathscr{F})$ such that*

$$\langle[\gamma^2\tau - (\gamma\tau)^2] + [\gamma\tau - \gamma\sigma]^2,\ Q_\theta\rangle < \varepsilon$$

for all $\gamma \in A$ and $\theta \in S$ is a weak neighborhood of σ in $\mathscr{B}(\mathscr{F})$.

Proof. Since σ is nonrandomized, one has $\gamma^2\sigma = (\gamma\sigma)^2$ and consequently

$$\gamma^2\tau - (\gamma\tau)^2 + (\gamma\tau - \gamma\sigma)^2 = \gamma^2\tau - \gamma^2\sigma - 2(\gamma\tau - \gamma\sigma)\,\gamma\sigma\,.$$

Let then $\mu_{t,\gamma}$ be the element of $L(\mathscr{F})$ which has density $\gamma\sigma$ with respect to Q_t. The set $K = \{\mu_{t,\gamma};\ t \in S,\ \gamma \in A\}$ is a strongly precompact subset of $L(\mathscr{F})$. The system of inequalities

$$(\gamma^2\tau - \gamma^2\sigma)\,Q_t - 2(\gamma\tau - \gamma\sigma)\,\mu < \varepsilon$$

for $t \in S$, $\gamma \in A$ and $\mu \in K$ defines a weakly open subset of $\mathscr{B}(\mathscr{F})$. This proves the desired result.

The following theorem represents an extension of Part 2 of Hájek's Theorem 4.1.

Theorem 3. *Let Z, Γ and W be fixed and let f be a function from Θ to $(-\infty, +\infty]$. Let $\mathscr{F} = \{Q_\theta; \theta \in \Theta\}$ be an experiment indexed by Θ. For the pair $(f, \mathscr{F})$ all three properties (A), (B) and (C) are equivalent.*

Proof. The concluding inequality in Property (C) involves a triplet $(A, S, \varepsilon) \in \mathscr{T}$ and a supremum $\sup_{\gamma \in A} |\gamma\varrho_1 - \gamma\varrho_2|$ taken in the dual $M(\mathscr{E})$ of $L(\mathscr{E})$. It is to this supremum that the P_θ are applied.

However, since A is precompact one can find a finite set $\{\gamma_j; j = 1, 2, \ldots, m\} \subset A$ such that for every $\gamma \in A$ one has $\|\gamma - \gamma_j\| < \varepsilon/2$ for at least one index j. This gives

$$\sup_{\gamma \in A} |\gamma\varrho_1 - \gamma\varrho_2| < \frac{\varepsilon}{2} + \sum_j |\gamma_j\varrho_1 - \gamma_j\varrho_2| .$$

Hence it will be enough to show that one can achieve the inequality $\langle|\gamma\varrho_1 - \gamma\varrho_2|, P_\theta\rangle < \varepsilon_0$ for an $\varepsilon_0 > 0$ such that $2m\varepsilon_0 < \varepsilon$.

Taking two elements ϱ_1 and ϱ_2 of $\mathscr{B}(\mathscr{E})$ introduce their average $\varrho = \frac{1}{2}(\varrho_1 + \varrho_2)$ and the difference $\varphi = \frac{1}{2}(\varrho_1 - \varrho_2)$. Then

$$(\gamma\varphi)^2 + (\gamma\varrho)^2 = \tfrac{1}{2}[(\gamma\varrho_1)^2 + (\gamma\varrho_2)^2] .$$

Also $(\gamma\varphi)^2 \leqq \gamma^2\varrho - (\gamma\varrho)^2$. Thus the desired inequality will hold if

$$\langle[\gamma^2\varrho - (\gamma\varrho)^2], P_\theta\rangle < \varepsilon_1 < \tfrac{1}{4}\varepsilon_0^2$$

for all $\gamma \in A$ and $\theta \in S$.

Let U be the set of elements $\tau \in \mathscr{B}(\mathscr{F})$ which satisfy the inequality

$$\langle[\gamma^2\tau - (\gamma\tau)^2], Q_\theta\rangle < \frac{\varepsilon_1}{2}$$

for all $\gamma \in A$ and $\theta \in S$. By Lemma 3 and Proposition 2 this is a weak neighborhood of the unique element σ of $D(f, \mathscr{F})$. Thus, by Proposition 1 (Section 3) there is a triplet $(S_1, V, \beta) \in \mathscr{L}$ such that $V_\theta\, \tau\, Q_\theta \leqq f(\theta) + \beta$, $\theta \in S_1$, implies $\tau \in U$.

Let $a = \sup\{\|\gamma\|; \gamma \in A\}$ and let $b = \sup\{|V_\theta(z)|; \theta \in S_1, z \in Z\}$. Select α and δ so that $\alpha < \beta/2$, $4b\delta < \beta$ and $a^2\delta < \varepsilon_1$. Let $S_2 = S \cup S_1$. Assume $\delta_{S_2}(\mathscr{F}, \mathscr{E}) \leqq \delta$ and let T be a transition which achieves the inequality $\|P_\theta - TQ_\theta\| \leqq 2\delta$ for all $\theta \in S_2$.

If two elements ϱ_1 and ϱ_2 both satisfy the relation $V_\theta\, \varrho_i\, P_\theta \leqq f(\theta) + \alpha$ for $\theta \in S_1$ the same is true of their average $\varrho = \frac{1}{2}(\varrho_1 + \varrho_2)$. Letting $\tau = \varrho T$ this gives

$$V_\theta\, \tau\, Q_\theta \leqq f(\theta) + \alpha + 2b\delta \leqq f(\theta) + \beta .$$

Therefore, $\tau = \varrho T$ belongs to the neighborhood U. This may be written $\langle[\lambda^2 \varrho T - (\gamma \varrho T)^2], Q_\theta\rangle < \varepsilon_1/2$. However, as in the proof of Lemma 3, one has $[(\gamma^2\varrho) - (\gamma\varrho)^2]\, T \leqq \gamma^2 \varrho T - (\gamma \varrho T)^2$. Thus

$$[\gamma^2\varrho - (\gamma\varrho)^2]\, TQ_\theta < \frac{\varepsilon_1}{2}$$

and

$$[\gamma^2\varrho - (\gamma\varrho)^2]\, P_\theta < \frac{\varepsilon_1}{2} + \tfrac{1}{2}\|\gamma\|^2\, \|P_\tau - TQ_\tau\| \leqq \frac{\varepsilon_1}{2} + a^2\delta < \varepsilon_1 .$$

This concludes the proof of the implications (A) ⇒ (B) ⇒ (C). As already noted the implication (C) ⇒ (A) is trivial and the theorem is completely proved.

Remark 1. The preceding Theorem 3 has some peculiarities of structure which deserve comment. Keeping Z, Γ and W fixed as usual, the property (B) can be broken into two pieces: (B_1) the set $D(f, \mathscr{F})$ has one and only one element σ, and (B_2) this σ is nonrandomized.

It should be noted that taken individually the properties (B_i) involve the experiment $\mathscr{F}$ itself and not only its equivalence class. It is a common occurrence that a procedure which is nonrandomized on a suitable version of $\mathscr{F}$ will be randomized on an equivalent but different version $\mathscr{F}'$.

In spite of this the combination (B) of (B_1) and (B_2) is a property of equivalence classes, as stated in (A) and emphasized by (C).

Remark 2. In the statement of (C) we have used sets S such that $\{Q_\theta;\ \theta \in \Theta\}$ is precompact. One could as well use finite sets. However, one cannot hope to obtain the final inequalities of property (C) by requiring that the procedures $W_\theta \varrho P_\theta \leqq f(\theta)$ on all of Θ or even by requiring in addition that $\Delta(\mathscr{F}, \mathscr{E})$ be very small. This is clear from the wording of property (A) but is also indicated by the fact that when the Q_θ are nonatomic the semistrong neighborhood system of $\sigma \in \mathscr{B}(\mathscr{F})$ is much finer than its weak neighborhood system (Lemma 1, Section 2).

Remark 3. There are cases where the weak and semistrong topologies coincide. For instance, this happens if $L(\mathscr{F})$ is finite dimensional. In such cases one could have conjectured that Theorem 3 would be a direct corollary of Theorem 2, without assuming nonrandomization properties for σ. The equivalence of (A) and (B) says that such is not the case.

In brief, there is no hope of obtaining property (C) for randomized procedures unless one is willing to place additional restrictions on the particular versions of the experiments $\mathscr{E}$ and $\mathscr{F}$ to be used. Since in the applications, the structure of the experiments called $\mathscr{E}$ here is usually nondescript, it is not clear how much can be achieved in this direction.

Remark 4. The foolowing feature, present in Theorems 1, 2 and 3, may be worth emphasizing. A look at property (C) shows that its structure is of the form "... given $(Z, \Gamma, W, \mathscr{F}, f, A, S, \varepsilon)$ there exists (δ, S_1, V, α) such that ... ". Thus the neighboring experiments $\mathscr{E}$ do not enter in the determination of the finite set S_1. The determination of (δ, S_1, V, α) is carried out entirely on the experiment $\mathscr{F}$ itself. However, the whole argument depends on keeping $\mathscr{F}$ fixed.

5. Some applications to the LAN case

Many of the assumptions customarily used in asymptotic arguments lead to limiting experiments of the Gaussian shift type. Since our purpose here is only to exemplify the use of Theorems 1, 2 and 3 we shall consider only the finite dimensional situation. It should however be noted that the infinite dimensional case arises very naturally. The book by Hájek and Šidák [5] contains numerous examples. Some others are treated in detail in a forthcoming paper of W. Moussatat.

Consider then a real vector space Θ equipped with a norm, noted $\|\theta\|$, of the Euclidean or Hilbertian type. By this is meant that $\|\cdot\|$ satisfies the median identity

$$\|\tfrac{1}{2}(\theta_1 + \theta_2)\|^2 + \|\tfrac{1}{2}(\theta_1 - \theta_2)\|^2 = \tfrac{1}{2}\{\|\theta_1\|^2 + \|\theta_2\|^2\}\,.$$

Such a norm can be used to define a Gaussian family of measures as follows. Assuming that the dimension $\dim\Theta$ is a finite integer k, let $\mathscr{X}$ be a copy of Θ and let λ be a Lebesgue measure on $\mathscr{X}$. Let $\mathscr{D} = \{G_\theta;\ \theta \in \Theta\}$ be the family of measures defined by the Radon-Nikodym densities

$$\frac{\mathrm{d}G_\theta}{\mathrm{d}\lambda} = C\exp\{-\tfrac{1}{2}\|x - \theta\|^2\}\,.$$

The family $\mathscr{D}$ will be called the standard Gaussian shift experiment attached to $(\Theta, \|\cdot\|)$.

The equivalence class, or type of $\mathscr{D}$ in $\mathbf{E}(\Theta)$ can be characterized in many ways. One possibility (which would extend to infinite dimensional Hilbert spaces) is as follows.

First, one can always define a log likelihood ratio

$$\Lambda(G_\theta, G_0) = \log\frac{\mathrm{d}G_\theta}{\mathrm{d}G_0}$$

as the logarithm of the Radon-Nikodym density $\mathrm{d}G_\theta/\mathrm{d}G_0$ of the part of G_θ which is dominated by G_0. Allowing infinite values, this is a random variable for the distributions induced by the denominator measure G_0.

Second, note that $\mathscr{D}$ is *homogeneous* in the sense that any two measures G_s, G_t, $(s, t) \in \Theta \times \Theta$ are always mutually absolutely continuous.

Letting $L(\theta)$ be the process $L(\theta) = \Lambda(G_\theta, G_0) + \frac{1}{2}\|\theta\|^2$, the type of $\mathcal{D}$ in $\mathbf{E}(\Theta)$ is characterized by the requirements that

(i) For every pair $(s, t) \in \Theta \times \Theta$, one has $L(s + t) = L(s) + L(t)$ almost surely.
(ii) $\mathcal{D}$ is homogeneous.

Let $\{\mathscr{E}_n = P_{\theta,n}; \theta \in \Theta\}$ be a sequence of experiments indexed by the same space $(\Theta, \|\cdot\|)$. Let $L_n(\theta) = \Lambda(P_{\theta,n}; P_{0,n}) + \frac{1}{2}\|\theta\|^2$.

Lemma 4. *The sequence $\{\mathscr{E}_n\}$ converges weakly in $\mathbf{E}(\Theta)$ to the Guassian experiment $\mathcal{D}$ if and only if for every pair $(s, t) \in \Theta \times \Theta$,*

(i) *the variables $L_n(s + t) - [L_n(s) + L_n(t)]$ tend to zero in $\{P_{0,n}\}$ probability,*
(ii) *the sequences $\{P_{s,n}\}$ and $\{P_{t,n}\}$ are contiguous.*

This is well known.

Definition. One says that a sequence $\{\mathscr{E}_n\}$, $\mathscr{E}_n = \{P_{\theta,n}; \theta \in \Theta\}$ indexed by the space $(\Theta, \|\cdot\|)$ satisfies condition (GS), or is *asymptotically standard Gaussian shift* if the conditions of Lemma 4 are satisfied.

It should be clear that the above condition (GS) differs from the LAN assumptions of [1] only by unessential notational conventions.

The present author [3] used somewhat stronger assumptions which could be phrased as follows.

Definition. The sequence $\{\mathscr{E}_n\}$ satisfies the condition (GS, K) if it satisfies (GS) and if the relation $\|t_n - t\| \to 0$ implies $L_n(t_n) - L_n(t) \to 0$, in probability.

The (GS, K) assumption is easily seen to be equivalent to the condition that for every compact set $K \subset \Theta$ the distance $\Delta_K(\mathscr{E}_n, \mathcal{D})$ tends to zero as $n \to \infty$. Even though this is stronger than (GS), it is definitely weaker than the assumptions used in [6]. The latter would involve Prohorov type convergence of stochastic processes, while both (GS) and (GS, K) are only concerned with convergences of finite dimensional (four dimensional in fact) distributions in these processes.

To exemplify the use of the results of Section 4 consider first a testing problem. Assume for this that H_0 is a hyperplane in Θ and that H_1 is one of the open half spaces generated by H_0. Take a decision space Z consisting of two points, $Z = \{0, 1\}$ and a loss function W such that $W_\theta(0) = 1 - W_\theta(1) = 0$ for $\theta \in H_0$ and $W_\theta(0) = 1 - W_\theta(1) = 1$ for $\theta \in H_1$.

The observed random vector X of $\mathcal{D}$ can be written $X = X_0 + \xi e_1$ for $X \in H_0$ equal to the orthogonal projection of X on H_0 and for a unit vector $e_1 \in H_1$ orthogonal to H_0. The customary test for $\mathcal{D}$ is to reject H_0 if ξ is larger than some constant b. Let φ be the corresponding test function. It is a function defined on $\mathscr{X}$ such that $\varphi(x) = 0$ if $\xi < b$ and $\varphi(x) = 1$ if $\xi \geq b$.

Suppose now that the sequence $\mathscr{E}_n = \{P_{\theta,n};\ \theta \in \Theta\}$ satisfies the condition (GS). One can then expand the logarithm of likelihood ratio $\Lambda(P_{\theta,n};\ P_{0,n})$ in the form

$$\Lambda(P_{\theta,n};\ P_{0,n}) = \theta X_n' - \tfrac{1}{2}\|\theta\|^2 + \varepsilon_n$$

with an ε_n which tends to zero in probability.

Proposition 3. *Let f be the risk function of φ for the Gaussian experiment $\mathscr{D}$. Assuming that $\{\mathscr{E}_n\}$ satisfies (GS), let φ_n be any test available for $\mathscr{E}_n$, subject to the restriction that the risk function of φ_n converges pointwise on $H_0 \cup H_1$ to the limit f. Then $\int|\varphi_n - \varphi(X_n)|\,\mathrm{d}P_{\theta,n}$ tends to zero as $n \to \infty$. Furthermore, if (GS, K) is satisfied the convergence to zero is uniform on the compacts of Θ, for suitable selection of X_n.*

Proof. For the experiment $\mathscr{D}$ the test φ is nonrandomized. It is also admissible. Since the linear space of $\{G_\theta;\ \theta \in H_0 \cup H_1\}$ is dense in $L(\mathscr{D})$ any admissible test is uniquely determined by its risk function. Thus, the pair $(f, \mathscr{D})$ satisfies property (B) of Section 4.

Note also that the expectations $\int\varphi(x_n)\,P_{\theta,n}(\mathrm{d}x_n)$ converge pointwise to $\int\varphi\,\mathrm{d}G_\theta$, because φ has a set of discontinuities of measure zero. Therefore, the tests φ_n^* defined by $\varphi_n^*(x) = \varphi(x_n)$ have limiting risk function equal to f.

This being the case, the result is an immediate consequence of Theorem 3.

Remark. The Gaussian character of $\mathscr{D}$ does not enter fully in the preceding argument. One can obtain similar results for many exponential families.

Even for the Gaussian experiment $\mathscr{D}$, the example selected here is just one of very many possibilities. Completeness of the family $\{G_\theta;\ \theta \in H_0 \cup H_1\}$ is readily available for many sets of the type $H_0 \cup H_1$. The one crucial feature which is not always easily checked is the admissibility of the test φ. However, admissibility of tests has been proved for a substantial variety of situations (see for instance [7], [8], [9] and especially the first theorem in [8]. The argument of Proposition 3 carries over naturally to these situations.

Leaving testing aside, let us consider some examples of estimation problems. For the purpose one may take a decision space Z which is a copy of Θ. The vector lattice Γ can be taken equal to the space of bounded continuous functions defined on Z.

To duplicate the first part of Hájek's Theorem 4.1, let ℓ be a function from Θ to $[0, \infty)$. Assume that ℓ is such that for every real u the set $\{\theta;\ \ell(\theta) \leqq u\}$ is a closed convex symmetric subset of Θ. Take for loss function the function $W_\theta(z) = \ell(z - \theta)$.

For such loss functions an application of T. W. Anderson's lemma [10] shows that, in the Gaussian case, a minimax estimator is the observed value $X \in \mathscr{X}$ itself.

Thus, the value of the minimax risk is simply $a = E\,\ell(Y)$, for a vector Y which is $\mathscr{N}(0, I)$ on $\mathscr{X}$.

Suppose then that $\{\mathscr{E}_n\}$, $\mathscr{E}_n = \{P_{\theta,n};\ \theta \in \Theta\}$ satisfies the (GS) condition. Let W^b be defined by $W_\theta^b(z) = \min\{b, W_\theta(z)\}$. Take any decision procedure ϱ_n available on $\mathscr{E}_n$ and let $R_n(\theta, \varrho_n, b)$ be its risk for the loss function W^b.

One of the consequences of Theorem 1 is that

$$\liminf_{b \to \infty} \sup_S \limsup_n \sup_{\theta \in S} R_n(\theta, \varrho_n, b) \geqq a\,,$$

S being allowed to run over finite subsets of Θ.

This particular statement does not involve any restriction on the dimension of Θ, except that, with the notation used here, $\dim \Theta = k$ is finite.

On the contrary, the second part of Hájek's Theorem 4.1 seems to involve a restriction on the dimension of Θ. In fact, Hájek uses very strongly the assumption that $\dim \Theta = 1$. This is directly related to the fact that the usual invariant estimate for the normal distribution is admissible only when $\dim \Theta \leqq 2$.

To obtain a simple example of application of Theorem 3 one can proceed as follows. Take $Z = \Theta$ and $\Gamma = C^b(Z)$ as before.

Consider a loss function W such that for each $\theta \in \Theta$ the function $z \to W_\theta(z)$ is a strictly convex function of z. Assume also that $\lim_{\|z\| \to \infty} W(z) = \infty$.

For such loss functions admissible estimates form, as usual, a complete class. However, it is easily seen that in this strictly convex case an admissible estimate cannot be randomized.

In brief, if f is an admissible risk function for $\mathscr{D}$ and for a strictly convex loss function W, as described, then the pair $(f, \mathscr{D})$ enjoys the properties (A), (B), (C) of Section 4.

The uniqueness of the procedure which achieves the risk f follows from the fact that if σ_1 and σ_2 are different procedures achieving f then $\sigma = \frac{1}{2}(\sigma_1 + \sigma_2)$ would be a randomized procedure with the same risk f. This is impossible, hence the uniqueness.

It follows from the foregoing that Theorem 3, Section 4, is applicable to the case described. To illustrate further, let us assume that the sequence $\{\mathscr{E}_n\}$, $\mathscr{E}_n = P_{\theta,n};\ \theta \in \Theta$ satisfies the requirement (GS). Assume that $P_{\theta,n}$ is a measure on a space $(\mathscr{Y}_n, \mathscr{B}_n)$ and consider (nonrandomized) estimates $\tau_{n,i}$ which are functions from $(\mathscr{Y}_n, \mathscr{B}_n)$ to Z.

Letting $W^b = \min(b, W)$ as before these estimates have risks

$$R_{n,i}(\theta, b) = \int W_\theta^b[\tau_{n,i}(y)]\, P_{\theta,n}(dy)\,.$$

For the limiting experiment $\mathscr{D}$, let τ be the function from $\mathscr{X}$ to Z which achieves the admissible risk $f(\theta) = \int W_\theta[\tau(x)]\, G_\theta(dx)$ for W itself. Then, according to Theorem 3, the relations

$$\lim_b \lim_n R_{n,i}(\theta, b) \leqq f(\theta)\,, \quad \theta \in \Theta\,, \quad i = 1, 2$$

always imply that $\tau_{n,1} - \tau_{n,2}$ tend to zero in $P_{\theta,n}$ probability, for all θ.

If τ is continuous and if X_n is the term in the expansion

$$\log \frac{dP_{\theta,n}}{dP_{0,n}} = \theta X_n' - \tfrac{1}{2}\|\theta\|^2 + \varepsilon_n$$

one could also say that $\tau_{n,i}(y) - \tau(X_n)$ converges in probability to zero.

(It is a result of [11] that for the Gaussian case considered here, the admissible estimates τ are always continuous.)

Note that the foregoing illustration does not actually use the full information provided by Theorem 3.

It is also very clear that the argument depends on the Gaussian character of $\mathscr{D}$ only very slightly. Thus, the result would be applicable in many other cases as well.

References

[1] Hájek, J. (1972). Local asymptotic minimax and admissibility in estimation. *Proc. of the Sixth Berkeley Symposium on Math. Stat. and Prob.* University of California Press, Vol. 1, 175—194.

[2] Fisher, R. A. (1925). Theory of statistical estimation. *Proc. Cambridge Philos. Soc.*, **22**, 700—725.

[3] Le Cam, L. (1974). Notes on asymptotic methods in statistical decision theory. *Centre de Recherches Mathématiques*, Université de Montréal, Chapter 1.

[4] Le Cam, L. (1972). Limits of experiments. *Proc. of the Sixth Berkeley Symposium on Math. Stat. and Prob.* University of California Press, Vol. 1, 245—261.

[5] Hájek, J. - Šidák, Z. (1967). "Theory of rank tests". Academia, Prague.

[6] Hájek, J. (1971). Limiting properties of likelihoods and inference. *Foundations of Statistical Inference*, V. P. Godambe and D. A. Sprott, eds. Holt Rinehart and Winston, Toronto, Montreal, 142—162.

[7] Birnbaum, A. (1955). Characterizations of complete classes of tests of some multiparametric hypotheses, with applications to likelihood ratio tests. *Ann. Math. Statist.*, **26**, 21—36.

[8] Stein, C. (1956), The admissibility of Hotelling's T^2 test. *Ann. Math. Statist.*, **27**, 616—623.

[9] Matthes, T. K. - Truax, D. R. (1967). Tests of composite hypotheses for the multivariate exponential family. *Ann. Math. Statist.*, **38**, 681—697.

[10] Anderson, T. W. (1955). The integral of a symmetric unimodal function over a symmetric convex set and some probability inequalities. *Proc. Amer. Math. Soc.*, **6**, 170—176.

[11] Brown, L. D. (1971). Admissible estimators, recurrent diffusions and insoluble boundary value problems. *Ann. Math. Statist.*, **42**, 855—903.

UNIVERSITY OF CALIFORNIA, BERKELEY, CALIFORNIA, U.S.A.

Received November 1974
Revised September 1975

ON AGGREGATING CONTROLLED MARKOV CHAINS

by

PETR MANDL

1. Introduction

Consider a system A, which is checked at times $0, \Delta, 2\Delta, \ldots$. Let there be only finitely many possible states of A. We label them by numbers and identify the state space with $I = \{1, 2, \ldots, r\}$. The observed states form a random sequence $\{X_n, n = 0, 1, \ldots\}$. A is a *controlled Markov chain* if the following holds. Whenever state i is observed, the probability distribution of the next observation is

$$[p(i, 1; z), \; p(i, 2; z), \ldots, p(i, r; z)] \, , \quad z \in \mathscr{Z}(i) \, ,$$

where z is a control parameter, $\mathscr{Z}(i)$ its range in state i. The parameter is selected in accordance with a control policy (briefly a control) depending only on events which occurred up to the present time (non-anticipativity). Let Z_n denote the control parameter value chosen at time $n\Delta$, $n = 0, 1, \ldots$. Thus

$$P(X_{n+1} = i \mid \text{events up to } n\Delta) = p(X_n, i; Z_n) \, . \tag{1}$$

The simplest controls are the stationary ones, under which

$$Z_n = z(X_n) \, , \quad n = 0, 1, \ldots \, , \tag{2}$$

i.e., the parameter depends on the actual state only. The elements of the set $\mathscr{U}$ of all stationary controls are therefore functions mapping $i \in I$ into $z(i) \in \mathscr{Z}(i)$. This will be expressed in symbols as $u \sim z(i)$. If (2) holds, then $\{X_n, n = 0, 1, \ldots\}$ is a Markov chain with transition probability matrix $\|p(i, j; z(i))\|_{i,j \in I}$.

With the trajectory of the system a sequence

$$S_n = S_0 + \sum_{m=0}^{n-1} \sigma(X_m, X_{m+1}; Z_m) \, , \quad n = 1, 2, \ldots \, ,$$

is associated. S_n will be interpreted here as the reward (gain) accumulated up to time $n\Delta$.

In the paper *aggregates* will be considered, consisting of independent units of a small number of types, say $A', A'', \ldots$. Independent means that the transitions obeying (1) occur independently. The number of units may be large or not, it is essential that Δ is small in the time scale of the entire aggregate. The aggregate is again a controlled Markov chain, the states of which are vectors

$$J_n = (j_1'(n), \ldots, j_{r'}'(n), j_1''(n), \ldots, j_{r''}''(n), \ldots),$$

giving the distribution of the states of the units. It is assumed that the rewards of the units are accumulated. The totals obtained are denoted by V_n, $n = 0, 1, \ldots$.

Optimization problems in which the criterion is additive with respect to the rewards of the units, like the expected discounted reward

$$\mathsf{E} \sum_{n=1}^{\infty} e^{-nh\Delta}(V_n - V_{n-1}), \quad h > 0, \tag{3}$$

have a rather straight-forward solution. Namely, each unit has to use the control maximizing its part of the criterion. Several convenient criteria do not fulfil this assumption. For example in the problem of keeping $\{V_n, n = 0, 1, \ldots\}$ within prescribed bounds the criterion could be

$$\mathsf{E} e^{-R h \Delta} \quad \text{where} \quad R = \inf\{n : V_n \notin (l_0, l_1)\}. \tag{4}$$

(One can imagine that units of some types are producers, and units of the remaining types are consumers, and that the problem is to maintain the stocked quantity of a commodity within desired limits.) In general the control minimizing $\mathsf{E} e^{-Rh\Delta}$ will be a function of (J_n, V_n), and therefore rather awkward to compute. In the paper we shall be concerned with controls of the form $\{u(V_n), n = 0, 1, \ldots\}$, where $u(x)$ maps $(-\infty, \infty)$ into $\mathcal{U}' \times \mathcal{U}'' \times \ldots$. The interpretation is as follows. The choice of $u = (u', u'', \ldots)$, $u' \sim z'(i)$, $u'' \sim z''(i)$, ..., means that the control parameter is made equal $z'(1), \ldots, z'(r')$ for all units of type A' in state $1, \ldots, r'$, respectively. Similarly for units of type A'', etc. First we shall give conditions, under which $\{V_n, n = 0,1,\ldots\}$ converges in distribution to a controlled difffusion process as $\Delta \to 0$. Then we shall show, how optimization methods for Markov processes with boundaries can be used to obtain approximately optimal controls in the mentioned class.

The aim of the paper is to suggest a procedure, how to control a large number of quickly unfolding units together with a method of computing the control policy. An approach developed previously by the author ([5]) is used. First steps in this direction of research were presented in [3]. Although the motivation of the paper does not come from a particular applied problem, it is supposed that the results have practical value.

2. Stationary controls

In this section we state assumptions about the controlled Markov chains (the units of the aggregate) we shall consider, and restate some facts about stationary controls.

Assumption 1. *$\mathscr{Z}(i)$, $i \in I$, are closed bounded subsets of R^s. $p(i, j; z)$, $\sigma(i, j; z)$, $i, j \in I$, are continuous functions on $\mathscr{Z}(i)$.*

Assumption 2. For each $u \in \mathscr{U}$, $u \sim z(i)$, *the matrix $\|p(i, j; z(i))\|_{i,j \in I}$ is not totally decomposable.*

Introduce

$$\varrho_1(i, z) = \sum_j p(i, j; z)\, \sigma(i, j; z)\,, \quad i \in I\,, \quad z \in \mathscr{Z}(i)\,,$$

$$\varrho_2(i, z) = \sum_j p(i, j; z)\, \sigma(i, j; z)^2\,, \quad i \in I\,, \quad z \in \mathscr{Z}(i)\,.$$

Take $u \in \mathscr{U}$, $u \sim z(i)$, and assume that (2) holds. Then, in virtue of Assumption 2, $\{X_n, n = 0, 1, \ldots\}$ is a Markov chain with only one class of recurrent states. Central limit theorem holds for

$$S_n = S_0 + \sum_{m=0}^{n-1} \sigma(X_m, X_{m+1}; z(X_m))\,, \quad n = 1, 2, \ldots\,,$$

$(S_n - \alpha(u))/\sqrt{n}$ has asymptotically normal distribution $N(0, \beta(u)^2)$, as $n \to \infty$. (The case $\beta(u) = 0$ is not excluded at this stage.) As known (see, e.g., [5]), the constants $\alpha(u)$, $\beta(u)^2$ are obtained from two systems of linear equations

$$(5) \qquad \varrho_1(i, z(i)) + \sum_j p(i, j; z(i))\, \Phi_{1j}(u) - \Phi_{1i}(u) - \alpha(u) = 0\,, \quad i \in I\,, \quad \Phi_{11}(u) = 0\,,$$

$$(6) \qquad \varrho_2(i, z(i)) + \sum_j p(i, j; z(i))\, [2(\sigma(i, j; z(i)) - \alpha(u))\, (\Phi_{1j}(u) - \alpha(u)) + \Phi_{2j}(u)]$$
$$- \Phi_{2i}(u) - \beta(u)^2 - \alpha(u)^2 = 0\,, \quad i \in I\,, \quad \Phi_{21}(u) = 0\,,$$

for unknowns $\Phi_{11}(u), \ldots, \Phi_{1r}(u)$, $\alpha(u)$, and $\Phi_{21}(u), \ldots, \Phi_{2r}(u)$, $\beta(u)^2$. The constants $\alpha(u)$, $\beta(u)^2$ contain essential information about the control u for our purposes. The auxiliary constants $\Phi_{11}(u), \ldots, \Phi_{2r}(u)$ will be used in the proofs.

If in addition to (5) holds

$$(7) \qquad \max_{z \in \mathscr{Z}(i)} \{\varrho_1(i, z) + \sum_j p(i, j; z)\, \Phi_{1j}(u) - \Phi_{1i}(u) - \alpha(u)\} = 0\,, \quad i \in I\,,$$

then

$$\alpha(u) = \max_{v \in \mathscr{U}} \alpha(v)\,.$$

3. Limit distribution of $\{V_n, n = 0, 1, \ldots\}$

We shall prove a limit theorem for $\{V_n, n = 0, 1, \ldots\}$ under the assumption that the transition time Δ tends to zero. We have therefore to consider a *family of aggregates* defined for Δ in a interval $(0, \Delta_0]$.

For the sake of simplicity, only two types of units A' and A'' will be assumed. Let their numbers be N' and N'', respectively. All components of the definition of A' and A'' are supposed to be constant with respect to Δ except the functions σ defining the reward. Quantities associated with the types and possibly depending on Δ will be denoted by Greek letters.

To avoid exceedingly complex denotation, let us make some conventions. Primes will be added to symbols to distinguish between the two types. Relations assumed or proved for both types of units will have symbols without primes. No indexes for individual units will be introduced, but symbol $\bar{\Sigma}$ will mean summation over all units. The indication of the dependence on Δ will be omitted in most cases. K will be used as a generic symbol denoting finite positive constants. (Thus, e.g., $2K = K$ etc.).

The basic assumptions concerning the dependence of the aggregate on Δ are the following.

$$(8) \qquad |\sigma(i, j; z)| \leqq K\Delta^{1/2}(N' + N'')^{-1/2}\,, \quad i, j \in I\,, \quad z \in \mathscr{Z}(i)\,, \quad \Delta \in (0, \Delta_0]\,,$$

$$(9) \qquad \lim_{\Delta \to 0} \Delta^{-1}(N' + N'')\, \varrho_1(i, z) = r_1(i, z)\,,$$

$$(10) \qquad \lim_{\Delta \to 0} \Delta^{-1}(N' + N'')\, \varrho_2(i, z) = r_2(i, z)\,,$$

uniformly in $z \in \mathscr{Z}(i)$, $i \in I$,

$$(11) \qquad \lim_{\Delta \to 0} N(N' + N'')^{-1} = q\,.$$

According to the denotational convention made, (8)–(11) are supposed to hold for both, A' and A''. N is assumed to be bounded for $\Delta \in [\varepsilon, \Delta_0]$ for each $\varepsilon > 0$.

In further discussion quantities $a(u)$, $w_{1i}(u)$, $b^2(u)$, $w_{2i}(u)$ will have important role. They are solutions of two systems of equations obtained from (5), (6) by letting $\Delta \to 0$ after multiplication by $\Delta^{-1}(N' + N'')$.

$$(12) \quad r_1(i, z(i)) + \sum_j p(i, j; z(i))\, w_{1j}(u) - w_{1i}(u) - a(u) = 0\,, \quad i \in I\,, \quad w_{11}(u) = 0\,,$$

$$(13) \quad r_2(i, z(i)) + \sum_j p(i, j; z(i))\, w_{2j}(u) - w_{2i}(u) - b(u)^2 = 0\,, \quad i \in I\,, \quad w_{21}(u) = 0\,,$$

$$u \sim z(i) \in U\,.$$

It is not difficult to verify that

$$\lim_{\Delta \to 0} \Delta^{-1}(N' + N'')\,\Phi_{ij}(u) = w_{ij}(u)\,, \quad i = 1, 2\,, \quad j \in I\,, \tag{14}$$

$$\lim_{\Delta \to 0} \Delta^{-1}(N' + N'')\,\alpha(u) = a(u)\,, \quad \lim_{\Delta \to 0} \Delta^{-1}(N' + N'')\,\beta(u)^2 = b(u)^2\,, \tag{15}$$

uniformly in $u \in \mathscr{U}$.

Let V_0 be fixed, not depending on Δ. Choose a control $u(x)$, $x \in (-\infty, \infty)$, as defined in Section 1, not depending on Δ as well. $u(x)$ together with the initial state, on which we do not impose any restrictions, specify the probability law of the evolution of the aggregate. With random sequence $\{V_n, n = 0, 1, \dots\}$ we shall associate a continuous random function

$$\hat{V}_t = V_{[t/\Delta]} + \{t/\Delta\}\,(V_{[t/\Delta]+1} - V_{[t/\Delta]})\,, \quad t \geqq 0\,, \tag{16}$$

where $[\;]$ and $\{\;\}$ mean the integer and the fractional parts, respectively. The same denotation will be used for other random sequences. E.g., $\hat{V}_t = \hat{\Sigma}\hat{S}_t$.

Definition (16) relates to $\{V_n, n = 0, 1, \dots\}$ a probability distribution in the space of continuous functions. Take $T > 0$. Let C_T be the space of all continuous functions on $[0, T]$ with the uniform metric. Further, for $t \in [0, T]$, let $\mathscr{C}_t$ be the σ-algebra on C_T generated by the sets

$$\{f \in C_T : f(s) \leqq x\}\,, \quad s \in [0, t]\,, \quad x \in (-\infty, \infty)\,.$$

Thus in particular, $\mathscr{C}_T$ are the Borel sets in C_T. The random function $\{x_t, t \in [0, T]\}$ is defined on $(C_T, \mathscr{C}_T)$ by the relation $x_t(f) = f(t)$, $t \in [0, T]$, $f \in C_T$. The probability distribution of $\{\hat{V}_t, t \in [0, T]\}$ is the probability measure $\mathscr{P}_T^\Delta$ induced on $(C_T, \mathscr{C}_T)$ by $\{\hat{V}_t, t \in [0, T]\}$.

Theorem 1. *Let the control $u(x)$ be continuous on $(-\infty, \infty)$, and such that $a(u(x))$ is Lipschitz continuous, $b(u(x)) > 0$ on $(-\infty, \infty)$. Then $\mathscr{P}_T^\Delta$ converges, as $\Delta \to 0$, weakly to the probability distribution $\mathscr{P}_T$ such that*

$$\begin{aligned} \mathrm{d}x_t &= [q'a(u'(x_t)) + q''a''(u''(x_t))]\,\mathrm{d}t \\ &\quad + [q'b'(u'(x_t))^2 + q''b''(u''(x_t))^2]^{1/2}\,\mathrm{d}w_t\,, \quad t \in [0, T]\,, \\ &\mathscr{P}_T(x_0 = V_0) = 1\,, \end{aligned} \tag{17}$$

where $\{w_t, t \in [0, T]\}$ is a standartized Wiener process on $(C_T, \mathscr{C}_T, \mathscr{P}_T)$.

Remark 1. According to the results of [7], $\mathscr{P}_T$ is unique. Let us have $u(x)$ satisfying the hypotheses of the theorem. The proof will be decomposed in a sequence of lemmas. The abbreviated denotation $\alpha(x)$, $\beta(x)$, $\Phi_{ij}(x)$, ..., will be used instead of $\alpha(u(x))$, $\beta(u(x))$, $\Phi_{ij}(u(x))$, ... We set also

$$c(x) = q'a'(x) + q''a''(x)\,, \tag{18}$$

$$e(x) = (q'b'(x)^2 + q''b''(x)^2)^{1/2}\,. \tag{19}$$

Consider the aggregate (precisely the family of aggregates) subjected to the control $u(x)$. Let for $n = 0, 1, \ldots$ be $\mathscr{F}_n$ the σ-algebra of random events defined on the history of the aggregate up to time $n\Delta$.

Lemma 1. *Let* $M_0 = V_0$,

$$M_n = V_n - \sum_{m=0}^{n-1}(N'\alpha'(V_m) + N''\alpha''(V_m)) + \tilde{\Sigma}\sum_{m=0}^{n-1}(\Phi_{1X_{m+1}}(V_m) - \Phi_{1X_m}(V_m))\,, \tag{20}$$

$$n = 1, 2, \ldots .$$

Then $\{M_n, n = 0, 1, \ldots\}$ *is a martingale with respect to* $\{\mathscr{F}_n, n = 0, 1, \ldots\}$.

Proof. Take an arbitrary unit of the aggregate. Set

$$Y_n = \sigma(X_n, X_{n+1}; Z_n) - \alpha(V_n) + \Phi_{1X_{n+1}}(V_n) - \Phi_{1X_n}(V_n)\,, \quad n = 0, 1, \ldots . \tag{21}$$

Recall that $Z_n = z(X_n, V_n)$ where $z(i, x) \sim u(x)$. Since

$$M_{n+1} - M_n = \tilde{\Sigma} Y_n\,, \quad n = 0, 1, \ldots,$$

to verify the martingale property it is sufficient to show that

$$\mathsf{E}\{Y_n \mid \mathscr{F}_n\} = 0\,, \quad n = 0, 1, \ldots . \tag{22}$$

In virtue of (1),

$$\mathsf{E}\{Y_n \mid \mathscr{F}_n\} = \varrho_1(X_n, z(X_n, V_n)) - \alpha(V_n)$$
$$+ \sum_j p(X_n, j; z(X_n, V_n))\,\Phi_{1j}(V_n) - \Phi_{1X_n}(V_n)\,.$$

The right-hand side is equal to 0 by (5). □

Lemma 2. *Let* $\mathscr{Q}_T^\Delta$ *be the probability distribution of* $\{\hat{M}_t, t \in [0, T]\}$, *and let* $\{\Delta_j, j = 1, 2, \ldots\}$ *be an arbitrary sequence in* $(0, \Delta_0]$ *such that* $\lim_{j\to\infty} \Delta_j = 0$. *Then the sequence* $\{\mathscr{Q}_T^{\Delta_j}, j = 1, 2, \ldots\}$ *is tight. Moreover, the integrals* $\int x_t^4 \,\mathrm{d}\mathscr{Q}_T^{\Delta_j}$, $t \in [0, T]$, $j = 1, 2, \ldots$, *are uniformly bounded.*

Proof. The tightness of $\{\mathcal{Q}_T^{\Delta_j}, j = 1, 2, \ldots\}$ will be established by demonstrating (see [1])

$$(23) \qquad \lim_{\delta \to 0} \overline{\lim_{j \to \infty}} \mathcal{Q}_T^{\Delta_j}\Big(\sup_{|s-t|<\delta} |x_s - x_t| > \varepsilon \Big) = 0$$

for arbitrary $\varepsilon > 0$.

Take $\varepsilon > 0$, $\delta > 0$, $\Delta \in (0, \Delta_0]$. Set $k = [\delta/\Delta] + 1$. Then

$$(24) \qquad \mathcal{Q}_T^{\Delta}\Big(\sup_{|s-t|<\delta} |x_s - x_t| < \varepsilon \Big) \leqq \sum_{l=0}^{[T/k\Delta]} \mathcal{Q}_T^{\Delta}\Big(\sup_{lk\Delta \leqq s \leqq (l+1)k\Delta} |x_s - x_{lk\Delta}| > \tfrac{1}{3}\varepsilon \Big)$$

$$= \sum_{l=0}^{[T/k\Delta]} P\Big(\sup_{lk \leqq m \leqq (l+1)k} |M_m - M_{lk}| > \tfrac{1}{3}\varepsilon \Big) \leqq \sum_{l=0}^{[T/k\Delta]} \left(\frac{3}{\varepsilon}\right)^4 \mathsf{E}(M_{(l+1)k} - M_{lk})^4 .$$

In the last step submartingale inequality was used.

We have

$$(25) \qquad \mathsf{E}(M_{(l+1)k} - M_{lk})^4 = \mathsf{E}\Big(\sum_{n=lk}^{(l+1)k-1} \tilde{\Sigma} Y_n \Big)^4 ,$$

where Y_n is defined as in (21). Using (22) one gets

$$(26) \qquad \mathsf{E}\Big(\sum_{n=lk}^{(l+1)k-1} \tilde{\Sigma} Y_n \Big)^4 = \mathsf{E}\sum_n (\tilde{\Sigma} Y_n)^4 + 4\mathsf{E}\sum_n \sum_{m<n} (\tilde{\Sigma} Y_m)(\tilde{\Sigma} Y_n)^3$$

$$+ 6\mathsf{E}\sum_n \Big(\sum_{m<n} \tilde{\Sigma} Y_m \Big)^2 (\tilde{\Sigma} Y_n)^2 = \mathsf{E}\sum_n (3\tilde{\Sigma}\tilde{\Sigma} Y_n^2 Y_n^2 - 2\tilde{\Sigma} Y_n^4)$$

$$+ \mathsf{E}\sum_n \sum_{m<n} \tilde{\Sigma}\tilde{\Sigma} Y_m Y_n^3 + 6\mathsf{E}\sum_n \Big(\sum_{m<n} \tilde{\Sigma} Y_m \Big)^2 \tilde{\Sigma} Y_n^2 .$$

From (8) it is not difficult to deduce

$$(27) \qquad |Y_n| \leqq K\Delta^{1/2}(N' + N'')^{-1/2}, \quad n = 0, 1, \ldots .$$

Hence,

$$\mathsf{E}\sum_n \Big(\sum_{m<n} \tilde{\Sigma} Y_m \Big)^2 \tilde{\Sigma} Y_n^2 \leqq K\Delta \sum_n \mathsf{E}\Big(\sum_{m<n} \tilde{\Sigma} Y_m \Big)^2 = K\Delta \sum_n \sum_{m<n} \tilde{\Sigma}\mathsf{E} Y_m^2 ,$$

and from (26)

$$(28) \qquad \mathsf{E}\Big(\sum_{n=lk}^{(l+1)k-1} \tilde{\Sigma} Y_n \Big)^4 \leqq K\Delta^2(3k + 10k(k-1)) \leqq K\Delta^2 k^2 .$$

(24), (25), and (28) give the inequality

$$\mathcal{Q}_T^{\Delta}\Big(\sup_{|s-t|<\delta} |x_s - x_t| > \varepsilon \Big) \leqq \varepsilon^{-4} KT(\delta + \Delta) ,$$

from which (23) is immediately obtained.

To demonstrate the second assertion, start with the inequalities

$$\int x_t^4 \, d\mathcal{Q}_T^\Delta \leqq (1 - \{t/\Delta\}) \, \mathsf{E} M_{[t/\Delta]}^4 + \{t/\Delta\} \, \mathsf{E} M_{[t/\Delta]+1}^4 \leqq \max_{0 \leqq m \leqq [T/\Delta]+1} \mathsf{E} M_m^4 = \mathsf{E} M_{[T/\Delta]+1}^4 ,$$

$$t \in [0, T] .$$

$$\mathsf{E} M_{[T/\Delta]+1}^4 \leqq 8V_0^4 + 8\mathsf{E}(M_{[T/\Delta]+1} - M_0)^4$$

$$\leqq 8V_0^4 + K\Delta^2([T/\Delta] + 1)^2 \leqq 8V_0^4 + K(T + \Delta_0)^2 .$$

From here the assertion follows. □

Corollary 1.

$$\mathsf{E}(V_{n+1} - V_n)^2 \leqq K\Delta , \quad n = 0, 1, \ldots .$$

Proof. We have

$$\mathsf{E}(V_{n+1} - V_n)^2 \leqq 3[\mathsf{E}(\tilde{\Sigma} Y_n)^2 + \mathsf{E}(N' \alpha'(V_n) + N'' \alpha''(V_n))^2$$

$$+ \mathsf{E}(\tilde{\Sigma}(\Phi_{1X_{n+1}}(V_n) - \Phi_{1X_n}(V_n)))^2] .$$

The last two terms in square brackets are less than $K\Delta^2$ in virtue of (14), (15). Further

$$\mathsf{E}(\tilde{\Sigma} Y_n)^2 = \tilde{\Sigma} \mathsf{E} Y_n^2 \leqq K\Delta$$

by (27). This establishes the corollary. □

Lemma 3. *Let* $\{F_t, t \in [0, T]\}$ *be the solution of*

(29) $$F_t = \int_0^t c(F_s) \, ds + \hat{M}_t , \quad t \in [0, T] .$$

Then for each $\varepsilon > 0$

(30) $$\lim_{\Delta \to 0} P(\sup_{t \in [0,T]} |F_t - \hat{V}_t| > \varepsilon) = 0 .$$

Proof. Note that $\{F_t, t \in [0, T]\}$ is well defined by (29), since $c(x)$ is Lipschitz continuous, and therefore (29) can be solved by successive approximations. To establish (30) we shall show first that

(31) $$\hat{M}_t = \hat{V}_t - \int_0^t c(\hat{V}_s) \, ds + R_t , \quad t \in [0, T] ,$$

where

(32) $$\lim_{\Delta \to 0} P(\max_{t \in [0,T]} |R_t| > \varepsilon) = 0 .$$

Consider (20). Let us establish the negligibility of its last term. It holds

$$\mathsf{E}\sup_{0\leq n\leq T/\Delta}\Big|\tilde{\Sigma}\sum_{m=0}^{n-1}(\Phi_{1X_{m+1}}(V_m)-\Phi_{1X_m}(V_m)\Big| = \mathsf{E}\sup_{0\leq n\leq T/\Delta}\Big|\tilde{\Sigma}\Big(\sum_{m=1}^{n-1}\Phi_{1X_m}(V_{m-1})-\Phi_{1X_m}(V_m)\Big)$$

$$+\Phi_{1X_n}(V_{n-1})-\Phi_{1X_0}(V_0)\Big)\Big|$$

$$\leqq \tilde{\Sigma}\Big(\sum_{m=1}^{[T/\Delta]}\mathsf{E}\big|\Phi_{1X_m}(V_{m-1})-\Phi_{1X_m}(V_m)\big| + \mathsf{E}\sup_{0\leq n\leq T/\Delta}\big|\Phi_{1X_n}(V_{n-1})\big| + \mathsf{E}\big|\Phi_{1X_0}(V_0)\big|\Big).$$

From (14) follows that to each $\varepsilon > 0$ there exist $\delta > 0$, $\bar{\Delta} > 0$ so that

$$|\Phi_{1j}(x)-\Phi_{1j}(y)| < \varepsilon\,\Delta(N'+N'')^{-1},\quad j\in I,\quad \text{for}\quad |x-y|<\delta,\quad \Delta\in(0,\bar{\Delta}).$$

Furthermore,

$$|\Phi_{1j}(x)| \leqq K\Delta(N'+N'')^{-1},\quad j\in I,\quad x\in(-\infty,\infty).$$

Hence, using Corollary 1 and Chebyshev's inequality,

$$\mathsf{E}\sup_{0\leq n\leq T/\Delta}\Big|\tilde{\Sigma}\sum_{m=0}^{n-1}(\Phi_{1X_{m+1}}(V_m)-\Phi_{1X_m}(V_m))\Big|$$

$$\leqq \tilde{\Sigma}\Big(\sum_{m=1}^{[T/\Delta]}\big[K\Delta(N'+N'')^{-1}P(|V_m-V_{m-1}|\geqq\delta)+\varepsilon\Delta(N'+N'')^{-1}\big]$$

$$+K\Delta(N'+N'')^{-1}\Big) \leqq K\Delta T\delta^{-2}+T\varepsilon+K\Delta,\quad \text{for}\quad \Delta\in(0,\bar{\Delta}).$$

The right-hand side tends to $T\varepsilon$ as $\Delta\to 0$, where ε is arbitrarily small.

The other sum in (20) is dealt with as follows.

$$\Big|\sum_{m=0}^{n-1}(N'\alpha'(V_m)+N''\alpha''(V_m)) - \Delta\sum_{m=0}^{n-1}c(V_m)\Big|$$

$$=\Delta\left|\sum_{m=0}^{n-1}\left(\frac{N'}{N'+N''}\Delta^{-1}(N'+N'')\,\alpha'(V_m)-q'\,a'(V_m)\right.\right.$$

$$\left.\left.+\frac{N''}{N'+N''}\Delta^{-1}(N''+N'')\,\alpha''(V_m)-q''\,a''(V_m)\right)\right|.$$

In virtue of (11), (15) this tends to zero uniformly in $n \leqq T/\Delta$. The Lipschitz continuity of $c(x)$ and Corollary 1 imply

$$\mathsf{E}\sup_{0\leq n\leq T/\Delta}\left|\int_0^{n\Delta}c(\hat{V}_s)\,ds-\Delta\sum_{m=0}^{n-1}c(V_m)\right| \leqq \mathsf{E}\sum_{m=0}^{[T/\Delta]}\Delta K|V_m-V_{m+1}| \leqq TK\sqrt{\Delta}.$$

From here (31) and (32) easily follow.

(29), (31) together with the Lipschitz continuity of $c(x)$ yield

(33) $$|F_t - \hat{V}_t| \leqq K \int_0^t |F_s - \hat{V}_s| \, ds + |R_t| , \quad t \in [0, T] .$$

(33) implies

(34) $$\sup_{t\in[0,T]} |F_t - \hat{V}_t| \leqq (\sup_{t\in[0,T]} |R_t|) \exp \{KT(\sup_{t\in[0,T]} |R_t|)\} .$$

(30) is an immediate consequence of (32) and (34). □

Let us summarize the estimates derived in the proof of Lemma 3.

Corollary 2.

(35) $$V_n = M_n + \int_0^{n\Delta} c(\hat{V}_s) \, ds + Q_n , \quad n = 0, 1, \dots ,$$

where the remainder terms Q_n *fulfil*

(36) $$\mathsf{E} \sup_{0\leqq m\leqq n} |Q_m| \leqq (K + \varepsilon n) \Delta , \quad n = 0, 1, \dots ,$$

with $\lim_{\Delta\to 0} \varepsilon = 0$.

Let $\{\Delta_j, j = 1, 2, \dots\}$ be an arbitrary sequence in $(0, \Delta_0]$ such that $\lim_{j\to\infty} \Delta_j = 0$, and $\{\mathcal{Q}_T^{\Delta_j}, j = 1, 2, \dots\}$ has a weak limit $\mathcal{Q}_T$, as $j \to \infty$.

Lemma 4. *Let* $t, s, s_1, \dots, s_k$ *be numbers from* $[0, T]$, *and let* $f(v_1, \dots, v_k)$ *be a bounded continuous function on* R^k. *Then*

(37)
$$\lim_{j\to\infty} \int (x_t - x_s)^i f(x_{s_1}, \dots, x_{s_k}) \, d\mathcal{Q}_T^{\Delta_j} = \int (x_t - x_s)^i f(x_{s_1}, \dots, x_{s_k}) \, d\mathcal{Q}_T , \quad i = 1, 2 .$$

Proof. By Lemma 2,

$$\int (x_t - x_s)^4 \, d\mathcal{Q}_T^{\Delta_j} \leqq K , \quad j = 1, 2, \dots .$$

This implies the uniform integrability of

$$(x_t - x_s)^i f(x_{s_1}, \dots, x_{s_k}) , \quad i = 1, 2 ,$$

which together with the weak convergence of $\{\mathcal{Q}_T^{\Delta_j}, j = 1, 2, \dots\}$ to $\mathcal{Q}_T$ gives (37). □

Lemma 5. $\{x_t, t \in [0, T]\}$ *is on* $(C_T, \mathscr{C}_T, \mathcal{Q}_T)$ *a quadratically integrable martingale with respect to* $\{\mathscr{C}_t, t \in [0, T]\}$.

Proof. Let $t > s > s_k > \ldots > s_1$, and $f(v_1, \ldots, v_k)$ be as in Lemma 4. From the martingale property of $\{M_n, n = 0, 1, \ldots\}$ follows for $\Delta_j < s - s_k$

$$\int (x_t - x_s) f(x_{s_1}, \ldots, x_{s_k}) \, d\mathcal{Q}_T^{\Delta_j} = \mathsf{E}(\hat{M}_t - \hat{M}_s) f(\hat{M}_{s_1}, \ldots, \hat{M}_{s_k}) = 0 .$$

Hence, (37) implies

$$\int (x_t - x_s) f(x_{s_1}, \ldots, x_{s_k}) \, d\mathcal{Q}_T = 0 . \tag{38}$$

Letting $s_k \to s$, it is seen that (38) holds for $t > s \geqq s_k > \ldots > s_1$.

(38) implies

$$\mathscr{E}\{x_t - x_s \mid \mathscr{C}_s\} = 0, \quad 0 \leqq s < t \leqq T .$$

Thus $\{x_t, t \in [0, T]\}$ is a martingale on $(C_T, \mathscr{C}_T, \mathcal{Q}_T)$. The integrability of its square follows from Lemma 4. □

Lemma 6. *On $(C_T, \mathscr{C}_T, \mathcal{Q}_T)$ holds for $0 \leqq s < t \leqq T$*

$$\mathscr{E}\{(x_t - x_s)^2 \mid \mathscr{C}_s\} = \mathscr{E}\left\{\int_s^t e(y_v)^2 \, dv \mid \mathscr{C}_s\right\}, \tag{39}$$

where $\{y_t, t \in [0, T]\}$ is the solution of

$$y_t = \int_0^t c(y_s) \, ds + x_t , \quad t \in [0, T] . \tag{40}$$

Proof. As in the preceding proof, to establish (39) it suffices to show that

$$\int (x_t - x_s)^2 f(x_{s}, \ldots, x_{s_k}) \, d\mathcal{Q}_T = \int \left(\int_s^t e(y_v)^2 \, dv \, f(x_{s_1}, \ldots, x_{s_k}) \right) d\mathcal{Q}_T \tag{41}$$

for $T \geqq t > s > s_k > \ldots > s_1 \geqq 0$, and f arbitrary continuous and bounded. By Lemma 4 the left-hand side of (41) equals

$$\lim_{j \to \infty} \mathsf{E}^{\Delta_j}(\hat{M}_t - \hat{M}_s)^2 f(\hat{M}_{s_1}, \ldots, \hat{M}_{s_k}) .$$

Let, for an arbitrary unit of the aggregate, Y_n be as in (21). Then

$$\begin{aligned}
\mathsf{E}\{Y_n^2 \mid \mathscr{F}_n\} &= \mathsf{E}\{\sigma(X_n, X_{n+1}; Z_n)^2 + 2(\sigma(X_n, X_{n+1}; Z_n) \\
&\quad - \alpha(V_n))(\Phi_{1X_{n+1}}(V_n) - \alpha(V_n)) - \alpha(V_n)^2 - 2(\sigma(X_n, X_{n+1}; Z_n) \\
&\quad - \alpha(V_n) + \Phi_{1X_{n+1}}(V_n) - \Phi_{1X_n}(V_n))\, \Phi_{1X_n}(V_n) + \Phi_{1X_{n+1}}(V_n)^2 \mid \mathscr{F}_n\} \\
&= \mathsf{E}\{\Phi_{2X_n}(V_n) - \Phi_{2X_{n+1}}(V_n) + \beta(V_n)^2 \\
&\quad + \Phi_{1X_{n+1}}(V_n)^2 - \Phi_{1X_n}(V_n)^2 \mid \mathscr{F}_n\}
\end{aligned}$$

in virtue of (1), (5), (6). Consequently, for $n > m$,

$$\text{(42)} \qquad \mathsf{E}\{(M_n - M_m)^2 \mid \mathscr{F}_m\} = \tilde{\Sigma}\sum_{l=m}^{n-1} \mathsf{E}\{Y_l^2 \mid \mathscr{F}_m\} = \mathsf{E}\{\sum_{l=m}^{n-1} (N'\beta'(V_l)^2$$

$$+ N''\beta''(V_l)^2 + \tilde{\Sigma}\sum_{l=m}^{n-1} (\Phi_{2X_l}(V_l) - \Phi_{2X_{l+1}}(V_l)$$

$$+ \Phi_{1X_{l+1}}(V_l)^2 - \Phi_{1X_l}(V_l)^2) \mid \mathscr{F}_m\}.$$

Set $n = [T/\Delta]$, $m = [s/\Delta]$, $m_1 = [s_1/\Delta], \ldots, m_k = [s_k/\Delta]$.

(42) implies

$$\mathsf{E}(M_n - M_m)^2 f(M_{m_1}, \ldots, M_{m_k}) = \mathsf{E}[\sum_{l=m}^{n-1} (N'\beta'(V_l)^2 + N''\beta''(V_l)^2)$$

$$+ \tilde{\Sigma}\sum_{l=m}^{n-1} (\Phi_{2X_l}(V_l) - \Phi_{2X_{l+1}}(V_l) + \Phi_{1X_{l+1}}(V_l)^2$$

$$- \Phi_{1X_l}(V_l)^2)] f(M_{m_1}, \ldots, M_{m_k}).$$

Using the same method as in the proof of Lemma 3 one shows that

$$\text{(43)} \quad \lim_{\Delta \to 0} \left|\mathsf{E}(\hat{M}_t - \hat{M}_s)^2 f(\hat{M}_{s_1}, \ldots, \hat{M}_{s_k}) - \mathsf{E}\int_s^t e(\hat{V}_v)^2 \, dv \, f(\hat{M}_{s_1}, \ldots, \hat{M}_{s_k})\right| = 0.$$

From Lemma 3 also follows

$$\text{(44)} \qquad \lim_{\Delta \to 0} \left| \mathsf{E}\int_s^t e(\hat{V}_v)^2 \, dv \, f(\hat{M}_{s_1}, \ldots, \hat{M}_{s_k}) - \mathsf{E}\int_s^t e(F_v)^2 \, dv \, f(\hat{M}_{s_1}, \ldots, \hat{M}_{s_k}) \right| = 0.$$

Finally, from (29) and (40),

$$\text{(45)} \qquad \mathsf{E}\int_s^t e(F_v)^2 \, dv \, f(\hat{M}_{s_1}, \ldots, \hat{M}_{s_k}) = \iint_s^t e(y_v)^2 \, dv \, f(x_{s_1}, \ldots, x_{s_k}) \, d\mathscr{Q}_T^{\Delta}.$$

Setting $\Delta = \Delta_j$, and letting $j \to \infty$, one obtains (39) from (43), (44), (45). □

Corollary 3.

$$x_t = x_0 + \int_0^t e(y_s) \, dw_s, \quad t \in [0, T],$$

where $\{w_t, t \in [0, T]\}$ is a Wiener process on $(C_T, \mathscr{C}_T, \mathscr{Q}_T)$.

Proof. $\{w_t = \int_0^t e(y_s)^{-1}\, dx_s,\ t \in [0, T]\}$ is a martingale, which satisfies

$$\mathscr{E}\{(w_t - w_s)^2 \mid \mathscr{C}_s\} = t - s \quad \text{for} \quad 0 \leqq s < t \leqq T.$$

This relation is a characteristic property a Wiener process. □

Proof of Theorem 1. The theorem will be established if we show that for any sequence $\{\bar{\Delta}_k,\ k = 1, 2, \ldots\}$ in $(0, \Delta_0]$, $\lim_{k\to\infty} \bar{\Delta}_k = 0$, the sequence of measures $\{\mathscr{P}^{\bar{\Delta}_k},\ k = 1, 2, \ldots\}$ contains a subsequence converging weakly to $\mathscr{P}_T$. Let $\{\Delta_j, j = 1, 2, \ldots\}$ be a subsequence of $\{\bar{\Delta}_k, k = 1, 2, \ldots\}$ such that $\{\mathscr{Q}_T^{\Delta_j}, j = 1, 2, \ldots\}$ has a weak limit $\mathscr{Q}_T$, as $j \to \infty$. Such subsequence exists by Lemma 2.

Denote by $\mathscr{R}_T^{\Delta}$ the measure induced on $(C_T, \mathscr{C}_T)$ by $\{F_t, t \in [0, T]\}$ defined in Lemma 3. From (29), (40) follows that the random process $\{y_t, t \in [0, T]\}$ on the probability space $(C_T, \mathscr{C}_T, \mathscr{Q}_T^{\Delta})$ has probability distribution $\mathscr{R}_T^{\Delta}$ as well. Hence, from $\mathscr{Q}_T^{\Delta_j} \to \mathscr{Q}_T$, from (40), and from Corollary 3 follows $\mathscr{R}_T^{\Delta_j} \to \mathscr{P}_T$. This and (30) imply $\mathscr{P}_T^{\Delta_j} \to \mathscr{P}_T$. □

4. Limit of the criterion

In Section 1 an example of a criterion which is not additive with respect to the rewards of the units was presented. Its definition (4) involves boundary conditions, i.e., restrictions on the behaviour of the process after reaching certain limits (absorption on the boundaries l_0, l_1). The theory of one-dimensional Markovian optimization problems with boundaries is presented in [4] and extended to singular boundaries in [6]. It is applicable to processes, satisfying the hypotheses of Theorem 1. This will be exhibited in the present section for the criterion

$$\text{(46)} \qquad \mathsf{E}D \quad \text{where} \quad D = \sum_{n=1}^{R} \mathrm{e}^{-nh\Delta}(g(V_n - V_{n-1}) + (1 - g)\,\Delta)\,,$$

$$R = \inf\{n : V_n \notin (l_0, l_1)\}\,, \quad g, h \text{ constants}\,, \quad 0 \leqq g \leqq 1\,, \quad h > 0\,.$$

(4) is a special case of (46) with $g = 0$. Infinite boundaries are not excluded. D is proportional to the discounted reward up to the first exit from (l_0, l_1) with a premium for staying within the boundaries.

Theorem 2. *Under the hypotheses of Theorem 1,*

$$\lim_{\Delta\to 0} \mathsf{E}D = v(V_0)\,,$$

where

$$\text{(47)} \qquad \tfrac{1}{2}e(x)^2 \frac{\mathrm{d}^2}{\mathrm{d}x^2} v(x) + c(x)\frac{\mathrm{d}}{\mathrm{d}x} v(x) - h\, v(x) + g\, c(x) + 1 - g = 0$$

for $x \in (l_0, l_1)$, *and on the boundaries* l_0, l_1

(48)
$$\overline{\lim_{x \to l_i}} |v(x)| = \begin{cases} 0 \text{ if } l_i \text{ is finite}, \\ \text{finite if } l_i \text{ is infinite}. \end{cases}$$

$c(x), e(x)$ *are defined as in* (18), (19).

Proof. Finite boundaries are regular boundaries of the second order differential operator in (47). Infinite boundaries are inaccessible, since $c(x)$ is bounded. Thus, (see, e.g., [4]) (47) together with (48) determine $v(x)$ uniquely.

Denote $n \wedge R = \min\{n, R\}$, and consider the sum

$$\sum_{n=1}^{R} e^{-nh\Delta}(V_n - V_{n-1}) = \sum_{n=1}^{\infty} e^{-nh\Delta}(V_{n\wedge R} - V_{(n-1)\wedge R}) = \sum_{n=0}^{\infty} (1 - e^{-h\Delta}) e^{-nh\Delta} V_{n\wedge R} - V_0 .$$

The series may be integrated termwise, since $V_n - V_{n-1}$, $n = 1, 2, \ldots$, are bounded for fixed Δ. Substituting for V_n from (35), and noting that $\mathsf{E} M_{n\wedge R} = V_0$ because $\{M_n, n = 1, 2, \ldots\}$ is a martingale, one gets

(49)
$$\mathsf{E}\sum_{n=1}^{R} e^{-nh\Delta}(V_n - V_{n-1}) = \mathsf{E}\sum_{n=0}^{\infty}(1 - e^{-h\Delta}) e^{-nh\Delta}\left(V_0 + \int_0^{(n\wedge R)\Delta} c(\hat{V}_s)\,ds + Q_{n\wedge R}\right)$$
$$- V_0 = \mathsf{E}\sum_{n=1}^{R} e^{-nh\Delta}\int_{(n-1)\Delta}^{n\Delta} c(\hat{V}_s)\,ds + \bar{Q} .$$

The remainder $\bar{Q}$ is estimated using (36).

$$|\bar{Q}| \leqq \sum_{n=0}^{\infty}(1 - e^{h\Delta}) e^{-nh\Delta}(K + \varepsilon n)\,\Delta \leqq \Delta K + \varepsilon\Delta(1 - e^{-h\Delta})^{-1} .$$

The last two terms tend to zero as $\Delta \to \infty$. Consequently, from (49) follows

$$\lim_{\Delta \to 0}\left|\mathsf{E}\sum_{n=1}^{R} e^{-nh\Delta}(V_n - V_{n-1}) - \mathsf{E}\int_0^{R\Delta} e^{-hs} c(\hat{V}_s)\,ds\right| = 0 ,$$

since

$$\lim_{\Delta \to 0}\left|\mathsf{E}\sum_{n=1}^{R} e^{-nh\Delta}\int_{(n-1)\Delta}^{n\Delta} c(\hat{V}_s)\,ds - \mathsf{E}\int_0^{R\Delta} e^{-hs} c(\hat{V}_s)\,ds\right| = 0 .$$

Next we shall show that

(50)
$$\lim_{\Delta \to 0} \mathsf{E} D = \lim_{\Delta \to 0} \mathsf{E}\int_0^{R\Delta} e^{-hs}(g\,c(\hat{V}_s) + 1 - g)\,ds = \mathscr{E}\int_0^{\varrho} e^{-hs}(g\,c(x_s) + 1 - g)\,ds ,$$

where $\{x_t, t \in [0, \infty)\}$ fulfils (17), i.e.,

(51) $$\mathrm{d}x_t = c(x_t)\,\mathrm{d}t + e(x_t)\,\mathrm{d}w_t\,, \quad t \in [0, \infty)\,,$$

and $\mathscr{P}(x_0 = V_0) = 1$. Further,

$$\varrho = \inf\{t : x_t \notin (l_0, l_1)\}\,.$$

For each $T > 0$ the probability distribution of $\{x_t, t \in [0, T]\}$ is $\mathscr{P}_T$ of Theorem 1. (50) is equivalent to

(52)
$$\lim_{\Delta \to 0} \int_0^\infty \mathrm{e}^{-hs} \mathsf{E}\chi\{R\Delta > s\}\,(g\,c(\hat{V}_s) + 1 - g)\,\mathrm{d}s = \int_0^\infty \mathrm{e}^{-hs} \mathscr{E}\chi\{\varrho > s\}\,(g\,c(x_s) + 1 - g)\,\mathrm{d}s,$$

where $\chi\{\,\}$ is the indicator function. For the validity of (52) it suffices that

(53) $$\lim_{\Delta \to 0} \mathsf{E}\chi\{\hat{V}_s \in (l_0, l_1),\ s \leqq t\}\,(g\,c(\hat{V}_t) + 1 - g)$$
$$= \mathscr{E}\chi\{x_s \in (l_0, l_1),\ s \leqq t\}\,(g\,c(x_t) + 1 - g) \quad \text{for} \quad t \in [0, \infty)\,.$$

The difference between $\{R_\Delta > s\}$ and $\{\hat{V}_s \in (l_0, l_1),\ s \leqq t\}$ is negligible in the limit. In the denotation of Theorem 1, (53) is

(54)
$$\lim_{\Delta \to 0} \int_{\{x_s \in (l_0, l_1), s \leqq t\}} (g\,c(x_t) + 1 - g)\,\mathrm{d}\mathscr{P}_t^\Delta = \int_{\{x_s \in (l_0, l_1), s \leqq t\}} (g\,c(x_t) + 1 - g)\,\mathrm{d}\mathscr{P}_t\,.$$

(54) follows from the weak convergence of $\mathscr{P}_t^\Delta$ to $\mathscr{P}_t$, and from the fact that the boundary of the integration domain

$$\{x_s \in [l_0, l_1],\ s \leqq t\} - \{x_s \in (l_0 - l_1),\ s \leqq t\}$$

has $\mathscr{P}_t$ measure zero.

Let, for the rest of the proof, $\{x_t, t \in [0, \infty)\}$ be a random process satisfying (51) with $x_0 = x$. This will be stressed in the symbol $\mathscr{E}_x$ for mathematical expectation. In virtue of (50), to establish the theorem we have to prove

(55) $$v(x) = \mathscr{E}_x \int_0^\varrho \mathrm{e}^{-hs}(g\,c(x_s) + 1 - g)\,\mathrm{d}s\,.$$

If ϱ is finite, then $x_\varrho = l_i$, where l_i is finite. Therefore $v(x_\varrho) = 0$. From the Itô formula and from (47), (51) we get

$$v(x) = -\int_0^\varrho d(e^{-hs} v(x_s)) = -\int_0^\varrho e^{-hs}\left(-hv(x_s)\, ds + \left(\frac{d}{dx} v(x_s)\right)(c(x_s)\, ds\right.$$

$$\left. + e(x_s)\, dw_s) + \frac{1}{2}\left(\frac{d^2}{dx^2} v(x_s)\right) e(x_s)^2\, ds\right) = \int_0^\varrho e^{-hs}(g\, c(x_s) + 1 - g)\, ds$$

$$+ \int_0^\varrho e^{-hs} e(x_s)\, dw_s\,.$$

From here (55) follows, since the mathematical expectation of the last term is zero. □

5. Optimization

In equation (47) $c(x)$, $e(x)^2$ are $c(u(x))$, $e(u(x))^2$, where

(56) $$c(u) = q'\, a'(u') + q''\, a''(u''),\quad e(u)^2 = q'\, b'(u')^2 + q''\, b''(u'')^2\,,$$

$$u \in \mathcal{U}' \times \mathcal{U}''\,.$$

Thus, the maximization of $v(V_0)$ is a one-dimensional Markovian optimization problem with compact parameter space $\mathcal{U}' \times \mathcal{U}''$ and continuous coefficients (56). The subsequent lemma gives a sufficient condition for optimality.

Lemma 7. *If in addition to* (47) *and* (48) *holds*

(57) $$\max_{u \in \mathcal{U}' \times \mathcal{U}''} \left\{\tfrac{1}{2} e(u)^2 \frac{d^2}{dx^2} v(x) + c(u) \frac{d}{dx} v(x) - h\, v(x) + g\, c(u) + 1 - g\right\} = 0$$

then the control $u(x)$ *maximizes* $v(x)$ *for all* $x \in (l_0, l_1)$.

Proof. Let (57) hold, and let $\bar{u}(x)$ be an arbitrary control satisfying the hypotheses of Theorem 1. Denote $\bar{c}(x) = c(\bar{u}(x))$, $\bar{e}(x) = e(\bar{u}(x))$. Let $\bar{v}(x)$ fulfill

(58) $$\tfrac{1}{2}\bar{e}(x)^2 \frac{d^2}{dx^2} \bar{v}(x) + \bar{c}(x) \frac{d}{dx} \bar{v}(x) - h\, \bar{v}(x) + g\, \bar{c}(x) + 1 - g = 0\,,$$

$$x \in (l_0, l_1)\,,$$

with boundary conditions (48). Inserting $\bar{u}(x)$ for u in the curly brackets in (57), and subtracting (58) one obtains

(59) $$\tfrac{1}{2}\, \bar{e}(x)^2 \frac{d^2}{dx^2} (v(x) - \bar{v}(x)) + \bar{c}(x) \frac{d}{dx} (v(x) - \bar{v}(x)) - h(v(x) - \bar{v}(x))$$

$$+ f(x) = 0\,,\quad x \in (l_0, l_1)\,,$$

where $f(x) \geqq 0$. Consequently, as in (55),

$$v(x) - \bar{v}(x) = \bar{\mathscr{E}}_x \int_0^{\varrho} e^{-hs} f(x_s)\, ds \geqq 0 . \quad \square \tag{60}$$

From Theorem 2 and Lemma 7 follows that we have to solve (57) with boundary conditions (48). This does not lead necessarily to a smooth control $u(x)$ as required in Theorem 1. However, $f(x)$ in (59) can be made small by appropriate choice of $\bar{u}(x)$ satisfying the hypotheses of Theorem 1. From (60) it is seen that the error in the criterion is then small as well.

Next we present some thoughts concerning the *numerical solution* of (57), (48). Introduce

$$\Xi = \{(c(u), \tfrac{1}{2} e(u)^2) : u \in \mathscr{U}' \times \mathscr{U}''\} .$$

(57) is then

$$\max_{(\vartheta,\zeta)\in\Xi} \left\{\zeta \frac{d^2}{dx^2} v(x) + \vartheta \left(\frac{d}{dx} v(x) + g\right) - h\, v(x) + 1 - g\right\} = 0 . \tag{61}$$

The procedure can be divided in two steps, namely in substituting a finite set Ξ_0 for Ξ, and in solving (61), (48) for Ξ_0.

i) ***Computation of*** Ξ_0. From (61) follows that only those elements of Ξ matter, which maximize

$$\vartheta \sin \psi - \zeta \cos \psi \tag{62}$$

for a $\psi \in [0, 2\pi)$. For given ψ, the pair (ϑ, ζ) maximizing (62), and the corresponding $u \in \mathscr{U}' \times \mathscr{U}''$ for which

$$\vartheta = c(u) = q'\, a'(u') + q''\, a''(u'') , \quad \zeta = \tfrac{1}{2} e(u)^2 = \tfrac{1}{2}(q'\, b'(u')^2 + q''\, b''(u'')^2) ,$$

can be obtained by solving (12), (13), and

$$\max_{z\in\mathscr{Z}(i)} \{r_1(i, z) \sin \psi - \tfrac{1}{2} r_2(i, z) \cos \psi + \sum_j p(i, j; z)\, w_j - w_i - \Theta\} = 0 , \tag{63}$$

$$i \in I ,$$

where

$$w_j = w_{1j}(u) \sin \psi - \tfrac{1}{2} w_{2j}(u) \cos \psi , \quad j \in I ,$$

$$\Theta = a(u) \sin \psi - \tfrac{1}{2} b(u)^2 \cos \psi .$$

This is done separately for A' and A''. (63) is an analogue of (7), and Howard's iteration method is one of the algorithms to solve it. Ξ_0 is constructed so that for a net $\{\psi_j, j = 0, 1, \ldots, n\}$ sufficiently dense in the relevant part of $[0, 2\pi)$, the maximizing pairs (ϑ_j, ζ_j), $j = 0, 1, \ldots, n$, are computed.

ii) ***Solution of***

$$\max_{(\vartheta,\zeta)\in\Xi_0} \left\{\zeta \frac{d^2}{dx^2} v(x) + \vartheta\left(\frac{d}{dx} v(x) + g\right) - h\, v(x) + 1 - g\right\} = 0\,. \tag{64}$$

Let $\psi_0 = \frac{1}{2}\pi$. Then $\vartheta_0 = \max\{\vartheta : (\vartheta, \zeta) \in \Xi\}$. For the sake of definiteness assume $\vartheta_0 > 0$, $l_0 = 0$, $l_1 = \infty$. Let $0 \leqq \psi_n < \psi_{n-1} < \ldots < \psi_1 < \psi_0 = \frac{1}{2}\pi$. From (55) follows that $v(x)$ fulfils

$$v(0) = 0\,, \quad \lim_{x\to\infty} v(x) = (g\vartheta_0 + 1 - g)\, h^{-1}\,, \tag{65}$$

since the influence of the boundary vanishes as $x \to \infty$. It will either hold

$$\zeta_0 \frac{d^2}{dx^2} v(x) + \vartheta_0\left(\frac{d}{dx} v(x) + g\right) - h\, v(x) + 1 - g = 0\,, \quad x \in (0, \infty)\,, \tag{66}$$

or there will be an m, $1 \leqq m \leqq n$, and $\xi_m = 0 < \xi_{m-1} < \ldots < \xi_0 < \xi_{-1} = \infty$ such that

$$\zeta_j \frac{d^2}{dx^2} v(x) + \vartheta_j\left(\frac{d}{dx} v(x) + g\right) - h\, v(x) + 1 - g = 0\,, \tag{67}$$

$$x \in (\xi_j, \xi_{j-1})\,, \quad j = 0, 1, \ldots, m\,,$$

$$v(\xi_{j-1} - 0) = v(\xi_{j-1} + 0)\,, \tag{68}$$

$$\frac{d}{dx} v(\xi_{j-1} - 0) = \frac{d}{dx} v(\xi_{j-1} + 0)\,, \quad j = 1, 2, \ldots, m\,.$$

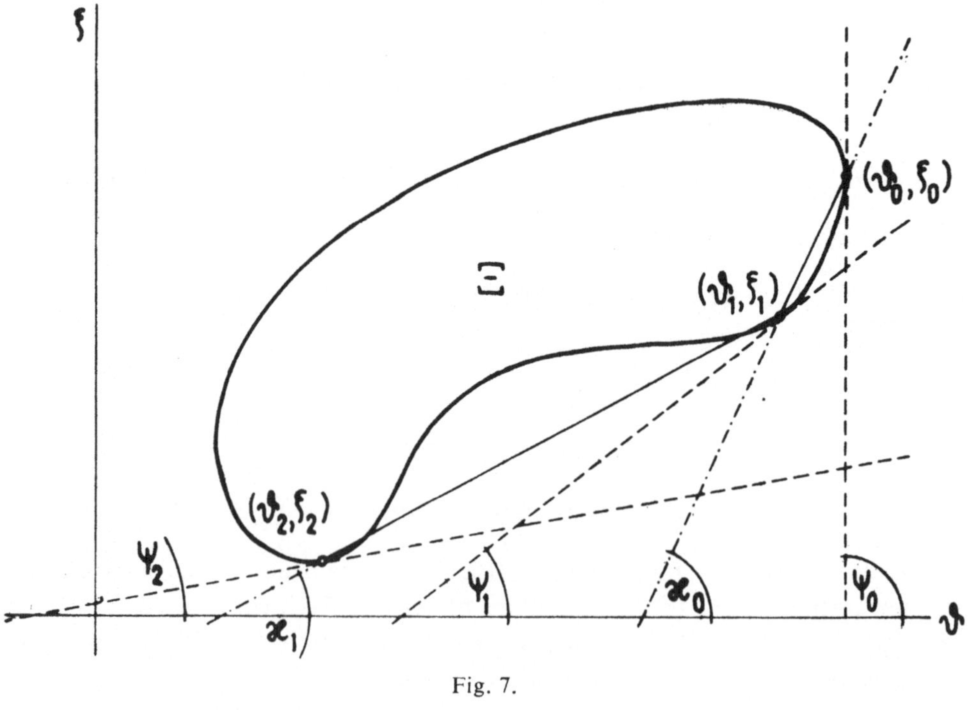

Fig. 7.

From Figure 7 it is seen that (64) implies

$$\cos \varkappa_j \left(\frac{d}{dx} v(x) + g\right) + \sin \varkappa_j \frac{d^2}{dx^2} v(x) \geqq 0 \geqq \cos \varkappa_{j-1} \left(\frac{d}{dx} v(x) + g\right)$$

$$+ \sin \varkappa_{j-1} \frac{d^2}{dx^2} v(x), \quad x \in (\xi_j, \xi_{j-1}), \quad j = 0, 1, \ldots, m.$$

Hence,

$$(69) \quad \cos \varkappa_j \left(\frac{d}{dx} v(\xi_j + 0) + g\right) + \sin \varkappa_j \frac{d^2}{dx^2} v(\xi_j + 0) = 0, \quad j = 0, 1, \ldots, m-1.$$

A consequence, of (69) is

$$\frac{d^2}{dx^2} v(\xi_j - 0) = \frac{d^2}{dx^2} v(\xi_j + 0), \quad j = 0, 1, \ldots, m-1.$$

The following iterative procedure suggests itself. Start by checking, whether (66) does solve the problem, i.e., whether for $v(x)$ satisfying (66), (65) holds

$$\cos \varkappa_0 \left(\frac{d}{dx} v(0) + g\right) + \sin \varkappa_0 \frac{d^2}{dx^2} v(0) \geqq 0.$$

If not, denote by $v_C^0(x)$ the solution of (66) satisfying the boundary condition at infinity, which depends on an undetermined constant C. Fix an initial value of C. Determine ξ_0 from

$$\cos \varkappa_0 \left(\frac{d}{dx} v_C^0(\xi_0) + g\right) + \sin \varkappa_0 \frac{d^2}{dx^2} v_C^0(\xi_0) = 0.$$

Then successively, for $j = 1, 2, \ldots$, solve (67) on $(0, \xi_{j-1})$ with initial conditions

$$v_C^j(\xi_{j-1}) = v_C^{j-1}(\xi_{j-1}), \quad \frac{d}{dx} v_C^j(\xi_{j-1}) = \frac{d}{dx} v_C^{j-1}(\xi_{j-1}),$$

and find ξ_j so that

$$\cos \varkappa_j \left(\frac{d}{dx} v_C^j(\xi_j) + g\right) + \sin \varkappa_j \frac{d^2}{dx^2} v_C^j(\xi_j) = 0.$$

If no $\xi_j > 0$ can be found, compute $v_C^j(0)$. As long as $v_C^j(0)$ differs substantially from 0, repeat the procedure with an improved value C.

References

[1] BILLINGSLEY, P. (1968). "Convergence of Probability Measures". Wiley, New York.

[2] HOWARD, R. A. (1960). "Dynamic Programming and Markov Processes". Technology Press, Cambridge (Mass.) — Wiley, New York.

[3] VAN KIEU, PHAM (1974). A diffusion approximation in the ruin problem for a controlled Markov chain. *Kybernetika* (Prague), **10**, 125—132.

[4] MANDL, P. (1968). "Analytical Treatment of One-dimensional Markov Processes". Academia, Prague — Springer, Heidelberg.

[5] MANDL, P. (1974). Estimation and control in Markov chains. *Adv. Appl. Prob.*, **6**, 40—60.

[6] MORTON, R. (1971). On the optimal control of stationary diffusion processes with natural boundaries and discounted cost. *J. Appl. Prob.*, **8**, 561—572.

[7] STROOCK, D. W. - VARADHAN, S. R. S. (1969). Diffusion processes with continuous coefficients. *Com. Pure and Appl. Math.* XXII, 345—400, 479—530.

CHARLES UNIVERSITY, PRAGUE, CZECHOSLOVAKIA

Received September 1975

A NOTE ON CONTIGUITY AND HELLINGER DISTANCE*

by

J. OOSTERHOFF (1), AND W. R. VAN ZWET (2)

1. Introduction

For $n = 1, 2, \dots$ let $(\mathscr{X}_{n1}, \mathscr{A}_{n1}), \dots, (\mathscr{X}_{nn}, \mathscr{A}_{nn})$ be arbitrary measurable spaces. Let P_{ni} and Q_{ni} be probability measures defined on $(\mathscr{X}_{ni}, \mathscr{A}_{ni})$, $i = 1, \dots, n$; $n = 1, 2, \dots$, and let $P_n^{(n)} = \prod_{i=1}^{n} P_{ni}$ and $Q_n^{(n)} = \prod_{i=1}^{n} Q_{ni}$ $(n = 1, 2, \dots)$ denote the product probability measures. For each i and n let X_{ni} be the identity map from $\mathscr{X}_{ni}$ onto $\mathscr{X}_{ni}$. Then P_{ni} and Q_{ni} represent the two possible distributions of the random element X_{ni} as well as the probability measures of the underlying probability space. Obviously $X_{n1}, \dots, X_{nn}$ are independent under both $P_n^{(n)}$ and $Q_n^{(n)}$ $(n = 1, 2, \dots)$.

The sequence $\{Q_n^{(n)}\}$ is said to be contiguous with respect to the sequence $\{P_n^{(n)}\}$ if $\lim_{n\to\infty} P_n^{(n)}(A_n) = 0$ implies $\lim_{n\to\infty} Q_n^{(n)}(A_n) = 0$ for any sequence of measurable sets A_n. This one-sided contiguity notion is denoted by $\{Q_n^{(n)}\} \lhd \{P_n^{(n)}\}$ (the notation is due to H. Witting & G. Nölle [7]). The sequences $\{P_n^{(n)}\}$ and $\{Q_n^{(n)}\}$ are said to be contiguous with respect to each other if both $\{Q_n^{(n)}\} \lhd \{P_n^{(n)}\}$ and $\{P_n^{(n)}\} \lhd \{Q_n^{(n)}\}$. This two-sided contiguity concept we denote by $\{P_n^{(n)}\} \lhd\rhd \{Q_n^{(n)}\}$.

The main purpose of this note is to characterize contiguity of product probability measures in terms of their marginals. To this end we introduce the Hellinger distance $H(P, Q)$ between two probability measures P and Q on the same σ-field, defined by

$$(1.1) \qquad H(P, Q) = \left\{ \int (p^{1/2} - q^{1/2})^2 \, d\mu \right\}^{1/2} = \left\{ 2 - 2 \int p^{1/2} q^{1/2} \, d\mu \right\}^{1/2},$$

where $p = dP/d\mu$, $q = dQ/d\mu$ and μ is any σ-finite measure dominating $P + Q$. This metric is independent of the choice of μ and satisfies $0 \leq H(P, Q) \leq 2^{1/2}$.

---

* Report SW 36/75 Mathematisch Centrum, Amsterdam

AMS (MOS) subject classification scheme (1970): 62E20

KEY WORDS & PHRASES: *asymptotic normality, contiguity, Hellinger distance, log likelihood ratio.*

Defining the total variation distance of P and Q by

$$\|P - Q\| = \sup |P(A) - Q(A)|, \tag{1.2}$$

where the supremum is taken over all measurable sets A, we have the following inequalities (Le Cam [4])

$$\tfrac{1}{2}H^2(P, Q) \leqq \|P - Q\| \leqq H(P, Q). \tag{1.3}$$

The Hellinger distances of the product measures and of their marginals are connected by the relationship

$$H^2(P_n^{(n)}, Q_n^{(n)}) = 2 - 2\prod_{i=1}^{n}\{1 - \tfrac{1}{2}H^2(P_{ni}, Q_{ni})\}. \tag{1.4}$$

For further reference we first mention two easy results, viz.

$$\sum_{i=1}^{n} H^2(P_{ni}, Q_{ni}) = o(1) \quad \text{for} \quad n \to \infty \Rightarrow \{P_n^{(n)}\} \lhd \rhd \{Q_n^{(n)}\}, \tag{1.5}$$

and

$$\{Q_n^{(n)}\} \lhd \{P_n^{(n)}\} \Rightarrow \sum_{i=1}^{n} H^2(P_{ni}, Q_{ni}) = O(1) \quad \text{for} \quad n \to \infty. \tag{1.6}$$

The proof of (1.5) is an immediate consequence of the string of implications

$$\sum_{i=1}^{n} H^2(P_{ni}, Q_{ni}) = o(1) \Rightarrow \sum_{i=1}^{n} \log\{1 - \tfrac{1}{2}H^2(P_{ni}, Q_{ni})\} = o(1)$$

$$\Rightarrow H^2(P_n^{(n)}, Q_n^{(n)}) = o(1) \Rightarrow \|P_n^{(n)} - Q_n^{(n)}\| = o(1) \Rightarrow \{P_n^{(n)}\} \lhd \rhd \{Q_n^{(n)}\}.$$

To prove (1.6) suppose that $\limsup_{n\to\infty} H(P_n^{(n)}, Q_n^{(n)}) = 2^{1/2}$. Then by (1.3) $\limsup_{n\to\infty} \|P_n^{(n)} - Q_n^{(n)}\| = 1$ in contradiction to $\{Q_n^{(n)}\} \lhd \{P_n^{(n)}\}$. Thus $\limsup_{n\to\infty} H^2(P_n^{(n)}, Q_n^{(n)}) < 2$, therefore $\liminf_{n\to\infty} \prod_{i=1}^{n}\{1 - \tfrac{1}{2}H^2(P_{ni}, Q_{ni})\} > 0$ and hence $\limsup_{n\to\infty} \sum_{i=1}^{n} H^2(P_{ni}, Q_{ni}) < \infty$ and the proof is complete.

It can be shown by counterexamples that in (1.5) the condition cannot be weakened to $\sum_{i=1}^{n} H^2(P_{ni}, Q_{ni}) = O(1)$, and that in (1.6) the conclusion cannot be strengthened to $\sum_{i=1}^{n} H^2(P_{ni}, Q_{ni}) = o(1)$, for $n \to \infty$. Hence there remains a gap between the sufficient condition and the necessary condition for contiguity in (1.5) and (1.6) respectively. In section 2 we obtain conditions which are both sufficient and necessary for contiguity of the product measures by adding another condition to $\sum_{i=1}^{n} H^2(P_{ni}, Q_{ni}) = O(1)$.

In many applications asymptotic normality of the log likelihood ratio statistic Λ_n (see (3.1)) plays an important part. Since

$$\mathscr{L}(\Lambda_n \mid P_n^{(n)}) \to_w \mathscr{N}(-\tfrac{1}{2}\sigma^2; \sigma^2) \quad \text{implies} \quad \{P_n^{(n)}\} \lhd\rhd \{Q_n^{(n)}\}$$

(cf. Hájek & Šidák [1], Le Cam [2], [3], [4], Roussas [6]), we have to impose stronger conditions on the marginals P_{ni} and Q_{ni} to ensure the asymptotic normality of Λ_n. Some sufficient (and almost necessary) conditions for the asymptotic normality of Λ_n, which are clearly stronger than those in section 2, are given in section 3. These conditions are closely related to some earlier results of Le Cam [3], [4].

2. Contiguity of product measures

We begin by noting the following useful implication:

$$\text{(2.1)} \qquad \{Q_n^{(n)}\} \lhd \{P_n^{(n)}\} \Rightarrow [\lim_{n\to\infty} \sum_{i=1}^{n} P_{ni}(A_{ni}) = 0 \Rightarrow \lim_{n\to\infty} \sum_{i=1}^{n} Q_{ni}(A_{ni}) = 0]$$

for any collection of measurable sets A_{ni}. For suppose $\lim_{n\to\infty} \sum_{i=1}^{n} P_{ni}(A_{ni}) = 0$. Then $\lim_{n\to\infty} P_n^{(n)}(\bigcup_{i=1}^{n} A_{ni}) = 0$, hence by contiguity $\lim_{n\to\infty} Q_n^{(n)}(\bigcup_{i=1}^{n} A_{ni}) = 1 - \lim_{n\to\infty} \prod_{i=1}^{n}(1 - Q_{ni}(A_{ni}))$ $= 0$ and therefore $\lim_{n\to\infty} \sum_{i=1}^{n} Q_{ni}(A_{ni}) = 0$.

Now let μ_{ni} be a σ-finite measure on $(\mathscr{X}_{ni}, \mathscr{A}_{ni})$ dominating $P_{ni} + Q_{ni}$ and write $p_{ni} = \mathrm{d}P_{ni}/\mathrm{d}\mu_{ni}$ and $q_{ni} = \mathrm{d}Q_{ni}/\mathrm{d}\mu_{ni}$ $(i = 1, \ldots, n;\ n = 1, 2, \ldots)$. The main result of this section is

Theorem 1. $\{Q_n^{(n)}\} \lhd \{P_n^{(n)}\}$ *iff*

$$\text{(2.2)} \qquad \limsup_{n\to\infty} \sum_{i=1}^{n} H^2(P_{ni}, Q_{ni}) < \infty$$

and

$$\text{(2.3)} \qquad \lim_{n\to\infty} \sum_{i=1}^{n} Q_{ni}(q_{ni}(X_{ni})/p_{ni}(X_{ni}) \geqq c_n) = 0 \quad \textit{whenever} \quad c_n \to \infty\,.$$

Proof. First assume that (2.2) and (2.3) are satisfied. Write

$$L_{ni} = q_{ni}(X_{ni})/p_{ni}(X_{ni})\,, \quad i = 1, \ldots, n\,; \quad n = 1, 2, \ldots\,,$$

and consider $\prod_{i=1}^{n} L_{ni}$. It is easily shown (cf. Le Cam [4], Roussas [6]) that $\{Q_n^{(n)}\} \lhd \{P_n^{(n)}\}$ is equivalent to tightness of the sequence of distributions $\{\mathscr{L}(\prod_{i=1}^{n} L_{ni} \mid Q_n^{(n)});$

$n = 1, 2, \ldots\}$. The tightness of this set of distributions can also be expressed in the more convenient form

$$\lim_{n\to\infty} Q_n^{(n)}\Big(\prod_{i=1}^{n} L_{ni} \geqq k_n\Big) = 0 \quad \text{whenever} \quad k_n \to \infty\,. \tag{2.4}$$

Hence we have to prove (2.4). Let $0 < k_n \to \infty$. Let $0 < c_n \to \infty$ be real numbers to be chosen in the sequel. If 1_A denotes the indicator function of the set A, we have by (2.3) and Markov's inequality for $n \to \infty$

$$Q_n^{(n)}\Big(\prod_{i=1}^{n} L_{ni} \geqq k_n\Big)$$

$$\leqq Q_n^{(n)}\Big(\prod_{i=1}^{n} L_{ni} \geqq k_n \wedge L_{ni} < c_n \quad \text{for} \quad i = 1, \ldots, n\Big) + Q_n^{(n)}\Big(\bigcup_{i=1}^{n} \{L_{ni} \geqq c_n\}\Big)$$

$$\leqq Q_n^{(n)}\Big(\prod_{i=1}^{n} L_{ni}^{1/2}\, 1_{(0,c_n)}(L_{ni}) \geqq k_n^{1/2}\Big) + \sum_{i=1}^{n} Q_{ni}(L_{ni} \geqq c_n)$$

$$\leqq k_n^{-1/2} \prod_{i=1}^{n} \int_{q_{ni}<c_n p_{ni}} q_{ni}^{3/2} p_{ni}^{-1/2}\, d\mu_{ni} + o(1)\,.$$

Since for all $c_n \geqq 1$

$$\int_{q_{ni}<c_n p_{ni}} q_{ni}^{3/2} p_{ni}^{-1/2}\, d\mu_{ni}$$

$$\leqq \int_{q_{ni}<c_n p_{ni}} q_{ni}\, d\mu_{ni} + \int_{q_{ni}<c_n p_{ni}} q_{ni} p_{ni}^{-1/2}(q_{ni}^{1/2} - p_{ni}^{1/2})\, d\mu_{ni}$$

$$\leqq 1 + \int_{q_{ni}<c_n p_{ni}} q_{ni}^{1/2} p_{ni}^{-1/2}(q_{ni}^{1/2} - p_{ni}^{1/2})^2\, d\mu_{ni} + \int_{q_{ni}<c_n p_{ni}} q_{ni}^{1/2}(q_{ni}^{1/2} - p_{ni}^{1/2})\, d\mu_{ni}$$

$$\leqq 1 + c_n^{1/2} \int (q_{ni}^{1/2} - p_{ni}^{1/2})^2\, d\mu_{ni} + 1 - \int q_{ni}^{1/2} p_{ni}^{1/2}\, d\mu_{ni}$$

$$- \int_{q_{ni}\geqq c_n p_{ni}} q_{ni}^{1/2}(q_{ni}^{1/2} - p_{ni}^{1/2})\, d\mu_{ni} \leqq 1 + (c_n^{1/2} + \tfrac{1}{2})\, H^2(P_{ni}, Q_{ni})\,,$$

it follows that

$$\limsup_{n\to\infty} Q_n^{(n)}\Big(\prod_{i=1}^{n} L_{ni} \geqq k_n\Big)$$

$$\leqq \limsup_{n\to\infty} k_n^{-1/2} \prod_{i=1}^{n} \{1 + (c_n^{1/2} + \tfrac{1}{2})\, H^2(P_{ni}, Q_{ni})\}$$

$$\leqq \limsup_{n\to\infty} k_n^{-1/2} \exp\Big\{(c_n^{1/2} + \tfrac{1}{2}) \sum_{i=1}^{n} H^2(P_{ni}, Q_{ni})\Big\}\,.$$

Choosing c_n in such a way that $c_n = o((\log k_n)^2)$ for $n \to \infty$, (2.2) implies $Q_n^{(n)}(\prod_{i=1}^{n} L_{ni} \geqq k_n) = o(1)$ for $n \to \infty$ and (2.4) is established.

Conversely, suppose that $\{Q_n^{(n)}\} \lhd \{P_n^{(n)}\}$. Since (1.6) implies that (2.2) is satisfied, it remains to prove (2.3). Let $0 < c_n \to \infty$ and consider the inequality, valid for $c_n \geqq 4$,

$$\sum_{i=1}^{n} \int_{q_{ni} \geqq c_n p_{ni}} p_{ni}\, d\mu_{ni} \leqq c_n^{-1/2} \sum_{i=1}^{n} \int_{q_{ni} \geqq c_n p_{ni}} p_{ni}^{1/2} q_{ni}^{1/2}\, d\mu_{ni}$$

$$= c_n^{-1/2} \left\{ \sum_{i=1}^{n} \int_{q_{ni} \geqq c_n p_{ni}} p_{ni}^{1/2}(q_{ni}^{1/2} - p_{ni}^{1/2})\, d\mu_{ni} + \sum_{i=1}^{n} \int_{q_{ni} \geqq c_n p_{ni}} p_{ni}\, d\mu_{ni} \right\}$$

$$\leqq c_n^{-1/2} \left\{ \sum_{i=1}^{n} \int_{q_{ni} \geqq c_n p_{ni}} (q_{ni}^{1/2} - p_{ni}^{1/2})^2\, d\mu_{ni} + \sum_{i=1}^{n} \int_{q_{ni} \geqq c_n p_{ni}} p_{ni}\, d\mu_{ni} \right\}$$

$$\leqq c_n^{-1/2} \left\{ \sum_{i=1}^{n} H^2(P_{ni}, Q_{ni}) + \sum_{i=1}^{n} \int_{q_{ni} \geqq c_n p_{ni}} p_{ni}\, d\mu_{ni} \right\}.$$

Since by (2.2) $c_n^{-1/2} \sum_{i=1}^{n} H^2(P_{ni}, Q_{ni}) \to 0$ for $n \to \infty$, it follows that $\lim_{n\to\infty} \sum_{i=1}^{n} P_n(L_{ni} \geqq c_n) = 0$. Hence (2.1) implies that $\lim_{n\to\infty} \sum_{i=1}^{n} Q_{ni}(L_{ni} \geqq c_n) = 0$ and the proof of the theorem is complete. □

Corollary 1. $\{P_n^{(n)}\} \lhd\rhd \{Q_n^{(n)}\}$ *iff* (2.2) *and* (2.3) *are satisfied and*

(2.5) $$\lim_{n\to\infty} \sum_{i=1}^{n} P_{ni}(p_{ni}(X_{ni})/q_{ni}(X_{ni}) \geqq c_n) = 0 \quad \textit{whenever} \quad c_n \to \infty .$$

In connection with contiguity Hellinger distance seems to be a more appropriate metric than total variation distance. Note that from (1.3) and (1.6) we immediately obtain the implication

(2.6) $$\{Q_n^{(n)}\} \lhd \{P_n^{(n)}\} \Rightarrow \sum_{i=1}^{n} \|P_{ni} - Q_{ni}\|^2 = O(1) \quad \text{for} \quad n \to \infty ,$$

where again the order term cannot be strenghtened to $o(1)$. However, $\sum_{i=1}^{n} \|P_{ni} - Q_{ni}\|^2 = O(1)$ is too weak a condition to replace (2.2) in Theorem 1. On the other hand we cannot strengthen this condition to $\sum_{i=1}^{n} \|P_{ni} - Q_{ni}\|^r = O(1)$ for some $r < 2$, since $\{Q_n^{(n)}\} \lhd \{P_n^{(n)}\}$ does not necessarily imply $\sum_{i=1}^{n} \|P_{ni} - Q_{ni}\|^r = O(1)$ for any positive $r < 2$. The following example serves to illustrate these points.

Example. Let μ_{ni} denote Lebesgue measure on (0,1), let $p_{ni} = 1_{(0,1)}$ and let $q_{ni} = (1 + n^{-1/2})\, 1_{(0,1-n^{-1/2})} + n^{-1/2}\, 1_{[1-n^{-1/2},1)}$, $i = 1, \ldots, n$; $n = 1, 2, \ldots$. Then $\sum_{i=1}^{n} \|P_{ni} - Q_{ni}\|^2 = (1 - n^{-1/2})^2 \leqq 1$ and (2.3) is trivially satisfied since q_{ni}/p_{ni} is uniformly bounded. But $\{Q_n^{(n)}\} \lhd \{P_n^{(n)}\}$ does not hold because $\sum_{i=1}^{n} H^2(P_{ni}, Q_{ni}) = 2n\{1 - \int q_{ni}^{1/2}\, d\mu_{ni}\} = 2n\{1 - (1 + n^{-1/2})^{1/2}\,(1 - n^{-1/2}) - n^{-3/4}\} = n^{1/2}(1 + o(1))$ for $n \to \infty$.

Taking $q_{ni} = (1 + n^{-1/2})\, 1_{(0,1/2)} + (1 - n^{-1/2})\, 1_{[1/2,1)}$ for all i and n, we have $\{Q_n^{(n)}\} \lhd \{P_n^{(n)}\}$ since (2.3) is satisfied and $\sum_{i=1}^{n} H^2(P_{ni}, Q_{ni}) = 2n\{1 - \frac{1}{2}(1 + n^{-1/2})^{1/2} - \frac{1}{2}(1 - n^{-1/2})^{1/2}\} = \frac{1}{4} + o(1)$ for $n \to \infty$. However, in this case $\sum_{i=1}^{n} \|P_{ni} - Q_{ni}\|^r = n(\frac{1}{2}n^{-1/2})^r \to \infty$ for $n \to \infty$ if $r < 2$.

3. Asymptotic normality of Λ_n

Define

$$\Lambda_n = \sum_{i=1}^{n} \log \{q_{ni}(X_{ni})/p_{ni}(X_{ni})\}\,, \quad n = 1, 2, \ldots. \tag{3.1}$$

Note that, with probability one, Λ_n is well-defined under $P_n^{(n)}$, although Λ_n may assume the value $-\infty$ with positive probability under $P_n^{(n)}$.

In our search for necessary and sufficient conditions for the weak convergence $\mathscr{L}(\Lambda_n \mid P_n^{(n)}) \to_w \mathscr{N}(-\frac{1}{2}\sigma^2; \sigma^2)$ in terms of the marginal distributions of the X_{ni} we shall confine ourselves to the case where the summands in (3.1) satisfy the traditional u.a.n. condition (cf. Loève [5]).

Theorem 2. *For any* $\sigma \geqq 0$

$$\mathscr{L}(\Lambda_n \mid P_n^{(n)}) \to_w \mathscr{N}(-\tfrac{1}{2}\sigma^2; \sigma^2) \tag{3.2}$$

and

$$\lim_{n\to\infty} \max_{1 \leqq i \leqq n} P_{ni}(|\log \{q_{ni}(X_{ni})/p_{ni}(X_{ni})\}| \geqq \varepsilon) = 0 \tag{3.3}$$

for every $\varepsilon > 0$ *iff for every* $\varepsilon > 0$

$$\lim_{n\to\infty} \sum_{i=i}^{n} H^2(P_{ni}, Q_{ni}) = \tfrac{1}{4}\sigma^2\,, \tag{3.4}$$

(3.5) $$\lim_{n\to\infty} \sum_{i=1}^{n} Q_{ni}(q_{ni}(X_{ni})/p_{ni}(X_{ni}) \geqq 1 + \varepsilon) = 0\,,$$

(3.6) $$\lim_{n\to\infty} \sum_{i=1}^{n} P_{ni}(p_{ni}(X_{ni})/q_{ni}(X_{ni}) \geqq 1 + \varepsilon) = 0\,,$$

or equivalently, iff (3.4) *holds and for every* $\varepsilon > 0$

(3.7) $$\lim_{n\to\infty} \sum_{i=1}^{n} \int_{|q_{ni}-p_{ni}|\geqq \varepsilon p_{ni}} (q_{ni}^{1/2} - p_{ni}^{1/2})^2 \,d\mu_{ni} = 0\,.$$

Proof. To simplify the notation we write $r_{ni} = q_{ni}/p_{ni}$. We first show that (3.5) and (3.6) are equivalent to (3.7). From

$$\sum_{i=1}^{n} \int_{|q_{ni}-p_{ni}|\geqq \varepsilon p_{ni}} (q_{ni}^{1/2} - p_{ni}^{1/2})^2 \,d\mu_{ni}$$

$$= \sum_{i=1}^{n} \left\{ \int_{r_{ni}\geqq 1+\varepsilon} q_{ni}(1 - r_{ni}^{-1/2})^2 \,d\mu_{ni} + \int_{r_{ni}\leqq 1-\varepsilon} p_{ni}(1 - r_{ni}^{1/2})^2 \,d\mu_{ni} \right\}$$

we obtain the double inequality

$$\{1 - (1 + \varepsilon)^{-1/2}\}^2 \sum_{i=1}^{n} Q_{ni}(r_{ni}(X_{ni}) \geqq 1 + \varepsilon)$$

$$+ \{1 - (1 - \varepsilon)^{1/2}\}^2 \sum_{i=1}^{n} P_{ni}(r_{ni}^{-1}(X_{ni}) \geqq (1 - \varepsilon)^{-1})$$

$$\leqq \sum_{i=1}^{n} \int_{|q_{ni}-p_{ni}|\geqq \varepsilon p_{ni}} (q_{ni}^{1/2} - p_{ni}^{1/2})^2 \,d\mu_{ni}$$

$$\leqq \sum_{i=1}^{n} Q_{ni}(r_{ni}(X_{ni}) \geqq 1 + \varepsilon) + \sum_{i=1}^{n} P_{ni}(r_{ni}^{-1}(X_{ni}) \geqq (1 - \varepsilon)^{-1})$$

and the equivalence of (3.5) and (3.6) to (3.7) is immediate.

Next we note that both (3.2), (3.3) and (3.4), (3.5), (3.6) imply $\{P_n^{(n)}\} \lhd\rhd \{Q_n^{(n)}\}$ (cf. Corollary 1).

The remainder of the proof relies on the normal convergence theorem (cf. Loève [5]). According to an equivalent form of this theorem (3.2) and (3.3) are equivalent to

(3.8) $$\lim_{n\to\infty} \sum_{i=1}^{n} P_{ni}(|\log r_{ni}(X_{ni})| \geqq \delta) = 0 \quad \text{for every} \quad \delta > 0\,,$$

(3.9) $$\lim_{\delta\downarrow 0} \lim_{n\to\infty} \sum_{i=1}^{n} \int_{|\log r_{ni}|\leqq\delta} (\log r_{ni}) \,dP_{ni} = -\tfrac{1}{2}\sigma^2\,,$$

$$\lim_{\delta\downarrow 0} \lim_{n\to\infty} \sum_{i=1}^{n} \left\{ \int_{|\log r_{ni}|\leqq\delta} (\log r_{ni})^2 \,dP_{ni} - \left(\int_{|\log r_{ni}|\leqq\delta} (\log r_{ni}) \,dP_{ni} \right)^2 \right\} = \sigma^2\,.$$

(3.10)

By the contiguity of $\{P_n^{(n)}\}$ and $\{Q_n^{(n)}\}$ and (2.1) the condition (3.8) is equivalent to (3.5) and (3.6) and hence to (3.7). Henceforth we assume (3.7), (3.8) and $\{P_n^{(n)}\} \lhd \rhd \lhd \rhd \{Q_n^{(n)}\}$. We still have to show that (3.4) is equivalent to (3.9) and (3.10).

Let $0 < \delta < 1$. For $|\log r_{ni}| \leqq \delta$ we have the expansion

$$\begin{aligned}(3.11)\qquad \log r_{ni} &= 2 \log \{1 + (q_{ni}^{1/2} - p_{ni}^{1/2})\, p_{ni}^{-1/2}\} \\ &= 2(q_{ni}^{1/2} - p_{ni}^{1/2})\, p_{ni}^{-1/2} - (q_{ni}^{1/2} - p_{ni}^{1/2})^2\, p_{ni}^{-1}(1 + \varrho_{ni\delta})\end{aligned}$$

with $|\varrho_{ni\delta}| < 2\delta$. Thus

$$\begin{aligned}&\int_{|\log r_{ni}| \leqq \delta} (\log r_{ni})\, p_{ni}\, d\mu_{ni} \\ &= -2 \int_{|\log r_{ni}| \leqq \delta} (q_{ni}^{1/2} - p_{ni}^{1/2})^2\, d\mu_{ni} + \int_{|\log r_{ni}| \leqq \delta} (q_{ni} - p_{ni})\, d\mu_{ni} \\ &\quad - \int_{|\log r_{ni}| \leqq \delta} \varrho_{ni\delta}(q_{ni}^{1/2} - p_{ni}^{1/2})^2\, d\mu_{ni}\,.\end{aligned}$$

Since by (3.7)

$$\lim_{n\to\infty} \left\{\sum_{i=1}^{n} \int_{|\log r_{ni}| \leqq \delta} (q_{ni}^{1/2} - p_{ni}^{1/2})^2\, d\mu_{ni} - \sum_{i=1}^{n} H^2(P_{ni}, Q_{ni})\right\} = 0$$

and by (3.8), $\{P_n^{(n)}\} \lhd \rhd \{Q_n^{(n)}\}$ and (2.1)

$$\sum_{i=1}^{n} \int_{|\log r_{ni}| \leqq \delta} (q_{ni} - p_{ni})\, d\mu_{ni} = -\sum_{i=1}^{n} \int_{|\log r_{ni}| > \delta} (q_{ni} - p_{ni})\, d\mu_{ni} \to 0$$

for $n \to \infty$, we have

$$\begin{aligned}(3.12)\qquad &\lim_{\delta\downarrow 0} \limsup_{n\to\infty} \left|\sum_{i=1}^{n} \int_{|\log r_{ni}| \leqq \delta} (\log r_{ni})\, dP_{ni} + 2\sum_{i=1}^{n} H^2(P_{ni}, Q_{ni})\right| \\ &\leqq \lim_{\delta\downarrow 0} \limsup_{n\to\infty} 2\delta \sum_{i=1}^{n} H^2(P_{ni}, Q_{ni}) = 0\,,\end{aligned}$$

where we have used (1.6). Similarly,

$$\begin{aligned}(3.13)\qquad &\lim_{\delta\downarrow 0} \limsup_{n\to\infty} \sum_{i=1}^{n} \left\{\int_{|\log r_{ni}| \leqq \delta} (\log r_{ni})\, dP_{ni}\right\}^2 \\ &\leqq \lim_{\delta\downarrow 0} \limsup_{n\to\infty} \delta \sum_{i=1}^{n} \left|\int_{|\log r_{ni}| \leqq \delta} (\log r_{ni})\, dP_{ni}\right| \\ &\leqq \lim_{\delta\downarrow 0} \limsup_{n\to\infty} \delta(2 + 2\delta) \sum_{i=1}^{n} H^2(P_{ni}, Q_{ni}) = 0\,.\end{aligned}$$

Finally (3.11) implies that for $|\log r_{ni}| \leqq \delta < 1$

$$(\log r_{ni})^2 = 4(q_{ni}^{1/2} - p_{ni}^{1/2})^2 p_{ni}^{-1} + \bar{\varrho}_{ni\delta}(q_{ni}^{1/2} - p_{ni}^{1/2})^2 p_{ni}^{-1}$$

with $|\bar{\varrho}_{ni\delta}| < 10\delta$. Hence, in view of (3.7) and (1.6),

$$\lim_{\delta \downarrow 0} \limsup_{n\to\infty} \left| \sum_{i=1}^{n} \int_{|\log r_{ni}| \leqq \delta} (\log r_{ni})^2 \, dP_{ni} - 4\sum_{i=1}^{n} H^2(P_{ni}, Q_{ni}) \right| = 0 . \tag{3.14}$$

The equivalence of (3.4) to (3.9) and (3.10) is now an immediate consequence of (3.12), (3.13) and (3.14). The theorem is proved. □

In the one sample case where, for each n, $X_{n1}, \ldots, X_{nn}$ are identically distributed, condition (3.3) is implied by (3.2) and Theorem 2 slightly simplifies. This remains true in the k sample case ($k \geqq 2$) provided all sample sizes tend to infinity.

The first part of the proof of Theorem 2 also shows that the conditions (2.3) and (2.5) in Corollary 1 may be replaced by the single condition

$$\lim_{n\to\infty} \sum_{i=1}^{n} \int_{|q_{ni}-p_{ni}| \geqq c_n p_{ni}} (q_{ni}^{1/2} - p_{ni}^{1/2})^2 \, d\mu_{ni} = 0 \quad \text{whenever} \quad c_n \to \infty .$$

The proof of Theorem 2 could also be given in a more roundabout way. Introducing the r.v.'s

$$W_{ni} = 2\{q_{ni}(X_{ni})/p_{ni}(X_{ni})\}^{1/2} - 2 , \quad i = 1, \ldots, n ; \quad n = 1, 2, \ldots ,$$

one shows that $\mathscr{L}(\sum_{i=1}^{n} W_{ni} \mid P_n^{(n)}) \to_w \mathscr{N}(-\frac{1}{4}\sigma^2; \sigma^2)$ iff $\mathscr{L}(\Lambda_n \mid P_n^{(n)}) \to_w \mathscr{N}(-\frac{1}{2}\sigma^2; \sigma^2)$, provided the respective u.a.n. conditions are satisfied. It is then not difficult to prove that the weak convergence of $\sum_{i=1}^{n} W_{ni}$ and the u.a.n. condition on the summands are equivalent to (3.4) and (3.7). In this proof (3.7) appears as the Lindeberg condition in the central limit theorem applied to $\sum_{i=1}^{n} W_{ni}$.

The equivalence of both weak convergence results has first been proved by Le Cam ([3], [4]). The initial assumptions $\lim_{n\to\infty} \sup_{1\leqq i\leqq n} H^2(P_{ni}, Q_{ni}) = 0$ and $\limsup_{n\to\infty} \|P_n^{(n)} - Q_n^{(n)}\| < 1$ made by Le Cam are not restrictive since they are implied by our condition (3.7) and the contiguity of $\{P_n^{(n)}\}$ and $\{Q_n^{(n)}\}$, respectively. One part of this proof is also contained in Hájek & Šidák [1].

References

[1] HÁJEK, J. - ŠIDÁK, Z. (1967). "Theory of rank tests". Academic Press, New York.

[2] LE CAM, L. (1960). Locally asymptotically normal families of distributions. *Univ. California Publ. Statist.*, **3**, 37—98, University of California Press.

[3] LE CAM, L. (1966). Likelihood functions for large numbers of independent observations. *Research papers in statistics (Festschrift for J. Neyman)*, 167—187, F. N. David (ed.), Wiley, New York.

[4] LE CAM, L. (1969). Théorie asymptotique de la décision statistique. *Les Presses de l'Université de Montréal.*

[5] LOÈVE, M. (1963). "Probability theory (3rd ed.)". Van Nostrand, New York.

[6] ROUSSAS, G. G. (1972). Contiguity of probability measures: some applications in statistics, *Cambridge University Press.*

[7] WITTING, H. - NÖLLE G. (1970). "Angewandte mathematische Statistik". Teubner, Stuttgart.

(1) CATH. UNIV. NYMEGEN, NYMEGEN, THE NETHERLANDS

(2) CENTRAAL REKENINSTITUUT DER RIJKSUNIVERSITEIT, LEIDEN, THE NETHERLANDS

Received October 1975

FIRST ORDER EFFICIENCY IMPLIES SECOND ORDER EFFICIENCY*)

by

J. PFANZAGL

1. Notations

Let $(X, \mathscr{A})$ be a measurable space and $P_\theta \mid \mathscr{A}$, $\theta \in \Theta$, a family of p-measures with open $\Theta \subset \mathbf{R}^{p+1}$, where $\Theta = \Theta_0 \times \mathbf{T}$ with $\Theta_0 \subset \mathbf{R}$, $\mathbf{T} \subset \mathbf{R}^p$. The elements of Θ will be denoted by θ or (θ_0, τ) or $(\theta_0, \theta_1, \ldots, \theta_p)$. By $P_\theta^n \mid \mathscr{A}^n$ we denote the n-fold independent product of identical components $P_\theta \mid \mathscr{A}$. For notational convenience we shall consider X^n as a subspace of $X^{\mathbf{N}}$ and $\mathscr{A}^n$ as a sub-σ-field of $\mathscr{A}^{\mathbf{N}}$. This will enable us to consider $\mathscr{A}^n$-measurable functions on X^n as $\mathscr{A}^n$-measurable functions on $X^{\mathbf{N}}$ and to write the argument as $\mathbf{x} = (x_i)_{i \in \mathbf{N}}$ instead of $(x_1, \ldots, x_n)$.

We shall use the following *convention*: if in a product an index occurs at least twice, this means summation over this index from 0 to p in case of a Roman index and from 1 to p in case of a Greek index.

Given a vector $v = (v_0, \ldots, v_q)$ let $v_{(0)}$ denote the vector $(v_1, \ldots, v_q)$.

For a function $f : X \times \Theta \to \mathbf{R}$ denote the partial derivatives with respect to θ by

$$f^{(i_1 \ldots i_k)}(x, \theta) := (\partial^k / \partial\theta_{i_1} \ldots \partial\theta_{i_k}) f(x, \theta) .$$

Denote the integral with respect to $P_\theta \mid \mathscr{A}$ by

$$P_\theta(f(\cdot, \tau)) := \int f(x, \tau)\, P_\theta(\mathrm{d}x) .$$

Suppressing the dependence on n and t, we define for $n \in \mathbf{N}$ and $t \in \mathbf{R}$

$$\tilde{f}(\mathbf{x}, \theta) := n^{-1/2} \sum_{\nu=1}^{n} \big(f(x_\nu, \theta) - P_\theta(f(\cdot, \theta))\big) ,$$

$$\tilde{f}'(\mathbf{x}, \theta) := n^{-1/2} \sum_{\nu=1}^{n} \big(f(x_\nu, \theta) - P_{(\theta_0 + n^{-1/2}t, \tau)}(f(\cdot, \theta))\big) .$$

* The results of this paper have first been presented at the Meeting of Dutch Statisticians at Lunteren, November 1975.

(For typographical reasons, we write $\tilde{f}^{(\alpha)}$ instead of $\widetilde{f^{(\alpha)}}$ etc.)

Given a σ-finite measure $\mu \mid \mathscr{A}$ dominating $P_\theta \mid \mathscr{A}$, $\theta \in \Theta$, denote by $p(\cdot, \theta)$ a density of $P_\theta \mid \mathscr{A}$ with respect to $\mu \mid \mathscr{A}$, and let $l := \log p$. For $i, j = 0, \ldots, p$ let

$$L_{(i)(j)}(\theta) := P_\theta(l^{(i)}(\cdot, \theta)\, l^{(j)}(\cdot, \theta)),$$

$$L_{(ij)}(\theta) := P_\theta(l^{(ij)}(\cdot, \theta)), \quad \text{etc.}$$

Furthermore, let

$$\Lambda := ((L_{(i)(j)})_{i,j=0,\ldots,p})^{-1}.$$

We define

$$\lambda_i(\cdot, \theta) := \Lambda_{ij}(\theta)\, l^{(j)}(\cdot, \theta) \quad \text{for} \quad i = 0, \ldots, p.$$

Finally,

$$\varphi(v) := (2\pi)^{-1/2} \exp\left[-\tfrac{1}{2}v^2\right],$$

$$\Phi(v) := \int_{r<v} \varphi(r)\, dr,$$

$$N_\alpha := \Phi^{-1}(\alpha).$$

For a q-dimensional covariance matrix $\Sigma = (\sigma_{ij})_{i,j=1,\ldots,q}$ let φ_Σ denote the Lebesgue density of the normal distribution with mean vector zero and covariance matrix Σ.

1.1. Definition. Let $f_n : X^{\mathbf{N}} \times \Theta \to \mathbf{R}$, $n \in \mathbf{N}$, be such that $f_n(\cdot, \theta)$ is $\mathscr{A}^n$-measurable for $\theta \in \Theta$ and $n \in \mathbf{N}$.

We shall write $f_n(\cdot, \theta) = l_n(r)$ if for every $\theta \in \Theta$ there exist a neighborhood U_θ of θ and a constant $a > 0$ such that uniformly for $\sigma \in U_\theta$ and $\|\delta - \sigma\| \leqq n^{-1/2} \log n$

$$P_\delta^n\{|f_n(\cdot, \sigma)| > (\log n)^a\} = o(n^{-r}).$$

Let $\theta_0 \in \Theta_0$ be fixed and assume that $f_n : X^{\mathbf{N}} \times \mathbf{T} \to \mathbf{R}$ for $n \in \mathbf{N}$. Then we shall write $f_n(\cdot, \tau) = l_n(r, \theta_0)$ if the above condition holds for every $\theta = (\theta_0, \tau)$ with $\tau \in \mathbf{T}$.

Furthermore, $f_n(\cdot, \theta) = g_n(\cdot, \theta) + n^{-s}\, l_n(r)$ is defined as $n^s(f_n(\cdot, \theta) - g_n(\cdot, \theta)) = l_n(r)$.

2. Regularity conditions

The following Conditions M_r, L_r and D refer to a function $f : X \times \Theta \to \mathbf{R}$ such that $f(\cdot, \theta)$ is $\mathscr{A}$-measurable for $\theta \in \Theta$.

Condition M_r. For every $\theta \in \Theta$ there exists a neighborhood U_θ of θ such that

$$\sup_{\sigma, \delta \in U_\theta} P_\delta(|f(\cdot, \sigma)|^r) < \infty.$$

Condition L_r. For every $\theta \in \Theta$ there exist a neighborhood U_θ of θ and an $\mathscr{A}$-measurable function $k(\cdot, \theta) : X \to \mathbf{R}$ such that

(a) $|f(x, \sigma) - f(x, \delta)| \leqq \|\sigma - \delta\| \, k(x, \theta)$ for $x \in X$ and $\sigma, \delta \in U_\theta$;

(b) k fulfills Condition M_r.

Condition D. The order of integration with respect to $\mu \mid \mathscr{A}$ and differentiation with respect to $\theta \in \Theta$ of the function $p(\cdot, \theta) f(\cdot, \theta)$ may be interchanged.

The following Condition C refers to a vector-valued function $f = (f_1, \ldots, f_q) : X \times \Theta \to \mathbf{R}^q$ such that $f(\cdot, \theta)$ is $\mathscr{A}$-measurable and square-integrable with respect to $P_\theta \mid \mathscr{A}$ for $\theta \in \Theta$.

Condition C. For every $\theta \in \Theta$ there exist a neighborhood U_θ and a subset $\{j_1, \ldots, j_m\}$ of $\{1, \ldots, q\}$ such that

(a) for $\delta \in U_\theta$, the functions $f_{j_k}(\cdot, \delta)$, $k = 1, \ldots, m$, are P_δ-a.e. linearly independent and constitute a base for the subspace spanned by $f_j(\cdot, \delta)$, $j = 1, \ldots, q$, in $L^2(P_\delta \mid \mathscr{A})$.

(b) $\limsup\limits_{\|v\| \to \infty} \sup\limits_{\delta \in U_\theta} \left|P_\delta(\exp [i \sum\limits_{k=1}^{m} v_k f_{j_k}(\cdot, \delta)])\right| < 1.$

Let $\theta_0 \in \Theta$ be fixed. If Conditions M_r, L_r, C, D are to hold for every $\theta = (\theta_0, \tau)$ with $\tau \in \mathbf{T}$, then we shall call them Conditions $M_r(\theta_0)$, etc. Notice that the Conditions $M_r(\theta_0)$, etc., refer to neighborhoods U_θ of $\theta = (\theta_0, \tau)$ in Θ, not in $\{\theta_0\} \times \mathbf{T}$.

Condition B. A function $Q : \mathbf{R}^q \times \Theta \to \mathbf{R}$ fulfills Condition B if

(a) $Q(\cdot, \theta)$ admits continuous partial derivatives, bounded by polynomials locally uniformly in $\theta \in \Theta$;

(b) for every $\theta \in \Theta$ there exist a neighborhood U_θ of θ and constants $a, b > 0$ such that

$$|Q(v, \sigma) - Q(v, \delta)| \leqq \|\sigma - \delta\| \, (a + \|v\|^b) \quad \text{for} \quad v \in \mathbf{R}^q \quad \text{and} \quad \sigma, \delta \in U_\theta .$$

Finally, we shall need the following regularity conditions.

(i) $P_\theta \mid \mathscr{A}$, $\theta \in \Theta$, are mutually absolutely continuous.

(ii) $l(x, \cdot)$ admits continuous partial derivatives up to the second order on Θ for every $x \in X$, and $l^{(i)}$ and $l^{(ij)}$, $i, j = 0, \ldots, p$, fulfil Conditions L_3 and M_3.

(iii) $L_{(i)} = 0$ for $i = 0, \ldots, p$.

(iv) $L_{(i)(j)} = -L_{(ij)}$ for $i, j = 0, \ldots, p$.

(v) $(L_{(i)(j)})_{i,j=0,\ldots,p}$ is positive definite on Θ.

(vi) For $i, j = 0, \ldots, p$, $L_{(i)(j)}$ and $L_{(ij)}$ admit continuous partial derivatives on Θ, and $l^{(i)(j)}$ and $l^{(ij)}$ fulfill Condition D.

3. Auxiliary results

In Lemma 3.1, the parameter space Θ is an open subset of $\mathbf{R}^p$. Its elements will be denoted by $\theta = (\theta_1, \ldots, \theta_p)$. Correspondingly, Λ now denotes the inverse of $(L_{(\alpha)(\beta)})_{\alpha,\beta=1,\ldots,p}$, and $\lambda_\alpha := \Lambda_{\alpha\beta} l^{(\beta)}$, $\alpha = 1, \ldots, p$.

3.1. Lemma. Let $T^{(n)} = (T_\alpha^{(n)})_{\alpha=1,\ldots,p}$, $n \in \mathbf{N}$, *be an estimator sequence admitting the following asymptotic expansion for* $\alpha = 1, \ldots, p$:

$$n^{1/2}(T_\alpha^{(n)} - \theta_\alpha) = \tilde{g}_\alpha(\cdot, \theta) + n^{-1/2} l_n(0)\,, \tag{3.2}$$

where $P_\theta(g_\alpha(\cdot, \theta)) = 0$ *for* $\theta \in \Theta$, *and* g_α *fulfills Conditions* $\mathbf{M}_3$ *and* $\mathbf{L}_3$. *Assume that Conditions* (ii)–(v) *are fulfilled.*

Then for $\theta \in \Theta$ *and* $\alpha, \beta = 1, \ldots, p$,

$$P_\theta(l^{(\alpha)}(\cdot, \theta)\, g_\beta(\cdot, \theta)) = \delta_{\alpha\beta}\,.$$

Proof. To simplify our notations, we shall omit the parameter whenever this is convenient.

Starting from $T^{(n)}$, $n \in \mathbf{N}$, an asymptotic m.l. estimator sequence $\theta^{(n)}$, $n \in \mathbf{N}$, of order $o(n^0)$ for the nuisance parameters may be constructed by Lemma 6 in Pfanzagl (1973b), p. 249. By Lemma 1 in Michel (1975), p. 77, we have for $\alpha = 1, \ldots, p$,

$$n^{1/2}(\theta_\alpha^{(n)} - \theta_\alpha) = \lambda_\alpha + n^{-1/2} l_n(0)\,.$$

Fix $\beta \in \{1, \ldots, p\}$, and for $a \in \mathbf{R}$ let

$$S_\alpha^{(n)} := \theta_\alpha^{(n)} \quad \text{for} \quad \alpha \neq \beta\,; \quad S_\beta^{(n)} := (1 - a)\, \theta_\beta^{(n)} + a T_\beta^{(n)}\,.$$

Then $n^{1/2}(S^{(n)} - \theta)$, $n \in \mathbf{N}$, has a limiting normal distribution under P with mean vector zero and continuous covariance matrix $P(hh')$, where

$$h_\alpha := \lambda_\alpha \quad \text{for} \quad \alpha \neq \beta\,; \quad h_\beta := (1 - a)\, \lambda_\beta + a g_\beta\,. \tag{3.3}$$

Since the covariance matrix $P(\lambda\lambda') = \Lambda$ is continuous, too, we obtain from Bahadur (1964), p. 1550, that the matrix

$$P(hh') - P(\lambda\lambda')$$

is positive semidefinite. This implies in particular that for every $\alpha \neq \beta$ the submatrix corresponding to the indices (α, β) is positive semidefinite. By (3.3) this implies that the following matrix is positive semidefinite.

$$\begin{pmatrix} 0 & aP(\lambda_\alpha(g_\beta - \lambda_\beta)) \\ aP(\lambda_\alpha(g_\beta - \lambda_\beta)) & a^2 P((\lambda_\beta + g_\beta)^2) + 2aP(\lambda_\beta(g_\beta - \lambda_\beta)) \end{pmatrix}.$$

Since the elements of the main diagonal have to be nonnegative for all $a \in \mathbf{R}$,

$$P(\lambda_\beta(g_\beta - \lambda_\beta)) = 0 .$$

Moreover, the determinant has to be nonnegative, hence

$$P(\lambda_\alpha(g_\beta - \lambda_\beta)) = 0 .$$

Therefore, for $\alpha, \beta = 1, \ldots, p$,

$$P(\lambda_\alpha g_\beta) = P(\lambda_\alpha \lambda_\beta) .$$

Since $\lambda_\alpha = \Lambda_{\alpha\gamma} l^{(\gamma)}$, this implies

$$\Lambda_{\alpha\gamma}\, P(l^{(\gamma)} g_\beta) = \Lambda_{\alpha\beta} .$$

Multiplying by $L_{(\varkappa)(\alpha)}$ and taking the sum over α we obtain for $\varkappa = 1, \ldots, p$

$$P(l^{(\varkappa)} g_\beta) = \delta_{\varkappa\beta} ,$$

the assertion.

The following considerations refer to a fixed p-measure $P \mid \mathscr{A}$ and an open neighborhood U_0 of 0 in $\mathbf{R}^p$.

3.4 Definition. Let $f_n : x^{\mathbf{N}} \times U_0 \to \mathbf{R}$, $n \in \mathbf{N}$, be a sequence of functions such that $f_n(\cdot, s)$ is $\mathscr{A}^n$-measurable for $s \in U_0$ and $n \in \mathbf{N}$.

We shall write $f_n(\cdot, s) = l_n^0(r)$ if there exist constants $a > 0$ and $c > 0$ such that for every $s \in [-c, c]$

$$P^n\{|f_n(\cdot, s)| > (\log n)^a\} = o(n^{-r}) .$$

This concept is weaker than that defined in 1.1.

3.5 Assumption. Let $F_n : X^{\mathbf{N}} \to \mathbf{R}$, $n \in \mathbf{N}$, be a sequence of $\mathscr{A}^n$-measurable functions admitting an asymptotic expansion

$$F_n(\mathbf{x}) = n^{-1/2} \sum_{\nu=1}^{n} f_0(x_\nu, n^{-1/2}s) \tag{3.6}$$

$$+ n^{-1/2} Q(n^{-1/2} \sum_{\nu=1}^{n} f(x_\nu, n^{-1/2}s), n^{-1/2}s) + n^{-1}\, l_n^0(0) ,$$

where $f = (f_0, \ldots, f_m)$ with $P(f(\cdot, 0)) = 0$.

Assume that there exist constants a, b, $\varepsilon > 0$ such that

$$|Q(v, \sigma) - Q(v, \delta)| \leqq \|\sigma - \delta\| (a + \|v\|^b) \quad \text{for} \quad v \in \mathbf{R}^{m+1} \quad \text{and} \quad \sigma, \delta \in U_0 , \tag{3.7}$$

$$|Q(v, 0) - Q(w, 0)| \leqq \|v - w\| (a + \|v\|^b) \quad \text{for} \quad v, w \in \mathbf{R}^{m+1} \quad \text{with} \tag{3.8}$$

$$\|v - w\| \leqq \varepsilon .$$

Assume, furthermore, that $f_0(x, \cdot)$ admits second order partial derivatives and $f_j(x, \cdot)$, $j = 1, \ldots, m$, admit first order partial derivatives on U_0 for every $x \in X$. Assume that $P(h(\cdot, 0)^2)$ is finite for $h = f_j, f_j^{(\alpha)}, f_0^{(\alpha\beta)}$, $j = 0, \ldots, m$; $\alpha, \beta = 1, \ldots, p$, and that for $h = f_j^{(\alpha)}, f_0^{(\alpha\beta)}$, $j = 0, \ldots, m$; $\alpha, \beta = 1, \ldots, p$, there exists an $\mathscr{A}^n$-measurable $k : X \to \mathbf{R}$ with $P(k^2)$ finite such that

$$(3.9) \qquad |h(x, \sigma) - h(x, \delta)| \leqq \|\sigma - \delta\| \, k(x) \quad \text{for} \quad x \in X \quad \text{and} \quad \sigma, \delta \in U_0 .$$

3.10 Lemma. *Let F_n, $n \in \mathbf{N}$, be a sequence of functions fulfilling Assumption* 3.5. *Then the following holds true.*

(a) $P(f_0^{(\alpha)}(\cdot, 0)) = 0$ *for* $\alpha = 1, \ldots, p$.

(b) *Arrange f in such a way that $f_\alpha = f_0^{(\alpha)}$ for $\alpha = 1, \ldots, p$. Let k be a q-dimensional vector whose components constitute a base of P-a.e. linearly independent functions for the subspace spanned by the components of f in $L^2(P \mid \mathscr{A})$. Let M denote a $(m + 1, q)$-matrix fulfilling $f = Mk$. Then the function Q occurring in the representation* (3.6) *fulfills the following relation identically in $(s, v) \in \mathbf{R}^p \times \mathbf{R}^q$*:

$$(3.11) \qquad s_\alpha (Mv)_\alpha + \tfrac{1}{2} s_\alpha s_\beta P(f_0^{(\alpha\beta)}(\cdot, 0)) + Q(Mv + s_\alpha P(f^{(\alpha)}(\cdot, 0)), 0) = Q(Mv, 0) .$$

Proof. For notational convenience we shall write f_0 for $f_0(\cdot, 0)$ etc.

(a) Let $s \in \mathbf{R}^p$ be arbitrary. Then

$$(3.12) \qquad n^{-1/2} \sum_{\nu=1}^{n} f_0(x_\nu, n^{-1/2} s) = \tilde{f}_0(\mathbf{x}) + s_\alpha \, P(f_0^{(\alpha)})$$
$$+ \, n^{-1/2} [s_\alpha \tilde{f}_0^{(\alpha)}(\mathbf{x}) + \tfrac{1}{2} s_\alpha s_\beta \, P(f_0^{(\alpha\beta)})] + n^{-1} \, l_n^0(0) .$$

For $j = 0, \ldots, m$,

$$(3.13) \qquad n^{-1/2} \sum_{\nu=1}^{n} f_j(x_\nu, n^{-1/2} s) = \tilde{f}_j(\mathbf{x}) + s_\alpha \, P(f_j^{(\alpha)}) + n^{-1/2} \, l_n^0(0) .$$

For notational convenience we assume w.l.g. that

$$(3.14) \qquad f_j = f_0^{(j)} - P(f_0^{(j)}) \quad \text{for} \quad j = 1, \ldots, p ,$$

and that f_j, $j = 0, \ldots, m$, are P-a.e. linearly independent. For $(s, v) \in \mathbf{R}^p \times \mathbf{R}^{m+1}$ define

$$(3.15) \qquad R(v, s) := s_\alpha v_\alpha + \tfrac{1}{2} s_\alpha s_\beta \, P(f_0^{(\alpha\beta)}) + Q(v + s_\alpha \, P(f^{(\alpha)})) .$$

Using (3.12), (3.13), and (3.15) we obtain from (3.6), applied for $\delta = n^{-1/2} s$, and from (3.7) and (3.8) that for every $s \in \mathbf{R}^p$

$$(3.16) \qquad F_n = \tilde{f}_0 + s_\alpha \, P(f_0^{(\alpha)}) + n^{-1/2} \, R(\tilde{f}, s) + n^{-1} \, l_n^0(0) .$$

The same reasoning, applied for $\delta = 0$, leads to

(3.17) $$F_n = \tilde{f}_0 + n^{-1/2} R(\tilde{f}, 0) + n^{-1} l_n^0(0).$$

Since $R(\tilde{f}, s) = l_n^0(0)$ for $s \in \mathbf{R}^p$, this implies

(3.18) $$P(f_0^{(\alpha)}) = 0 \quad \text{for} \quad \alpha = 1, \dots, p.$$

Hence, according to (3.14),

(3.19) $$f_j = f_0^{(j)} \quad \text{for} \quad j = 1, \dots, p.$$

(b) From (3.16) and (3.17) we obtain that for an appropriately chosen $a > 0$

$$P^n * \tilde{f}\{|R(\cdot, s) - R(\cdot, 0)| > n^{-1/2}(\log n)^a\} = o(n^0).$$

Since $P^n * \tilde{f}$, $n \in \mathbf{N}$, converges weakly to a normal distribution Q_0 with positive definite covariance matrix $P(ff')$, we have for every $s \in \mathbf{R}^p$ and $k \in \mathbf{N}$

$$Q_0\{|R(\cdot, s) - R(\cdot, 0)| > 1/k\}$$
$$\leqq \liminf_{n \to \infty} P^n * \tilde{f}\{|R(\cdot, s) - R(\cdot, 0)| > 1/k\} = 0.$$

Hence $R(\cdot, s) = R(\cdot, 0)$ Q_0-a.e. Since Q_0 is nondegenerate and $R(\cdot, s)$ is continuous for every $s \in \mathbf{R}^p$, this implies

$$R(v, s) = R(v, 0) \quad \text{for} \quad (s, v) \in \mathbf{R}^p \times \mathbf{R}^{m+1},$$

the assertion.

4. The asymptotic studentization procedure

We shall consider test-sequences for the hypothesis $\{(\theta_0, \tau) : \tau \in \mathbf{T}\}$ against alternatives δ with $\delta_0 > \theta_0$. We assume that we are given a sequence of $\mathscr{A}^n$-measurable test statistics $F_n : X^{\mathbf{N}} \to \mathbf{R}$, $n \in \mathbf{N}$, and an estimator sequence $T^{(n)} = (T_\alpha^{(n)})_{\alpha = 1, \dots, p}$, $n \in \mathbf{N}$, for the nuisance parameters which can be suitably approximated by asymptotic expansions.

4.1 Assumption. The sequence of test statistics F_n, $n \in \mathbf{N}$, admits an asymptotic expansion

(4.2) $$F_n = \tilde{f}_0(\cdot, \tau) + n^{-1/2} Q(\tilde{f}(\cdot, \tau), \tau) + n^{-1} l_n(\tfrac{1}{2}, \theta_0),$$

where $f = (f_0, \dots, f_m)$ and $P_{(\theta_0, \tau)}(f(\cdot, \tau)) = 0$ for $\tau \in \mathbf{T}$. The function Q fulfills Condition B. Furthermore, f_j, $j = 0, \dots, m$, fulfill Condition $M_4(\theta_0)$, and $f_j^{(\alpha)}, f_0^{(\alpha\beta)}$, $i = 0, \dots, m$; $\alpha, \beta = 1, \dots, p$, fulfill Conditions $\mathrm{M}_4(\theta_0)$ and $\mathrm{L}_4(\theta_0)$.

In Section 5 we shall give a few examples of test statistics fulfilling Assumption 4.1.

4.3 Assumption. The sequence of estimators $T^{(n)} = (T_\alpha^{(n)})_{\alpha=1,\dots,p}$, $n \in \mathbf{N}$, for the nuisance parameters admits the following asymptotic expansion for $\alpha = 1, \dots, p$:

$$n^{1/2}(T_\alpha^{(n)} - \tau_\alpha) = \tilde{g}_\alpha(\cdot, \tau) + n^{-1/2}\, l_n(\tfrac{1}{2}, \theta_0)\,, \tag{4.4}$$

where $P_{(\theta_0,\tau)}(g_\alpha(\cdot, \tau)) = 0$ for $\alpha = 1, \dots, p$, and $\tau \in \mathbf{T}$. Furthermore, g_α, $\alpha = 1, \dots, p$, fulfill Condition $M_4(\theta_0)$, and $g_\alpha^{(\beta)}$, $\alpha, \beta = 1, \dots, p$, fulfill Conditions $M_4(\theta_0)$ and $L_4(\theta_0)$.

In most special cases the test statistic F_n will be defined for all possible values θ_0, hence the functions f_j and Q occurring in the asymptotic expansion (4.2) will depend on θ_0. This is, however, irrelevant for our general considerations. Here, θ_0 is fixed by the hypothesis, and test statistics for other values do not occur in our considerations.

Moreover, the estimators for the nuisance parameters (hence also the functions g_α occurring in their asymptotic expansion (4.4)) may, in particular applications, depend on the hypothetical value θ_0, i.e. they may be estimators for the family $P_{(\theta_0,\tau)}$, $\tau \in \mathbf{T}$.

Relation (3.11) can be used to recast the representation (4.2) into a certain canonical form, distinguished by the fact that the arguments of Q are expressions $\tilde{f}_i^*$ with the following properties:

For a function $f : X \times \Theta \to \mathbf{R}$ with $P_\theta(f(\cdot, \theta)) = 0$ define

$$f^*(\cdot, \theta) := f(\cdot, \theta) + P_\theta(f^{(\beta)}(\cdot, \theta))\, \lambda_\beta(\cdot, \theta)\,.$$

Then $L_{(i)}(\theta) = 0$ for $i = 0, \dots, p$ implies $P(f^*(\cdot, \theta)) = 0$. If f fulfills Condition D, then for $i = 0, \dots, p$

$$0 = P_\theta(f(\cdot, \theta))^{(i)} = P_\theta(f^{(i)}(\cdot, \theta)) + P_\theta(l^{(i)}(\cdot, \theta)\, f(\cdot, \theta))\,,$$

and hence for $\alpha = 1, \dots, p$

$$P_\theta(l^{(\alpha)}(\cdot, \theta)\, f^*(\cdot, \theta)) = P_\theta(l^{(\alpha)}(\cdot, \theta)\, f(\cdot, \theta))$$
$$- P_\theta(l^{(\beta)}(\cdot, \theta)\, f(\cdot, \theta))\, \Lambda_{\beta i}(\theta)\, L_{(i)(\alpha)}(\theta) = 0\,.$$

Furthermore,

$$P_\theta(l^{(0)}(\cdot, \theta)\, f^*(\cdot, \theta)) = P_\theta(l^{(0)}(\cdot, \vartheta)\, f(\cdot, \theta))\,.$$

4.5 Lemma. *Under Assumption* 4.1 *and Condition* (v), *the sequence of test statistics* F_n, $n \in \mathbf{N}$, *admits the following asymptotic expansion:*

$$F_n = \tilde{f}_0(\cdot, \tau) + n^{-1/2}[\tilde{\lambda}_\alpha\cdot(, \theta_0, \tau)\, \tilde{f}_0^{(\alpha)*}(\cdot, \tau)$$
$$- \tfrac{1}{2}\tilde{\lambda}_\alpha(\cdot, \theta_0, \tau)\, \tilde{\lambda}_\beta(\cdot, \theta_0, \tau)\, P_{(\theta_0,\tau)}(f_0^{(\alpha\beta)}(\cdot, \tau)) + Q(\tilde{f}^*(\cdot, \tau), \tau)] + n^{-1} l_n(\tfrac{1}{2}, \theta_0)\,.$$

4.6 Definition. A test-sequence φ_n, $n \in \mathbf{N}$, is *of level* $\alpha + o(n^{-r})$ for the hypothesis $\{(\theta_0, \tau) : \tau \in \mathbf{T}\}$ if locally uniformly in $\tau \in \mathbf{T}$

$$P^n_{(\theta_0,\tau)}(\varphi_n) \leqq \alpha + o(n^{-r}) .$$

If this holds with an equality sign then the test-sequence is *asymptotically similar of level* $\alpha + o(n^{-r})$.

Starting from a sequence of test statistics F_n, $n \in \mathbf{N}$, we apply an asymptotic studentization procedure, based on an estimator-sequence $T^{(n)}$, $n \in \mathbf{N}$, for the nuisance parameters, to obtain a sequence of critical regions for the hypothesis $\{(\theta_0, \tau) : \tau \in \mathbf{T}\}$ which is asymptotically similar of level $\alpha + o(n^{-1/2})$.

For a positive definite covariance matrix $\Sigma = (\sigma_{ij})_{i,j=0,\ldots,q}$ and a function $R : \mathbf{R}^{q+1} \to \mathbf{R}$ for which the following integral exists, define

$$(4.7) \qquad H_\Sigma[R]\,(u_0, a) := \varphi(\sigma_{00}^{-1/2}(u_0 - a_0))^{-1} \int \varphi_\Sigma(u - a)\, R(u)\, du_{(0)} .$$

4.8 Theorem. *Let* F_n, $n \in \mathbf{N}$, *be a sequence of test statistics fulfilling Assumption* 4.1, *and let* $T^{(n)}$, $n \in \mathbf{N}$, *be a sequence of estimators for the nuisance parameters fulfilling Assumption* 4.3. *Assume that Conditions* (i)–(v) *are fulfilled, and that the vector*

$$h := (f_0, \lambda_1, \ldots, \lambda_p, g_1, \ldots, g_p, f_0^{(1)*}, \ldots, f_0^{(p)*}, f^*_{p+1}, \ldots, f^*_m)$$

fulfills Condition $C(\theta_0)$. *Assume, furthermore, that*

$$\sigma_{00}(\theta_0, \tau) := P_{(\theta_0,\tau)}(f_0(\cdot, \tau)^2) > 0 \quad \text{for} \quad \tau \in \mathbf{T} .$$

For $\tau \in \mathbf{T}$ *let* $k(\cdot, \tau)$ *denote a subvector of* $h(\cdot, \tau)$ *whose components constitute a base of* $P_{(\theta_0,\tau)}$*-a.e. linearly independent functions for the subspace spanned by the components of* $h(\cdot, \tau)$ *in* $L^2(P_{(\theta_0,\tau)} \mid \mathscr{A})$, *and assume w.l.g. that* $k_0(\cdot, \tau) = f_0(\cdot, \tau)$. *Let* $\Sigma(\theta_0, \tau) := P_{(\theta_0,\tau)}(k(\cdot, \tau)\, k(\cdot, \tau)')$, *and consider* $Q(f^*(\cdot, \tau), \tau)$ *as a function of* $k(\cdot, \tau)$ *in computing* $H_\Sigma[Q]$.

Then the following is true:

(*a*) *For all* $\alpha \in (0, 1)$, *the sequence of critical regions*

$$(4.9) \qquad C_{n,\alpha} := \{F_n > -N_\alpha \sigma_{00}(\theta_0, T^{(n)})^{1/2} + n^{-1/2}[d_1(N_\alpha, \theta_0, T^{(n)}) + d_2(N_\alpha, \theta_0, T^{(n)})]\}$$

with

$$d_1(N, \cdot) := P(\lambda_\alpha f_0^{(\alpha)}) + \tfrac{1}{2}\Lambda_{\alpha\beta}\, P(f_0^{(\alpha\beta)}) + (N^2 - 1)\, \sigma_{00}^{-1}[\tfrac{1}{6} P(f_0^3) + \Lambda_{0\alpha} P(f_0 f_0^{(\alpha)})\, P(l^{(0)} f_0) + \tfrac{1}{2}\Lambda_{0\alpha}\Lambda_{0\beta}\, P(f_0^{(\alpha\beta)})\, (P(l^{(0)} f_0))^2] + \sigma_{00}^{1/2}\, H_\Sigma[Q]\,(-N\sigma_{00}^{1/2}, 0) ,$$

$$d_2(N, \cdot) := -N^2\, \tfrac{1}{2}\sigma_{00}^{-1}\sigma_{00}^{(\alpha)}\, P(f_0 g_\alpha) ,$$

is asymptotically similar of level $\alpha + o(n^{-1/2})$ *for the hypothesis* $\{(\theta_0, \tau) : \tau \in \mathbf{T}\}$. *Its power function admits the following asymptotic expansion locally uniformly in* $\tau \in \mathbf{T}$ *and uniformly in* $|t| \leqq \log n$:

$$
\begin{aligned}
(4.10) \qquad P^n_{(\theta_0 + n^{-1/2}t, \tau)}(C_{n,\alpha}) &= \Phi\big(N + t\sigma_{00}^{-1/2} P(l^{(0)} f_0) \\
&\quad + n^{-1/2}\big[t^2 \tfrac{1}{2}\sigma_{00}^{-1/2}\big(P(l^{(0)2} f_0) + P(l^{(00)} f_0)\big) \\
&\quad - t\big(N + t\sigma_{00}^{-1/2} P(l^{(0)} f_0)\big)\, \sigma_{00}^{-1}\big(\tfrac{1}{2} P(l^{(0)} f_0^2) + \Lambda_{0\alpha} P(l^{(0)} f_0^{(\alpha)})\, P(l^{(0)} f_0)\big) \\
&\quad + t\big(2N + t\sigma_{00}^{-1/2} P(l^{(0)} f_0)\big)\, \sigma_{00}^{-2}\big(\tfrac{1}{6} P(f_0^3)\, P(l^{(0)} f_0) \\
&\quad + \Lambda_{0\alpha} P(f_0 f_0^{(\alpha)})\, (P(l^{(0)} f_0))^2 + \tfrac{1}{2}\Lambda_{0\alpha}\Lambda_{0\beta} P(f_0^{(\alpha\beta)})\, (P(l^{(0)} f_0))^3\big) \\
&\quad + tN\, \tfrac{1}{2}\sigma_{00}^{-2}\sigma_{00}^{(\alpha)}\big(\sigma_{00} P(l^{(0)} g_\alpha) - P(f_0 g_\alpha)\, P(l^{(0)} f_0)\big) \\
&\quad + H_\Sigma[Q]\,(-N\sigma_{00}^{1/2}, t\, P(l^{(0)} f)) \\
&\quad - H_\Sigma[Q]\,(-N\sigma_{00}^{1/2}, 0)\big]\big)\Big|_{\theta = (\theta_0, \tau), N = N_\alpha} + o(n^{-1/2}) .
\end{aligned}
$$

(b) *The distribution function of the sequence of test statistics* F_n, $n \in \mathbf{N}$, *admits the following asymptotic expansion locally uniformly in* $\tau \in \mathbf{T}$ *and uniformly in* $u \in \mathbf{R}$:

$$
(4.11) \qquad P^n_{(\theta_0, \tau)}\{F_n < u\} = \Phi\big(\sigma_{00}^{-1/2} u - n^{-1/2}\sigma_{00}^{-1/2} d_1(-\sigma_{00}^{-1/2} u, \cdot)\big)\Big|_{\vartheta = (\theta_0, \tau)} .
$$

In general, the power function of the sequence of critical regions $C_{n,\alpha}$, $n \in \mathbf{N}$, depends on Q and on the studentizing estimator sequence through terms of order $n^{-1/2}$. The dependence on g_α is such that a better covariance matrix does not necessarily lead to a better power. If, however, $C_{n,\alpha}$, $n \in \mathbf{N}$, is asymptotically efficient then its power function depends on Q and the studentizing estimator sequence only through terms of order $o(n^{-1/2})$. This is the content of Theorem 6.5.

5. Examples of test statistics

Theorems 4.8 and 6.5 refer to sequences of tests based on a test statistic with an asymptotic expansion (4.2). It is, therefore, of interest to show that the test statistics obtained by the usual methods for the construction of tests are, in fact, of this type. (It would be more fruitful, of course, to prove Theorem 6.5 without this restriction.)

A. Tests based on estimators

Assume that we are given an estimator sequence $(T_0^{(n)}, T_1^{(n)}, \ldots, T_p^{(n)})$, $n \in \mathbf{N}$, for $(\theta_0, \theta_1, \ldots, \theta_p)$ such that

$$n^{1/2}(T_i^{(n)} - \theta_i) = \tilde{g}_i(\cdot, \theta) + n^{-1/2} Q_i(\tilde{g}(\cdot, \theta), \theta) + n^{-1} l_n(\tfrac{1}{2}, \theta_0), \quad i = 0, \ldots, p,$$

with $g = (g_0, \ldots, g_m)$, $m \geqq p$. (For instance all minimum contrast estimators are of this type; see Chibisov (1973c), p. 298, Theorem 5.) Under certain regularity conditions, the test statistic $F_n := n^{1/2}(T_0^{(n)} - \theta_0)$ fulfills Assumptions 4.1, and $(T_1^{(n)}, \ldots \ldots, T_p^{(n)})$, considered as an estimator sequence for the nuisance parameters, fulfills Assumption 4.3. We remark that the test obtained from this test statistic will be efficient iff $g_0 = \lambda_0$.

If, in particular, $T^{(n)}$, $n \in \mathbf{N}$, is a sequence of asymptotic m.l. estimators of order $o(n^{-1/2})$, then by Lemma 8 in Pfanzagl (1973b), p. 253,

$$Q_i(\tilde{g}, \cdot) = \tfrac{1}{2}\Lambda_{ij} L_{(jkq)} \tilde{g}_k \tilde{g}_q + \tilde{g}_j \tilde{g}_{(i+1)(p+1)+j}$$

with

$$g = (\lambda_0, \ldots, \lambda_p, \Lambda_{0i} l^{(0i)}, \ldots, \Lambda_{0i} l^{(pi)}, \ldots, \Lambda_{pi} l^{(0i)}, \ldots, \Lambda_{pi} l^{(pi)}).$$

B. Tests obtained by desensitization

Assume that we are given a family of sequences of $\mathscr{A}^n$-measurable functions $K_n(\cdot, \theta_0, \tau) : X^{\mathbf{N}} \to \mathbf{R}$, $n \in \mathbf{N}$, $\tau \in \mathbf{T}$, which are asymptotically efficient if used as test statistics for the hypothesis $\delta_0 = \theta_0$ against alternatives $\delta_0 > \theta_0$, if the nuisance parameter τ is known. If the nuisance parameter is unknown, $K_n(\cdot, \theta_0, \tau)$ cannot be used immediately as a test statistic, but it suggests itself to replace τ in this case by an estimator, i.e. to use $K_n(\cdot, \theta_0, T^{(n)})$.

As far as we know there is no technical term for this procedure (i.e. for obtaining a test statistic for the hypothesis $\{(\theta_0, \tau) : \tau \in \mathbf{T}\}$ from a test statistic for the hypothesis (θ_0, τ), τ being known, and replacing τ by an estimator). Hence we suggest the term „desensitization". The procedure itself is so suggestive that it has been applied earlier, e.g. by Neyman (1954, 1959) in connection with his so-called $C(\alpha)$-tests.

In the following we shall give conditions on $K_n(\cdot, \theta_0, \tau)$ under which the test statistic $K_n(\cdot, \theta_0, T^{(n)})$ fulfills Assumption 4.1, and we shall show that these conditions are, for instance, fulfilled if $K_n(\cdot, \theta_0, \tau)$ is m.p. for testing the hypothesis (θ_0, τ) against an alternative $(\theta_0 + n^{-1/2}t, \tau)$.

Since θ_0 is fixed in our considerations, we shall write $K_n(\cdot, \tau)$ instead of $K_n(\cdot, \theta_0, \tau)$.

5.1 Definition. Let $f_n : X^{\mathbf{N}} \times \mathbf{T} \to \mathbf{R}$ be such that $f_n(\cdot, \tau)$ is $\mathscr{A}^n$-measurable for $\tau \in \mathbf{T}$ and $n \in \mathrm{N}$. We shall write $f_n(\cdot, \tau) = l_n(r, \theta_0)$ if for every $\tau \in \mathbf{T}$ there exist a neighbourhood U_τ of τ and a constant $a > 0$ such that uniformly for $\gamma \in U_\tau$ and $|\delta_0 - \theta_0| \leqq n^{-1/2} \log n$

$$P^n_{(\delta_0,\gamma)}\{\sup_{\|\varkappa - \gamma\| \leqq n^{-1/2}(\log n)^2} |f_n(\cdot, \varkappa)| > (\log n)^a\} = o(n^{-r}) .$$

Notice that this strengthens the definition of l_n given in 1.1.

5.2 Assumption. The sequence of functions $K_n : X^{\mathbf{N}} \times \mathbf{T} \to \mathbf{R}$, $n \in \mathrm{N}$, admits an asymptotic expansion

$$K_n(\cdot, \tau) = \tilde{k}_0(\cdot, \tau) + n^{-1/2} R(\tilde{k}(\cdot, \tau), \tau) + n^{-1} l_n(\tfrac{1}{2}, \theta_0) , \tag{5.3}$$

where $k = (k_0, \ldots, k_m)$ and $P_{(\theta_0,\tau)}(k(\cdot, \tau)) = 0$ for $\tau \in \mathbf{T}$. The function R fulfills Condition B and the functions $\tau \to P_{(\theta_0,\tau)}(k_0^{(\gamma)}(\cdot, \tau))^{(\alpha\beta)}$, $\alpha, \beta, \gamma = 1, \ldots, p$, fulfill locally uniform Lipschitz conditions on $\mathbf{T}$. Furthermore, k_j fulfill Condition $M_4(\theta_0)$ and $k_j^{(\alpha)}$ and $k_0^{(\alpha\beta)}$ fulfill Conditions $M_4(\theta_0)$ and $L_4(\theta_0)$ for $j = 0, \ldots, m$; $\alpha, \beta = 1, \ldots, p$.

5.4 Assumption. The sequence of estimators $T^{(n)} = (T_\alpha^{(n)})_{\alpha = 1,\ldots,p}$, $n \in \mathrm{N}$, for the nuisance parameters admits the following asymptotic expansion for $\alpha = 1, \ldots, p$:

$$n^{1/2}(T_\alpha^{(n)} - \theta_\alpha) = \tilde{g}_\alpha(\cdot, \tau) + n^{-1/2} Q_\alpha(\tilde{g}(\cdot, \tau), \tau) + n^{-1} l_n(\tfrac{1}{2}, \theta_0) , \tag{5.5}$$

where $g = (g_1, \ldots, g_q)$ and $P_{(\theta_0,\tau)}(g(\cdot, \tau)) = 0$ for $\tau \in \mathbf{T}$. The functions Q_α, $\alpha = 1, \ldots, p$, fulfill Condition B. Furthermore, g_i fulfill Condition $M_4(\theta_0)$, and $g_i^{(\alpha)}$ and $g_\gamma^{(\alpha\beta)}$ fulfill Conditions $M_4(\theta_0)$ and $L_4(\theta_0)$ for $i = 1, \ldots, q$; $\alpha, \beta, \gamma = 1, \ldots, p$.

5.6 Lemma. *Let $K_n : X^{\mathbf{N}} \times \mathbf{T} \to \mathbf{R}$, $n \in \mathrm{N}$, be a sequence of functions fulfilling Assumption* 5.2, *and let $T^{(n)}$, $n \in \mathrm{N}$, be a sequence of estimators for the nuisance parameters fulfilling Assumption* 5.4. *Assume that $l(x, \cdot)$ admits continuous first order derivatives on Θ for $x \in X$, and that $L_{(i)(i)}$, $i = 0, \ldots, p$, are locally uniformly bounded on Θ.*

Then $K_n(\cdot, T^{(n)})$, $n \in \mathrm{N}$, is a sequence of test statistics for the hypothesis $\{(\theta_0, \tau) : \tau \in \mathbf{T}\}$ fulfilling Assumption 4.1.

Proof. We need only observe that locally uniformly in $\tau \in \mathbf{T}$ and uniformly in $\|\delta_0 - \theta_0\| \leqq n^{-1/2} \log n$

$$P^n_{(\delta_0,\tau)}\{\|T^{(n)} - \tau\| > n^{-1/2}(\log n)^2\} = o(n^{-1}) .$$

(See e.g. Corollary 2 in Nagaev (1965), p. 215.) Then (5.3) implies

$$K_n(\cdot, T^{(n)}) = \tilde{k}_0(\cdot, T^{(n)}) + n^{-1/2} R(\tilde{k}(\cdot, T^{(n)}), T^{(n)}) + n^{-1} l_n(\tfrac{1}{2}, \theta_0) .$$

Expanding about $\tau = (\theta_1, \ldots, \theta_p)$ we obtain from (5.5)

$$K_n(\cdot, T^{(n)}) = \tilde{k}_0(\cdot, \tau) + \tilde{g}_\alpha(\cdot, \tau)\, P_{(\theta_0,\tau)}(k_0^{(\alpha)}(\cdot, \tau))$$

$$+ n^{-1/2}[P_{(\theta_0,\tau)}(k_0^{(\alpha)}(\cdot, \tau))\, Q_\alpha(\tilde{g}(\cdot, \tau), \tau) + \tilde{g}_\alpha(\cdot, \tau)\, \tilde{k}_0^{(\alpha)}(\cdot, \tau)$$

$$+ \tfrac{1}{2}\tilde{g}_\alpha(\cdot, \tau)\, \tilde{g}_\beta(\cdot \tau)\, P_{(\theta_0,\tau)}(k_0^{(\alpha\beta)}),$$

$$+ R(\tilde{k}(\cdot, \tau) + \tilde{g}_\alpha(\cdot, \tau)\, P_{(\theta_0,\tau)}(k^{(\alpha)}(\cdot, \tau)), \tau)] + n^{-1}\, l_n(\tfrac{1}{2}, \theta_0).$$

This is an asymptotic expansion of the type (4.2), and it is easily seen that Assumption 4.1 is fulfilled.

If $T^{(n)}$ consists of the components $(\theta_1^{(n)}, \ldots, \theta_p^{(n)})$ of an asymptotic m.l. estimator $(\theta_0^{(n)}, \theta_1^{(n)}, \ldots, \theta_p^{(n)})$ of order $o(n^{-1/2})$ for θ, then this asymptotic expansion is already of the canonical form introduced in Lemma 4.5.

Now we shall show that under certain conditions the m.p. test is equivalent to one fulfilling Assumption 5.2.

Assume that $l(x, \cdot)$ admits continuous partial derivatives up to the fourth order on Θ for $x \in X$, and that $l^{(i)}, l^{(ij)}, l^{(ijk)}, l^{(ijkq)}$, $i, j, k, q = 0, \ldots, p$, fulfill Conditions L_4 and M_4. Assume, furthermore, that $L_{(i)(j)}$, $i, j = 0, \ldots, p$, admit second order partial derivatives fulfilling locally uniform Lipschitz conditions on Θ. Then the m.p. test based on

$$\prod_{\nu=1}^{n} \frac{p(x_\nu, \theta_0 + n^{-1/2}t, \tau)}{p(x_\nu, \theta_0, \tau)}$$

is for $t > 0$ up to $o(n^{-1/2})$ equivalent to the test based on

$$\tilde{l}^{(0)}(\cdot, \theta_0, \tau) + n^{-1/2}\, \tfrac{1}{2}t\, \tilde{l}^{(00)}(\cdot, \theta_0, \tau) \tag{5.7}$$

(for which Assumption 5.2 is trivially fulfilled). The asserted equivalence follow from

$$t^{-1} \sum_{\nu=1}^{n} (l(x_\nu, \theta_0 + n^{-1/2}t, \tau) - l(x_\nu, \theta_0, \tau))$$

$$= \tilde{l}^{(0)}(\mathbf{x}, \theta_0, \tau) + \tfrac{1}{2}t\, L_{(00)}(\theta_0, \tau)$$

$$+ n^{-1/2}[\tfrac{1}{2}t\, \tilde{l}^{(00)}(\mathbf{x}, \theta_0, \tau) + \tfrac{1}{6}t^2\, L_{(000)}(\theta_0, \tau)] + n^{-1}\, l_n(\tfrac{1}{2}, \theta_0).$$

The efficient version of the $C(\alpha)$-test may be considered as the limiting case of (5.7) for $t \downarrow 0$.

6. Results on tests

6.1 Definition. (i) A test sequence φ_n, $n \in \mathbf{N}$, of level $\alpha + o(n^0)$ for the hypothesis $\{(\theta_0, \tau) : \tau \in \mathbf{T}\}$ is *first order efficient* (or *asymptotically efficient*) if locally uniformly in $\tau \in \mathbf{T}$ and uniformly in $|t| \leq \log n$

$$(6.2) \qquad P^n_{(\theta_0+n^{-1/2}t,\tau)}(\varphi_n) = \Phi(N_\alpha + t\,\Lambda_{00}(\theta_0,\tau)^{-1/2}) + o(n^0)\,.$$

(ii) A test sequence φ_n, $n \in \mathbf{N}$, of level $\alpha + o(n^{-1/2})$ for the hypothesis $\{(\theta_0, \tau) : : \tau \in \mathbf{T}\}$ is *second order efficient* if locally uniformly in $\tau \in \mathbf{T}$ and uniformly in $|t| \leq \leq \log n$

$$(6.3) \qquad P^n_{(\theta_0+n^{-1/2}t,\tau)}(\varphi_n) = \Phi(N_\alpha + t\,\Lambda_{00}(\theta_0,\tau)^{-1/2} + n^{-1/2}[tN_\alpha\, h_1(\theta_0,\tau) + t^2\, h_2(\theta_0,\tau)]) + o(n^{-1/2})$$

with

$$(6.4) \qquad h_1 := -\tfrac{1}{6}\Lambda_{00}^{-2}\Lambda_{0i}\Lambda_{0j}\Lambda_{0k}L_{(i)(j)(k)}\,,$$

$$h_2 := -\Lambda_{00}^{-5/2}\Lambda_{0i}\Lambda_{0j}\Lambda_{0k}(\tfrac{1}{3}L_{(i)(j)(k)} + \tfrac{1}{2}L_{(ij)(k)}) + \Lambda_{00}^{-3/2}\Lambda_{0i}\Lambda_{0j}(\tfrac{1}{2}L_{(0)(i)(j)} + L_{(0i)(j)})\,.$$

This definition is justified by the fact that for $t > 0$ $[t < 0]$ the right side of (6.3) is an upper [lower] bound for the power function of any sequence of tests φ_n, $n \in \mathbf{N}$, of level $\alpha + o(n^{-1/2})$ (see Chibisov (1973a), p. 40, Theorem 9.1; see also the Proposition in Pfanzagl (1973b), p. 260), and that there exist sequences of tests of level $\alpha + o(n^{-1/2})$ for which this bound is assumed. (For C(α)-tests see Chibisov (1973a), p. 38, Theorem 8.1; for the test based on the m.l. estimator see Pfanzagl (1973b), p. 213, Theorem (ii) and (iv).)

With the aid of Theorem 4.8 we obtain the following result.

6.5 Theorem. *Let F_n, $n \in \mathbf{N}$, be a sequence of test statistics fulfilling Assumption 4.1, and let $T^{(n)}$, $n \in \mathbf{N}$, be a sequence of estimators for the nuisance parameters fulfilling Assumption 4.3. Assume that Conditions (i)–(vi) are fulfilled, that the vector*

$$h := (f_0, \lambda_1, \ldots, \lambda_p, g_1, \ldots, g_p, f_0^{(1)*}, \ldots, f_0^{(p)*}, f_{p+1}, \ldots, f_m)$$

fulfills Condition C(θ_0), and that the components of h fulfill Condition D(θ_0). Assume, furthermore, that $P_{(\theta_0,\tau)}(f_0(\cdot,\tau)^2) > 0$ for $\tau \in \mathbf{T}$.

Then the sequence of asymptotically similar critical regions of level $\alpha + o(n^{-1/2})$ for the hypothesis $\{(\theta_0, \tau) : \tau \in \mathbf{T}\}$ obtained from F_n, $n \in \mathbf{N}$, by asymptotic studentization with $T^{(n)}$, $n \in \mathbf{N}$, as defined by (4.9), is second order efficient if it is first order efficient.

More explicitly: The sequence of critical regions is asymptotically efficient iff $f_0(\cdot,\tau) = c(\tau)\,\lambda_0(\cdot,\theta_0,\tau)$ $P_{(\theta_0,\tau)}$ – a.e. with $c(\tau) > 0$ for $\tau \in \mathbf{T}$. In this case, the dependence of the power on the function Q occurring in the expansion (4.2) as well as the dependence on the estimator sequence $T^{(n)}$, $n \in \mathbf{N}$, is of order $o(n^{-1/2})$.

6.6 Remark. Under suitable regularity conditions this means that the efficiency of the estimator sequence for the nuisance parameters influences the asymptotic expansion of the power function in the term n^{-1} (and higher). This implies in particular that the gain which might be achieved by estimating $(\theta_1, \ldots, \theta_p)$ under the condition that the true value of θ_0 is the one specified by the hypothesis and working with estimators $(T_1^{(n)}(\cdot, \theta_0), \ldots, T_p^{(n)}(\cdot, \theta_0))$ (instead of estimating $(\theta_0, \theta_1, \ldots, \theta_p)$ by $(T_0^{(n)}, T_1^{(n)}, \ldots, T_p^{(n)})$ and taking the subvector $(T_1^{(n)}, \ldots, T_p^{(n)})$ as estimator for $(\theta_1, \ldots, \theta_p)$) will influence the power function of order n^{-1}, at most.

6.7 Remark. If $f_0(\cdot, \tau) = c(\tau)\, \lambda_0(\cdot, \theta_0, \tau)$ $P_{(\theta_0,\tau)}$-a.e. with $c(\tau) > 0$ for $\tau \in \mathbf{T}$ then the functions d_1 and d_2 (defined in Theorem 4.8(a)) which determine the bound of the critical region $C_{n,\alpha}$ (defined in (4.9)) reduce to

$$d_1(N, \cdot) = cN^2[\Lambda_{00}^{-1} \Lambda_{0i}\Lambda_{0j}\Lambda_{0k}(\tfrac{1}{6}L_{(i)(j)(k)} + \tfrac{1}{2}L_{(ij)(k)})$$

$$- \Lambda_{0i}\Lambda_{0j}(\tfrac{1}{2}L_{(i)(j)(0)} + L_{(i)(j0)}) + \Lambda_{00}\Lambda_{0i}(\tfrac{1}{2}L_{(i)(0)(0)} + \tfrac{1}{2}L_{(i)(00)})]$$

$$+ c[\Lambda_{00}^{-1} \Lambda_{0i}\Lambda_{0j}\Lambda_{0k}(\tfrac{1}{3}L_{(i)(j)(k)} + \tfrac{1}{2}L_{(ij)(k)}) - \Lambda_{0i}\Lambda_{jk}(\tfrac{1}{2}L_{(i)(j)(k)} + \tfrac{1}{2}L_{(i)(jk)})]$$

$$+ \frac{1}{c} \Lambda_{00}^{-1/2} H_\Sigma[Q]\,(-cN\Lambda_{00}^{1/2}, 0)\,,$$

$$d_2(N, \cdot) = N^2[c\Lambda_{0i}\Lambda_{0j}(\tfrac{1}{2}L_{(i)(j)(\alpha)} + L_{(i)(j\alpha)}) - c^{(\alpha)}\Lambda_{00}]\, P(l^{(0)}g_\alpha)\,.$$

7. Results on estimators

In the sections on hypothesis testing it was appropriate to denote the parameter by $(\theta_0, \theta_1, \ldots, \theta_p)$ in order to stress the basically different rôle played by the parameter θ_0, specified by the hypothesis, and the nuisance parameters $(\theta_1, \ldots, \theta_p)$. In this section we consider the problem of estimating a vector parameter. Hence it is appropriate to denote the whole parameter by $(\theta_1, \ldots, \theta_p)$. Correspondingly, Λ now denotes the inverse of $(L_{(\alpha)(\beta)})_{\alpha,\beta=1,\ldots,p}$, and $\lambda_\alpha := \Lambda_{\alpha\beta} l^{(\beta)}$ for $\alpha = 1, \ldots, p$.

The results in this section refer to sequences of estimators admitting an asymptotic expansion of the following form.

7.1 Assumption. The sequence of estimators $T^{(n)} = (T_\alpha^{(n)})_{\alpha=1,\ldots,p}$, $n \in \mathbf{N}$, admits the following expansion for $\alpha = 1, \ldots, p$.

$$n^{1/2}(T_\alpha^{(n)} - \theta_\alpha) = \tilde{g}_\alpha + n^{-1/2}\, Q_\alpha(\tilde{g}(\cdot, \theta), \theta) + n^{-1}\, l_n(\tfrac{1}{2})\,, \tag{7.2}$$

where $P_\theta(g(\cdot, \theta)) = 0$ for $g = (g_1, \ldots, g_q)$ and $\theta \in \Theta$. The functions Q_α, $\alpha = 1, \ldots, p$, fulfill Condition B. Furthermore, g_i, $i = 1, \ldots, q$, fulfill Condition M_4 and $g_i^{(\beta)}$, $g_\alpha^{(\beta\gamma)}$, $\alpha, \beta, \gamma = 1, \ldots, p$, $i = 1, \ldots, q$, fulfill Conditions M_4 and L_4.

7.3 Definition. A sequence $Q_\theta^{(n)} \mid \mathscr{B}^q$, $n \in \mathbf{N}$, of p-measures admits an *E-expansion* (*Edgeworth-expansion*) $\varphi_\Sigma(1 + n^{-1/2}K)$ of order $o(n^{-1/2})$ locally uniformly in θ if, for a function $K : \mathbf{R}^q \times \Theta \to \mathbf{R}$, we have uniformly for $E \in \mathscr{E}$ and locally uniformly in θ

$$Q_\theta^{(n)}(E) = \int_E \varphi_{\Sigma(\theta)}(v)\,(1 + n^{-1/2}\,K(v, \theta))\,dv + o(n^{-1/2}),$$

where $\mathscr{E}$ denotes the class of all measurable convex subsets of $\mathbf{R}^q$.

If $T^{(n)}$, $n \in \mathbf{N}$, is any estimator sequence for which $n^{1/2}(T^{(n)} - \theta)$ is asymptotically normally distributed with continuous covariance matrix Σ, then $\Sigma - \Lambda$ is positive semidefinite (see e.g. Bahadur (1964), p. 1550). Hence in a certain sense, Λ is asymptotically the best possible covariance matrix. It is attained, for instance, by all asymptotic m.l. estimators (see e.g. Schmetterer (1974), p. 316, Theorem 3.9, for m.l. estimators and Pfanzagl (1972), p. 186, Theorem 2, for asymptotic m.l. estimators).

If an estimator sequence $T^{(n)}$, $n \in \mathbf{N}$, admits an expansion (7.2), then its distribution $P_\theta^n * n^{1/2}(T^{(n)} - \theta)$ admits an E-expansion $\varphi_\Sigma(1 + n^{-1/2}K)$ of order $o(n^{-1/2})$. We shall show that $\Sigma = \Lambda$ implies that the function K is uniquely determined up to a linear term.

7.4 Theorem. *Let $T^{(n)}$, $n \in \mathbf{N}$, be an estimator sequence fulfilling Assumption* 7.1. *Assume that Conditions* (i)–(vi) *are fulfilled, that the vector h consisting of the functions g_α, λ_α, $g_\alpha^{(\beta)*}$, $\alpha, \beta = 1, \ldots, p$, and g_j^*, $j = 1, \ldots, q$, fulfills Condition* C, *and that the components of h fulfill Condition* D. *Assume, furthermore, that the partial derivatives of Q_α, $\alpha = 1, \ldots, p$, (with respect to the first variable) fulfill Condition* B(*b*).

Then $P_\theta^n * n^{1/2}(T^{(n)} - \theta)$, $n \in \mathbf{N}$, admits locally uniformly in θ an E-expansion of order $o(n^{-1/2})$, say $\varphi_\Sigma(1 + n^{-1/2}K)$, with $\Sigma(\theta) = (P_\theta(g_\alpha(\cdot, \theta)\, g_\beta(\cdot, \theta)))_{\alpha,\beta=1,\ldots,p}$.

If the estimator sequence is asymptotically efficient, i.e. if $\Sigma = \Lambda$, then

$$K(v, \cdot) = a_\alpha v_\alpha - (\tfrac{1}{3}L_{(\alpha)(\beta)(\gamma)} + \tfrac{1}{2}L_{(\alpha)(\beta\gamma)})\, v_\alpha v_\beta v_\gamma\,, \tag{7.5}$$

where a_α, $\alpha = 1, \ldots, p$, are functions of θ.

7.6 Remark. It is clear in advance that K cannot be unique, since the sequence of transformations $\hat{T}^{(n)} := T^{(n)} + n^{-1}\,c(T^{(n)})$, $n \in \mathbf{N}$, with continuously differentiable functions $c_1, \ldots, c_p$, applied to an estimator sequence $T^{(n)}$, $n \in \mathbf{N}$, with E-expansion $\varphi_\Sigma(1 + n^{-1/2}K)$ leads to an estimator sequence $\hat{T}^{(n)}$, $n \in \mathbf{N}$, with E-expansion

$$v \to \varphi_\Sigma(v)\,(1 + n^{-1/2}(K(v, \cdot) + c_\alpha(\Sigma^{-1})_{\alpha\beta} v_\beta))\,.$$

K becomes unique if we require the estimator sequence to be componentwise asymptotically median unbiased $o(n^{-1/2})$.

7.7 Definition. An estimator sequence $T^{(n)}$, $n \in \mathbf{N}$, is *componentwise asymptotically median unbiased* $o(n^{-r})$ if locally uniformly in $\theta \in \Theta$, for $\alpha = 1, \ldots, p$,

$$P_\theta^n\{T_\alpha^{(n)} \geqq \theta_\alpha\} \geqq \tfrac{1}{2} + o(n^{-r})$$

and

$$P_\theta^n\{T_\alpha^{(n)} \leqq \theta_\alpha\} \geqq \tfrac{1}{2} + o(n^{-r}).$$

7.8 Corollary. *Assume that the regularity conditions of Theorem 7.4 are fulfilled. If the estimator sequence is asymptotically efficient and componentwise asymptotically median unbiased* $o(n^{-1/2})$ *then*

$$\begin{aligned}(7.9)\qquad K(v, \cdot) = \Big(\big[L_{(\alpha)(\beta)(\gamma)} + L_{(\alpha\beta)(\gamma)} + \tfrac{1}{2}L_{(\alpha)(\beta\gamma)}\big]\,\Lambda_{\beta\gamma} \\ - L_{(\delta)(\alpha)}\big[\tfrac{1}{3}L_{(\beta)(\gamma)(\rho)} + \tfrac{1}{2}L_{(\beta)(\gamma\rho)}\big]\,\Lambda_{\delta\beta}\Lambda_{\delta\gamma}\Lambda_{\delta\rho}\Lambda_{\delta\delta}^{-1}\Big)\,v_\alpha \\ - \big(\tfrac{1}{3}L_{(\alpha)(\beta)(\gamma)} + \tfrac{1}{2}L_{(\alpha)(\beta\gamma)}\big)\,v_\alpha v_\beta v_\gamma\,.\end{aligned}$$

7.10 Remark. Asymptotically normal estimators are asymptotically median unbiased $o(n^0)$ not only componentwise, but in the stronger sense that for each half-space H containing the origin

$$(7.11)\qquad P_\theta^n\{n^{1/2}(T^{(n)} - \theta) \in H\} \geqq \tfrac{1}{2} + o(n^0)\,.$$

Our Corollary implies that it is impossible in the general case to obtain an efficient estimator sequence for which (7.11) holds with $o(n^0)$ replaced by $o(n^{-1/2})$: By 7.8, the function K is uniquely determined if the estimator sequence is asymptotically efficient and componentwise asymptotically median unbiased $o(n^{-1/2})$ (i.e. if (7.11) holds with $o(n^{-1/2})$ for each half-space H containing the origin and having a bounding plane normal to one of the coordinate vectors). Hence it suffices to give an example where for the polynomial K given by (7.9),

$$\int_{H_0} \varphi_\Sigma(v)\,K(v)\,dv \neq 0$$

for some half-space H_0 with its boundary plane passing through the origin, for this implies that for any asymptotically efficient estimator sequence $T^{(n)}$, $n \in \mathbf{N}$, which is asymptotically median unbiased $o(n^{-1/2})$,

$$P_\theta^n\{n^{1/2}(T^{(n)} - \theta) \in H_0\} - \tfrac{1}{2}\,, \quad n \in \mathbf{N}\,,$$

is of order $n^{-1/2}$.

There are plenty of such examples, one being the 1-dimensional normal distribution with $\theta = (\mu, \sigma)$. In this case we have

$$\Sigma = \begin{pmatrix} \sigma^2 & 0 \\ 0 & \sigma^2/2 \end{pmatrix}$$

and from (7.9)

$$K(v,(\mu,\sigma)) = \tfrac{1}{3}\sigma^{-1}(v_2^3 - v_2).$$

For $H_0 = \{(v_1, v_2) : v_1 \leqq v_2\}$,

$$\int_{H_0} \varphi_\Sigma(v)\, K(v)\, dv = -1/(18\sqrt{(3\pi)}).$$

Hence in this case no efficient estimator sequence fulfills (7.11) with $o(n^0)$ replaced by $o(n^{-1/2})$.

7.12 Remark. It was mentioned above that $\Sigma - \Lambda$ is positive semidefinite and that, therefore, the covariance matrix Λ is optimal in a certain sense. A very intuitive interpretation of this optimality can be based on the theorem stating that "$\Sigma_2 - \Sigma_1$ positive semidefinite" implies "$\varphi_{\Sigma_2}(E) \leqq \varphi_{\Sigma_1}(E)$ for $E \in \mathscr{E}_0$", where $\mathscr{E}_0$ is the class of all measurable convex sets symmetric about the origin (see T. W. Anderson (1955), p. 173, Corollary 3). Furthermore, weak convergence to a normal distribution implies uniform convergence in the class of all measurable convex sets. (See e.g. R. R. Rao (1962), p. 665, Theorem 2.) Hence any sequence of asymptotic m.l. estimators $\theta^{(n)}$, $n \in \mathbb{N}$, is asymptotically optimal in the following sense. If $T^{(n)}$, $n \in \mathbb{N}$, is any sequence of asymptotically normal estimators with continuous covariance matrix, then uniformly for $E \in \mathscr{E}_0$,

$$P_\theta^n\{n^{1/2}(T^{(n)} - \theta) \in E\} \leqq P_\theta^n\{n^{1/2}(\theta^{(n)} - \theta) \in E\} + o(n^0).$$

This optimum property can now be strengthened as follows: Define a sequence of *adjusted asymptotic m.l. estimators* $\hat{\theta}^{(n)} := \theta^{(n)} + n^{-1} F(\theta^{(n)})$, $n \in \mathbb{N}$, where for $\varkappa = 1, \ldots, p$,

$$F_\varkappa := \tfrac{1}{2}\Lambda_{\alpha\beta}\Lambda_{\varkappa\gamma}(L_{(\alpha)(\beta)(\gamma)} + L_{(\alpha\beta)(\gamma)}) - \Lambda_{\varkappa\varkappa}^{-1}\Lambda_{\varkappa\alpha}\Lambda_{\varkappa\beta}\Lambda_{\varkappa\gamma}(\tfrac{1}{3}L_{(\alpha)(\beta)(\gamma)} + \tfrac{1}{2}L_{(\alpha\beta)(\gamma)}).$$

By Theorem (iv) in Pfanzagl (1973b), p. 213–215, this estimator sequence is componentwise asymptotically median unbiased $o(n^{-1})$. It is optimal in the sense that for any estimator sequence $T^{(n)}$, $n \in \mathbb{N}$, for which $P_\theta^n * n^{1/2}(T^{(n)} - \theta)$ admits an E-expansion of order $o(n^{-1/2})$, and which is componentwise asymptotically median unbiased $o(n^{-1/2})$, we have uniformly for $E \in \mathscr{E}_0$

$$P_\theta^n\{n^{1/2}(T^{(n)} - \theta) \in E\} \leqq P_\theta^n\{n^{1/2}(\hat{\theta}^{(n)} - \theta) \in E\} + o(n^{-1/2}). \tag{7.13}$$

Let Σ denote the asymptotic covariance matrix of $n^{1/2}(T^{(n)} - \theta)$. If $\Sigma - \Lambda$ is positive definite then the difference

$$\sup_{E \in \mathscr{E}_0} (P_\theta^n\{n^{1/2}(T^{(n)} - \theta) \in E\} - P_\theta^n\{n^{1/2}(\theta^{(n)} - \theta) \in E\})$$

is of order n^0 and negative. If $\Sigma = \Lambda$ then the $n^{-1/2}$-terms of the E-expansions also agree and the difference is of order $o(n^{-1/2})$.

In the one-dimensional case, the sequence of adjusted asymptotic m.l. estimators is even optimal in the stronger sense that (7.13) holds with $o(n^{-1/2})$ replaced by $o(n^{-1})$ if $P_\theta^n * n^{1/2}(T^{(n)} - \theta)$ admits an E-expansion of order $o(n^{-1})$ and if $T^{(n)}$, $n \in \mathbf{N}$, is asymptotically median unbiased $o(n^{-1})$. (See Pfanzagl (1975), *p*. 40, Theorem 7, and the remark following it.) It is unknown whether this stronger optimum property holds true also in the multidimensional case.

8. Proofs

To simplify our notations we shall omit the parameter $\theta = (\theta_0, \tau)$ whenever it is convenient. In particular, $P(f)$ means $P_{(\theta_0,\tau)}(f(\cdot, \theta_0, \tau))$ or $P_{(\theta_0,\tau)}(f(\cdot, \tau))$.

Proof of Theorem 4.8

(a) The first step of the asymptotic studentization procedure is as follows. Using $F_n = \tilde{f}_0 + n^{-1/2}\, l_n(\frac{1}{2}, \theta_0)$ we obtain that locally uniformly in $\tau \in \mathbf{T}$

$$P^n_{(\theta_0,\tau)}\{F_n > -N\, \sigma_{00}(\theta_0, \tau)^{1/2}\} = \alpha + o(n^0)\,.$$

Replacing τ by $T^{(n)}$ in $\sigma_{00}(\theta_0, \tau)$ we obtain that

$$\{F_n > -N\, \sigma_{00}(\theta_0, T^{(n)})^{1/2}\}$$

is a sequence of asymptotically similar critical regions of level $\alpha + o(n^0)$.

(b) If we refine the approximation to F_n, $n \in \mathbf{N}$, and add a correction term of order $n^{-1/2}$ to the bound $-N\, \sigma_{00}(\theta_0, T^{(n)})^{1/2}$ we obtain a sequence of critical regions which is asymptotically similar of level $\alpha + o(n^{-1/2})$. This second step of the asymptotic studentization procedure will be combined with an asymptotic expansion of the power function.

For the components of the vector $h = (h_0, \ldots, h_r)$ defined as

$$h := (f_0, \lambda_1, \ldots, \lambda_p, g_1, \ldots, g_p, f_0^{(1)*}, \ldots, f_0^{(p)*}, f_{p+1}^*, \ldots, f_m^*)$$

the following expansions hold uniformly for $|t| \leqq \log n$.

(8.1) $$\tilde{h}_j = \tilde{h}'_j + t\, P(l^{(0)}h_j) + n^{-1/2}\, l_n(\tfrac{1}{2}, \theta_0) \quad \text{for} \quad j = 0, \ldots, r\,,$$

(8.2) $$\tilde{f}_0 = \tilde{f}'_0 + t\, P(l^{(0)}f_0) + n^{-1/2}\, \tfrac{1}{2}P((l^{(0)2} + l^{(00)})\, f_0) + n^{-1}\, l_n(\tfrac{1}{2}, \theta_0)\,.$$

For $v = (v_0, \ldots, v_r) \in \mathbf{R}^{r+1}$ define

(8.3) $$R(v) := v_\alpha v_{2p+\alpha} - v_\alpha v_\beta\, \tfrac{1}{2}P(f_0^{(\alpha\beta)})\,.$$

Writing $Q(\tilde{f}_0^*, \ldots, \tilde{f}_m^*)$ as a function of $\tilde{h}$, we obtain from Lemma 4.5 on the canonical representation of F_n and from (8.1), (8.2) that uniformly for $|t| \leqq \log n$

(8.4)
$$\begin{aligned} F_n &= \tilde{f}_0 + n^{-1/2}(R + Q)(\tilde{h}) + n^{-1}\, l_n(\tfrac{1}{2}, \theta_0) \\ &= \tilde{f}'_0 + t\, P(l^{(0)} f_0) + n^{-1/2}[(R + Q)(\tilde{h}' + t\, P(l^{(0)} h)) \\ &\quad + t^2\, \tfrac{1}{2} P((l^{(0)2} + l^{(00)}) f_0)] + n^{-1}\, l_n(\tfrac{1}{2}, \theta_0)\,. \end{aligned}$$

By means of the canonical representation of F_n we have thus separated the term of order $n^{-1/2}$ of its asymptotic expansion into two functions with basically different arguments: R depends on h only through f_0, whereas the arguments of Q are f_j^*, $j = 0, \ldots, m$, with $P(l^{(\alpha)} f_j^*) = 0$ for $\alpha = 1, \ldots, p$; $j = 0, \ldots, m$. (For notational simplicity we write R and Q as functions of the full vector h.)

For $v = (v_0, \ldots, v_r) \in \mathbf{R}^{r+1}$ define

(8.5)
$$U(v) := v_{p+\alpha}\, \tfrac{1}{2} \sigma_{00}^{-1/2} \sigma_{00}^{(\alpha)}\,.$$

Making use of the asymptotic expansion (4.4) of the estimator sequence we have uniformly for $|t| \leqq \log n$

(8.6)
$$\begin{aligned} \sigma_{00}(\theta_0, T^{(n)})^{1/2} &= \sigma_{00}^{1/2} + n^{-1/2} \tilde{g}_\alpha\, \tfrac{1}{2} \sigma_{00}^{-1/2} \sigma_{00}^{(\alpha)} + n^{-1}\, l_n(\tfrac{1}{2}, \theta_0) \\ &= \sigma_{00}^{1/2} + n^{-1/2}\, U(\tilde{h}) + n^{-1}\, l_n(\tfrac{1}{2}, \theta_0) \\ &= \sigma_{00}^{1/2} + n^{-1/2}\, U(\tilde{h}' + t\, P(l^{(0)} h)) + n^{-1}\, l_n(\tfrac{1}{2}, \theta_0)\,. \end{aligned}$$

(c) Let $\gamma \in \mathbf{T}$ be fixed. For some neighborhood U_γ of γ the following is true. There exists a subvector $k = (k_0, \ldots, k_g)$ of h such that for every $\tau \in U_\gamma$ the components of $k(\cdot, \tau)$ constitute a base of $P_{(\theta_0,\tau)}$-a.e. linearly independent functions for the subspace spanned by the components of $h(\cdot, \tau)$ in $L^2(P_{(\theta_0,\tau)} \mid \mathscr{A})$. Since $P_{(\theta_0,\tau)}(f_0(\cdot, \tau)^2) > 0$ we may assume w.l.g. that $k_0(\cdot, \tau) = f_0(\cdot, \tau)$.

For $\tau \in U_\gamma$, let

$$h(\cdot, \tau) = M(\tau)\, k(\cdot, \tau)\,,$$

where $M(\tau)$ is an appropriately chosen $(r + 1,\ q + 1)$-matrix with $M_{0j}(\tau) = \delta_{0j}$ for $j = 0, \ldots, q$. There exists a neighborhood U_0 of θ_0 such that for $\tau \in U_\gamma$ and $\delta_0 \in U_0$ the components of $k(\cdot, \tau) - P_{(\delta_0,\tau)}(k(\cdot, \tau))$ constitute a base of $P_{(\delta_0,\tau)}$-a.e. linearly independent functions for the subspace spanned by the components of $h(\cdot, \tau) - P_{(\delta_0,\tau)}(h(\cdot, \tau))$ in $L^2(P_{(\delta_0,\tau)} \mid \mathscr{A})$. (Recall that $P_\theta \mid \mathscr{A}$, $\theta \in \Theta$, are equivalent.) Moreover,

$$h(\cdot, \tau) - P_{(\delta_0,\tau)}(h(\cdot, \tau)) = M(\tau)\, (k(\cdot, \tau) - P_{(\delta_0,\tau)}(k(\cdot, \tau)))\,.$$

From (8.4) and (8.6) we obtain uniformly for $|t| \leqq \log n$

$$(8.7) \qquad F_n = \tilde{f}'_0 + t\,P(l^{(0)}f_0) + n^{-1/2}[(R+Q)(M(\tilde{k}' + t\,P(l^{(0)}k))) + t^2\,\tfrac{1}{2}P((l^{(0)2} + l^{(00)})f_0)] + n^{-1}\,l_n(\tfrac{1}{2}, \theta_0)\,,$$

$$(8.8) \qquad \sigma_{00}(\theta_0, T^{(n)})^{1/2} = \sigma_{00}^{1/2} + n^{-1/2}\,U(M(\tilde{k}' + t\,P(l^{(0)}k))) + n^{-1}\,l_n(\tfrac{1}{2}, \theta_0)\,.$$

For appropriately chosen U_γ and U_0 we obtain from Lemma 2 in Pfanzagl (1973b), p. 242, that the sequence of measures induced by $P^n_{(\delta_0,\tau)}$ and the vector-valued functions

$$\mathbf{x} \to n^{-1/2} \sum_{\nu=1}^{n} (k(x_\nu, \tau) - P_{(\delta_0,\tau)}(k(\cdot, \tau)))$$

admits an E-expansion, say

$$(8.9) \qquad \varphi_{\Sigma(\delta_0,\tau)}(1 + n^{-1/2}\,A(\cdot, \delta_0, \tau))$$

of order $o(n^{-1/2})$ uniformly for $\tau \in U_\gamma$ and $\delta_0 \in U_0$, with positive definite covariance matrix

$$\Sigma(\delta_0, \tau) := P_{(\delta_0,\tau)}((k(\cdot, \tau) - P_{(\delta_0,\tau)}(k(\cdot, \tau)))\,(k(\cdot, \tau) - P_{(\delta_0,\tau)}(k(\cdot, \tau)))')\,.$$

Hence by the Corollary in Pfanzagl (1973b), p. 243, we have for $N \in \mathbf{R}$ and uniformly for $\tau \in U_\gamma$, $|t| \leqq \log n$, $u \in \mathbf{R}$,

$$(8.10) \qquad P^n_{(\theta_0+n^{-1/2}t,\tau)}\{\tilde{f}'_0 + n^{-1/2}(R + Q + NU)(M(\tilde{k}' + t\,P(l^{(0)}k))) > u\}$$

$$= \int_{v_0 > u} \varphi_{\Sigma(\theta_0+n^{-1/2}t,\tau)}(v)\,(1 + n^{-1/2}\,A(v, \theta_0 + n^{-1/2}t, \tau))\,dv$$

$$-n^{-1/2} \int_{v_0 > u} (\varphi_{\Sigma(\theta_0+n^{-1/2}t,\tau)}(v)\,(R + Q + NU)\,(M(v + tP(l^{(0)}k))))^{(0)}\,dv + o(n^{-1/2})\,.$$

The first integral on the right-hand side of (8.10) is an expression which depends only on the marginal distribution with respect to k_0 of $P^n_{(\theta_0+n^{-1/2}t,\tau)} * \tilde{k}'$ (see 8.9)) and reduces, therefore, to an asymptotic expansion of the distribution of $P^n_{(\theta_0+n^{-1/2}t,\tau)} * \tilde{k}'_0$. Recalling that $k_0 = f_0$, and using that uniformly for $\tau \in U_0$, $|t| \leqq \log n$,

$$\sigma_{00}(\theta_0 + n^{-1/2}t, \tau) = \sigma_{00} + n^{-1/2}t\,P(l^{(0)}f_0^2) + O(n^{-1}(\log n)^2)\,,$$

we have uniformly for $\tau \in U_\gamma$, $|t| \leqq \log n$, $u \in \mathbf{R}$

$$(8.11) \qquad \int_{v_0 > u} \varphi_{\Sigma(\theta_0+n^{-1/2}t,\tau)}(v)\,(1 + n^{-1/2}\,A(v, \theta_0 + n^{-1/2}t, \tau))\,dv$$

$$= P^n_{(\theta_0+n^{-1/2}t,\tau)}(\tilde{f}'_0 > u\} + o(n^{-1/2})$$

$$= \Phi(-\sigma_{00}^{-1/2}u + n^{-1/2}\sigma_{00}^{-3/2}[t\,\tfrac{1}{2}P(l^{(0)}f_0^2)\,u + \tfrac{1}{6}P(f_0^3)\,(\sigma_{00}^{-1}u^2 - 1)]) + o(n^{-1/2})\,.$$

(See Gnedenko-Kolmogorov (1954), p. 220, Theorem; for a uniform version see Pfanzagl (1973a), p. 1013, Lemma 4.) With the aid of (8.11) and of the operator H_Σ defined in (4.7), the asymptotic expansion (8.10) may be recast as follows. For $N \in \mathbf{R}$ and uniformly for $\tau \in U_\gamma$, $|t| \leqq \log n$, $u \in \mathbf{R}$,

$$(8.12) \quad P^n_{(\theta_0+n^{-1/2}t,\tau)}\{\tilde{f}'_0 + n^{-1/2}(R + Q + NU)(M(\tilde{k}' + tP(l^{(0)}k))) > u\}$$
$$= \Phi(-\sigma_{00}^{-1/2}u + n^{-1/2}(\sigma_{00}^{-3/2}(t\tfrac{1}{2}P(l^{(0)}f_0^2)u + \tfrac{1}{6}P(f_0^3)(\sigma_{00}^{-1}u^2 - 1))$$
$$+ H_\Sigma[(R + Q + NU)\circ M](u + t\,P(l^{(0)}f_0), t\,P(l^{(0)}k)))) + o(n^{-1/2}).$$

For $N = 0$ and $t = 0$ relation (8.12) reduces to an asymptotic expansion of the distribution function of F_n: Recalling (8.7) we have uniformly for $\tau \in U_\gamma$

$$(8.13) \quad P^n_{(\theta_0,\tau)}\{F_n < u\} = P^n_{(\theta_0,\tau)}\{\tilde{f}_0 + n^{-1/2}(R + Q)(M\tilde{k}) < u\}\, o(n^{-1/2})$$
$$= \Phi(\sigma_{00}^{-1/2}u - n^{-1/2}(\tfrac{1}{6}\sigma_{00}^{-3/2}P(f_0^3)(\sigma_{00}^{-1}u^2 - 1) + H_\Sigma[(R + Q)\circ M](u, 0)))$$
$$+ o(n^{-1/2}).$$

We are now ready to determine the correction term to the bound $-N\sigma_{00}(\theta_0, T^{(n)})^{1/2}$ of the critical region which leads to an asymptotically similar test of level $\alpha + o(n^{-1/2})$. From (8.7), (8.8), (8.12) we obtain for $N \in \mathbf{R}$ and uniformly for $\tau \in U_\gamma$, $|t| \leqq \log n$,

$$(8.14) \quad P^n_{(\theta_0+n^{-1/2}t,\tau)}\{F_n > -N\sigma_{00}(\theta_0, T^{(n)})^{1/2} + n^{-1/2}d\}$$
$$= P^n_{(\theta_0+n^{-1/2}t,\tau)}\{\tilde{f}'_0 + n^{-1/2}(R + Q + NU)(M(\tilde{k}' + t\,P(l^{(0)}k)))$$
$$> -N\sigma_{00}^{1/2} - t\,P(l^{(0)}f_0) + n^{-1/2}[d - t^2\tfrac{1}{2}P((l^{(0)2} + l^{(00)})f_0)]\} + o(n^{-1/2})$$
$$= \Phi(N + t\sigma_{00}^{-1/2}P(l^{(0)}f_0) + n^{-1/2}(-\sigma_{00}^{-1/2}d + (N^2 - 1)\tfrac{1}{6}\sigma_{00}^{-3/2}P(f_0^3)$$
$$+ t^2\tfrac{1}{2}\sigma_{00}^{-1/2}P((l^{(0)2} + l^{(00)})f_0) - t^2\tfrac{1}{2}\sigma_{00}^{-3/2}P(l^{(0)}f_0)P(l^{(0)}f_0^2)$$
$$+ t^2\tfrac{1}{6}\sigma_{00}^{-5/2}P(f_0^3)(P(l^{(0)}f_0))^2 - tN\tfrac{1}{2}\sigma_{00}^{-1}P(l^{(0)}f_0^2) + tN\tfrac{1}{3}\sigma_{00}^{-2}P(f_0^3)P(l^{(0)}f_0)$$
$$+ H_\Sigma[(R + Q + NU)\circ M](-N\sigma_{00}^{1/2}, t\,P(l^{(0)}k))) + o(n^{-1/2}).$$

Let

$$(8.15) \quad d_0 := (N^2 - 1)\tfrac{1}{6}\sigma_{00}^{-1}P(f_0^3) + \sigma_{00}^{1/2}H_\Sigma[(R + Q + NU)\circ M](-N\sigma_{00}^{1/2}, 0).$$

If we replace $d(\theta)$ by $d_0(\theta_0, T^{(n)})$ in the bound of the critical region (8.14) we obtain an asymptotically similar critical region of level $\alpha + o(n^{-1/2})$. Its power function admits the following expansion for $N \in \mathbf{R}$ and uniformly for $\tau \in U_\gamma$, $|t| \leqq \log n$,

$$(8.16)\qquad P^n_{(\theta_0+n^{-1/2}t,\tau)}\{F_n > -N\sigma_{00}(\theta_0, T^{(n)})^{1/2} + n^{-1/2}\, d(\theta_0, T^{(n)})\}$$

$$= \Phi(N + t\sigma_{00}^{-1/2} P(l^{(0)} f_0) + n^{-1/2}(t^2\, \tfrac{1}{2}\sigma_{00}^{-1/2}(P(l^{(0)2} + l^{(00)}) f_0)$$

$$- t(N + t\sigma_{00}^{-1/2} P(l^{(0)} f_0))\, \tfrac{1}{2}\sigma_{00}^{-1} P(l^{(0)} f_0^2)$$

$$+ t(2N + t\sigma_{00}^{-1/2} P(l^{(0)} f_0))\, \tfrac{1}{6}\sigma_{00}^{-2} P(f_0^3)\, P(l^{(0)} f_0)$$

$$+ H_\Sigma[(R + Q + NU) \circ M]\,(-N\sigma_{00}^{1/2}, t\, P(l^{(0)} k))$$

$$- H_\Sigma[(R + Q + NU) \circ M]\,(-N\sigma_{00}^{1/2}, 0))).$$

By an application of the relation

$$\varphi_\Sigma(u_0, v_{(0)}) = \sigma_{00}^{-1/2}\, \varphi(\sigma_{00}^{-1/2} u_0)\, \varphi_\Sigma(v_{(0)} \mid v_0 = u_0)$$

and of the formulas for the conditional moments (see e.g. Pfanzagl (1973b), p. 248), the asymptotic expansions (8.13) for the distribution function of the test statistic and (8.15) respectively (8.16) for the bound and the power function of the critical region are recast into the more explicit form given in (4.10). □

Proof of Theorem 6.5

(a) Assume that the sequence of critical regions (4.9) is asymptotically efficient. Then

$$(8.17)\qquad \sigma_{00}^{-1/2} P(l^{(0)} f_0) = \Lambda_{00}^{-1/2}.$$

By the Schwarz inequality,

$$(8.18)\qquad P(\lambda_0 f_0) \leqq P((\lambda_0)^2)^{1/2}\, P(f_0^2)^{1/2}.$$

Since $P(l^{(\alpha)} f_0) = -P(f_0^{(\alpha)}) = 0$ for $\alpha = 1, \ldots, p$ by Lemma 3.10(a), we have

$$P(\lambda_0 f_0) = \Lambda_{00}\, P(l^{(0)} f_0).$$

Furthermore,

$$P(\lambda_0^2) = \Lambda_{00} \quad \text{and} \quad P(f_0^2) = \sigma_{00}.$$

Hence (8.17) implies that equality holds in (8.18). Therefore, f_0 is P-a.e. proportional to λ_0 if the sequence of critical regions (4.9) is asymptotically efficient.

If, on the other hand, $f_0 = c\lambda_0$ P-a.e. with $c > 0$ then (8.17) is fulfilled, and hence the sequence of critical regions (4.9) is asymptotically efficient.

(b) Assume that $f_0 = c\lambda_0$ P-a.e. with $c > 0$. To show that the power function of the sequence of critical regions depends on Q and on the studentizing estimator sequence only through terms of order $o(n^{-1/2})$, we consider the asymptotic expansion of the power function as given in (4.10).

Let $f = (f_0, \ldots, f_m)$ denote the vector appearing in the asymptotic expansion (4.2) of the sequence of test statistics F_n, $n \in \mathbf{N}$. By Lemma 3.10(a),

$$f_0 = f_0^*$$

and, by convention,

$$f_\alpha = f_0^{(\alpha)} \quad \text{for} \quad \alpha = 1, \ldots, p\,.$$

We may therefore assume w.l.g. that the base of P-a.e. linearly independent functions for the subspace spanned by the components of

$$h = (f_0, \lambda_1, \ldots, \lambda_p, g_1, \ldots, g_p, f_0^{(1)*}, \ldots, f_0^{(p)*}, f_{p+1}^*, \ldots, f_m^*)$$

contains a base for the subspace spanned by $f_0^*, \ldots, f_m^*$ which consists of the components of a subvector, say $k^* = (k_0^*, \ldots, k_r^*)$, of $(f_0^*, \ldots, f_m^*)$. W.l.g. $k_0^* = f_0^* \,(= f_0)$. Let $\Sigma^* = P(k^* k^{*\prime})$ denote the pertaining covariance matrix. Note that

$$\sigma_{00}^* = P(k_0^{*2}) = P(f_0^2) = \sigma_{00}\,.$$

Since $f_0 = c\lambda_0$ P-a.e., we obtain the relations

$$P(l^{(0)} f_0) = c\,, \tag{8.19}$$

$$P(f_0 k_j^*) = \frac{1}{c}\,\sigma_{00}\, P(l^{(0)} k_j^*) \quad \text{for} \quad j = 0, \ldots, r\,. \tag{8.20}$$

Introducing $s := ct\sigma_{00}^{-1}$ and $u_0 := -N\sigma_{00}^{1/2}$ we may write the term of the asymptotic expansion (4.10) of the power function which depends on Q as follows:

$$H_{\Sigma^*}[Q]\,(u_0, t\, P(l^{(0)} k^*)) - H_{\Sigma^*}[Q]\,(u_0, 0) \tag{8.21}$$

$$= \varphi((\sigma_{00}^*)^{-1/2}(u_0 - s\sigma_{00}^*))^{-1} \int \varphi_{\Sigma^*}(u_0 - s\sigma_{00}^*, \ldots, u_r - s\sigma_{0r}^*)\, Q(u)\, du_{(0)}$$

$$- \varphi((\sigma_{00}^*)^{-1/2} u_0)^{-1} \int \varphi_{\Sigma^*}(u)\, Q(u)\, du_{(0)}\,.$$

Now it is easy to see that this difference vanishes, since

$$\varphi((\sigma_{00}^*)^{-1/2}(u_0 - s\sigma_{00}^*))^{-1}\, \varphi_{\Sigma^*}(u_0 - s\sigma_{00}^*, \ldots, u_r - s\sigma_{0r}^*)$$

is, in fact, independent of s, as can be checked in a straightforward way.

We remark that we could have separated the asymptotic expansion of $\sigma_{00}(\theta_0, T^{(n)})^{1/2}$ in the same way as that of F_n by using

$$g_\alpha = \lambda_\alpha + g_\alpha^* \quad \text{for} \quad \alpha = 1, \ldots, p\,.$$

(Compare (8.5) and (8.6).) Then g_α^* would have occurred in the asymptotic expansion of the power function through an expression analogous to (8.21), and by the same reasoning as above we could have concluded that g_α does not appear in the asymptotic expansion of order $o(n^{-1/2})$ of the power function if $f_0 = c\lambda_0$ P-a.e.

It can, however, be seen directly from (4.10) that g_α enters the asymptotic expansion of the power function only through the term

$$\sigma_{00}\, P(l^{(0)}g_\alpha) - P(l^{(0)}f_0)\, P(f_0 g_\alpha)$$

which reduces to $-c^2 \Lambda_{0\alpha}$ if $f_0 = c\lambda_0$ P-a.e.

It is now easily checked that for $f_0 = c\lambda_0$ P-a.e. the asymptotic expansion of the power function of the critical region defined in (4.9) is up to $o(n^{-1/2})$ equal to the envelope power function as given in Definition 5.1. □

Proof of Theorem 7.4

(a) Assumption 7.1 implies that $P_\theta^n * n^{1/2}(T^{(n)} - \theta)$, $n \in \mathbf{N}$, is asymptotically normally distributed with continuous covariance matrix $\Sigma := (P(g_\alpha g_\beta))_{\alpha,\beta=1,\dots,p}$. Since $\Sigma - \Lambda$ is positive semidefinite by Bahadur (1964), p. 1550, and since Λ is positive definite by assumption, Σ is positive definite. By the Corollary in Pfanzagl (1973b), p. 243, together with Lemma 1 in Sazonov (1968), p. 183, the sequence $P_\theta^n * n^{1/2}(T^{(n)} - \theta)$, $n \in \mathbf{N}$, admits locally uniformly in $\theta \in \Theta$ an E-expansion of order $o(n^{-1/2})$.

Assume that $\Sigma = \Lambda$. Then the E-expansion is of the form

(8.22) $$v \to \varphi_{\Lambda(\theta)}(v)\,(1 + n^{-1/2} K(v, \theta))\,.$$

(b) Let $a \in \mathbf{R}^p - \{0\}$. For notational convenience we assume w.l.g. that $a_1 \neq 0$. The matrix

$$C := \begin{bmatrix} a_1 & a_2 & \dots & a_p \\ & 1 & & \mathbf{0} \\ & & \ddots & \\ & \mathbf{0} & & 1 \end{bmatrix}$$

defines a one-one transformation of $\mathbf{R}^p$. Consider the family $P_{C^{-1}\eta}$, $\eta \in C\Theta$. Define $S^{(n)} := CT^{(n)}$, $n \in \mathbf{N}$, and $M := C\Lambda C'$. Then by (8.22) the sequence $P_{C^{-1}\eta}^n * * n^{1/2}(S^{(n)} - \eta)$, $n \in \mathbf{N}$, admits an E-expansion

(8.23) $$s \to \varphi_{M(C^{-1}\eta)}(s)\,(1 + n^{-1/2} K(C^{-1}s, C^{-1}\eta))$$

fulfilling the assumptions of Lemma 8.28.

For every $\alpha \in (0, 1)$, the asymptotically similar critical region (8.29) of level $\alpha + o(n^{-1/2})$ for the hypothesis η_1, obtained from the test statistic $n^{1/2}(S_1^{(n)} - \eta_1)$

by asymptotic studentization with $(S_2^{(n)}, \ldots, S_p^{(n)})$, is asymptotically efficient. By Theorem 6.5, applied to the transformed problem this implies that the power function (8.32) is up to $o(n^{-1/2})$ the same for every estimator sequence $T^{(n)}$, $n \in \mathbf{N}$, with asymptotic covariance matrix Λ, fulfilling Assumption 7.1. Since this holds true for every $\alpha \in (0, 1)$, $t \in \mathbf{R}$, we obtain that the function $u \to G(u) - G(0)$ is the same for every estimator sequence of such a kind.

Since by Lemma 8 in Pfanzagl (1973b), p. 273, the sequence of asymptotic m.l. estimators of order $o(n^{-1/2})$ is of such a kind, we have for all $u \in \mathbf{R}$ (see (8.31))

$$\int_{s_1 \leqq M_{11}(C^{-1}\eta)^{1/2}u} \varphi_{M(C^{-1}\eta)}(s)\,[K(C^{-1}s, C^{-1}\eta) - K_0(C^{-1}s, C^{-1}\eta)]\, ds = -c(C^{-1}\eta)\,\varphi(u),$$

where

$$(8.24) \quad K_0(v, \cdot) := \Lambda_{\alpha\beta}(\tfrac{1}{2}L_{(\alpha)(\beta)(\gamma)} + L_{(\alpha)(\beta\gamma)})\, v_\gamma - (\tfrac{1}{3}L_{(\alpha)(\beta)(\gamma)} + \tfrac{1}{2}L_{(\alpha)(\beta\gamma)})\, v_\alpha v_\beta v_\gamma$$

is the polynomial occurring in the E-expansion of the asymptotic m.l. estimator (see Michel (1975), p. 70). Applying the inverse transformation C^{-1}, and replacing $C^{-1}\eta$ by θ (and suppressing θ in the following) we obtain

$$\int_{a'v = M_{11}{}^{1/2}u} \varphi_\Lambda(v)\,[K(v) - K_0(v)]\, dv = -c\,\varphi(u).$$

Since the distribution of $a'v$ under φ_Λ is normal with mean zero and variance M_{11}, we have

$$\int_{a'v = M_{11}{}^{1/2}u} \varphi_\Lambda(v)\, a'v\, dv = \int_{r_1 \leqq M_{11}{}^{1/2}u} \varphi(M_{11}^{-1/2}r)\, M_{11}^{-1/2}r\, dr$$

$$= M_{11}^{1/2} \int_{-\infty}^{u} \varphi(w)\, w\, dw = -M_{11}^{1/2}\,\varphi(u).$$

Hence, with $c_0 = M_{11}^{-1/2}c$, for all $r \in \mathbf{R}$

$$(8.25) \qquad \int_{a'v \leqq r} \varphi_\Lambda(v)\,[K(v) - K_0(v)]\, dv = c_0 \int_{a'v \leqq r} \varphi_\Lambda(v)\, a'v\, dv.$$

(c) In drawing conclusions from the last relation we have to bear in mind that c_0 depends on a. Since (8.25) holds for every $r \in \mathbf{R}$, this implies for every function $g: \mathbf{R} \to \mathbf{R}$ for which the integral exists,

$$(8.26) \qquad \int \varphi_\Lambda(v)\, g(a'v)\,[K(v) - K_0(v)]\, dv = c_0(a) \int \varphi_\Lambda(v)\, g(a'v)\, a'v\, dv.$$

Applied for $g(u) = u$ relation (8.26) yields

(8.27) $$c_0(a) = (a'\Lambda a)^{-1}\, q'a$$

(with $q := \int \varphi_\Lambda(v)\, v[K(v) - K_0(v)]\, dv$).

Since

$$\int \varphi_\Lambda(v) \exp[a'v]\, v\, dv = \exp[\tfrac{1}{2}a'\Lambda a] \int \varphi_\Lambda(v - \Lambda a)\, v\, dv = \exp[\tfrac{1}{2}a'\Lambda a]\, \Lambda a\,,$$

we obtain from (8.26), applied for $g(u) = \exp[u]$, and (8.27) with

$L := ((L_{(\alpha)(\beta)})_{\alpha,\beta=1,\dots,p})$,

$$\int \varphi_\Lambda(v) \exp[a'v]\,(K(v) - K_0(v) - q'Lv)\, dv$$

$$= \int \varphi_\Lambda(v) \exp[a'v]\,[K(v) - K_0(v)]\, dv - q'L \int \varphi_\Lambda(v) \exp[a'v]\, v\, dv$$

$$= (a'\Lambda a)^{-1}\, q'a \int \varphi_\Lambda(v) \exp[a'v]\, a'v\, dv - q'L \int \varphi_\Lambda(v) \exp[a'v]\, v\, dv$$

$$= [(a'\Lambda a)^{-1}\, q'a(a'\Lambda a) - q'L\Lambda a] \exp[\tfrac{1}{2}a'\Lambda a] = 0\,.$$

Since this relation holds true for all $a \in \mathbf{R}^p - \{0\}$, the function $v \to K(v) - K_0(v) - q'Lv$ vanishes on $\mathbf{R}^p$ by Theorem 1 in Lehmann (1959), p. 132. Hence the assertion follows from the definition of K_0 given in (8.24). □

Proof of Corollary 7.8

If $P_\theta^n * n^{1/2}(T^{(n)} - \theta)$, $n \in \mathbf{N}$, admits an E-expansion

$$v \to \varphi_\Lambda(v)\,(1 + n^{-1/2}(a_\alpha v_\alpha + a_{\alpha\beta\gamma} v_\alpha v_\beta v_\gamma))$$

of order $o(n^{-1/2})$, then the distribution function of the marginal distribution $P_\theta^n * n^{1/2}(T_\varkappa^{(n)} - \theta_\varkappa)$, $n \in \mathbf{N}$, admits the following expansion (use the formulas for the conditional moments of a normal distribution):

$$P_\theta^n\{n^{1/2}(T_\varkappa^{(n)} - \theta_\varkappa) < u\}$$

$$= \Phi(\Lambda_{\varkappa\varkappa}^{-1/2} u) + n^{-1/2}\, \varphi(\Lambda_{\varkappa\varkappa}^{-1/2} u)\,[-\Lambda_{\varkappa\varkappa}^{-1/2} \Lambda_{\varkappa\alpha} a_\alpha$$

$$+ (-3\Lambda_{\varkappa\varkappa}^{-1/2} \Lambda_{\varkappa\alpha}\Lambda_{\beta\gamma} + \Lambda_{\varkappa\varkappa}^{-3/2} \Lambda_{\varkappa\alpha}\Lambda_{\varkappa\beta}\Lambda_{\varkappa\gamma} - \Lambda_{\varkappa\varkappa}^{-5/2} \Lambda_{\varkappa\alpha}\Lambda_{\varkappa\beta}\Lambda_{\varkappa\gamma} u^2)\, a_{\alpha\beta\gamma}] + o(n^{-1/2})\,.$$

Median unbiasedness of order $o(n^{-1/2})$ implies

$$\Lambda_{\varkappa\varkappa}^{-1/2} \Lambda_{\varkappa\alpha} a_\alpha = (\Lambda_{\varkappa\varkappa}^{-3/2} \Lambda_{\varkappa\alpha}\Lambda_{\varkappa\beta}\Lambda_{\varkappa\gamma} - 3\Lambda_{\varkappa\varkappa}^{-1/2} \Lambda_{\varkappa\alpha}\Lambda_{\beta\gamma})\, a_{\alpha\beta\gamma}\,.$$

If this holds true for $\varkappa = 1, \ldots, p$, we obtain

$$a_\delta = L_{(\delta)(\varkappa)} \Lambda_{\varkappa\varkappa}^{-1} \Lambda_{\varkappa\alpha} \Lambda_{\varkappa\beta} \Lambda_{\varkappa\gamma} a_{\alpha\beta\gamma} - 3\Lambda_{\beta\gamma} a_{\delta\beta\gamma}.$$

Since $a_{\alpha\beta\gamma} = -(\frac{1}{3}L_{(\alpha)(\beta)(\gamma)} + \frac{1}{2}L_{(\alpha)(\beta\gamma)})$ by Theorem 7.4, this implies the assertion. □

8.28 Lemma. *Let $T^{(n)}$, $n \in \mathbb{N}$, be an estimator sequence such that $P_\theta^n * n^{1/2}(T^{(n)} - \theta)$, $n \in \mathbb{N}$, admits an E-expansion $\varphi_\Sigma(1 + n^{-1/2}K)$ locally uniformly in $\theta \in \Theta$. Assume that K fulfills Condition B(b), that $K(\cdot, \theta)$ is bounded by polynomials, locally uniformly in $\theta \in \Theta$, and that Σ and $\sigma_{11}^{(\alpha)}$, $\alpha = 1, \ldots, p$, fulfill locally uniform Lipschitz conditions on Θ.*

For $n \in \mathbb{N}$ *and* $\alpha \in (0, 1)$ *let*

$$C_{n,\alpha} := \{T_1^{(n)} > \theta_1 - n^{-1/2} N_\alpha \sigma_{11}(\theta_1, T_2^{(n)}, \ldots, T_p^{(n)})^{1/2} - n^{-1} d(N_\alpha, \theta_1, T_2^{(n)}, \ldots, T_p^{(n)})\} \tag{8.29}$$

with

$$d(N, \cdot) := \sigma_{11}^{1/2} G(N, \cdot) + N^2 \tfrac{1}{2}\sigma_{11}^{-1} \sum_{j=2}^{p} \sigma_{ij}\sigma_{11}^{(j)}, \tag{8.30}$$

where

$$G(u, \cdot) := \varphi(u)^{-1} \int_{v_1 > \sigma_{11}^{1/2} u} \varphi_\Sigma(v) K(v, \cdot)\, dv. \tag{8.31}$$

Then uniformly for $|t| \leq \log n$ and locally uniformly in $\theta \in \Theta$

$$P^n_{(\theta_1 + n^{-1/2}t, \theta_2, \ldots, \theta_p)}(C_{n,\alpha}) = \Phi(N_\alpha + t\sigma_{11}(\theta)^{-1/2} + n^{-1/2}[G(N_\alpha, \theta) - G(N_\alpha + t\sigma_{11}(\theta)^{-1/2}, \theta) - tN_\alpha \tfrac{1}{2}\sigma_{11}(\theta)^{-2}\sigma_{1\alpha}(\theta)\sigma_{11}^{(\alpha)}(\theta) - t^2 \tfrac{1}{2}\sigma_{11}(\theta)^{-3/2}\sigma_{11}^{(1)}(\theta)]) + o(n^{-1/2}). \tag{8.32}$$

Proof of Lemma 8.28

We have locally uniformly in $\theta \in \Theta$

$$P_\theta^n\{n^{1/2}(T_1^{(n)} - \theta_1) > -N\sigma_{11}^{1/2}\} = \alpha + o(n^0).$$

Replacing $\sigma_{11}(\theta)^{1/2}$ by $\sigma_{11}(\theta_1, T_2^{(n)}, \ldots, T_p^{(n)})^{1/2}$ we obtain an asymptotically similar critical region of level $\alpha + o(n^0)$. This is the first step of the studentization procedure.

For the second step of the studentization procedure, observe that

$$\sigma_{11}(\theta_1, T_2^{(n)}, \ldots, T_p^{(n)})^{1/2} = \sigma_{11}^{1/2} + \tfrac{1}{2}\sigma_{11}^{-1/2} \sum_{j=2}^{p} (T_j^{(n)} - \theta_j)\sigma_{11}^{(j)} + n^{-1} l_n(\tfrac{1}{2}). \tag{8.33}$$

In much the same way as Lemma 1 in Pfanzagl (1973b), p. 236, the following can be proven: If $Q^{(n)} \mid \mathscr{B}^p$, $n \in \mathbb{N}$, admits an E-expansion $\varphi_\Sigma(1 + n^{-1/2}K)$, then

$$Q^{(n)} * (v \to (v_1 + n^{-1/2} \sum_{j=2}^{p} b_j v_j, v_2, \ldots, v_p)) \mid \mathscr{B}^p$$

admits an E-expansion

$$v \to \varphi_\Sigma(v)\,(1 + n^{-1/2} K(v)) - n^{-1/2}\, \varphi_\Sigma^{(1)}(v) \sum_{j=2}^{p} b_j v_j \,. \tag{8.34}$$

We remark that locally uniformly in $\theta \in \Theta$ and uniformly for $|t| \leqq \log n$

$$\sigma_{11}(\theta_1 + n^{-1/2}t, \theta_2, \ldots, \theta_p)^{-1/2} = \sigma_{11}^{-1/2} - n^{-1/2} t\, \tfrac{1}{2}\sigma_{11}^{-3/2}\sigma_{11}^{(1)} + O(n^{-1}(\log n)^2)\,. \tag{8.35}$$

Applying (8.34) and (8.35) we obtain locally uniformly in $\theta \in \Theta$ and uniformly for $|t| \leqq \log n$

$$\begin{aligned} & P^n_{(\theta_1 + n^{-1/2}t, \theta_2, \ldots, \theta_p)}\{n^{1/2}(T_1^{(n)} - \theta_1 - n^{-1/2}t) \\ & + N\, \tfrac{1}{2}\sigma_{11}^{-1/2} \sum_{j=2}^{p} (T_j^{(n)} - \theta_j)\, \sigma_{11}^{(j)} < u\} = \int_{v_1 < u} (\varphi_\Sigma(1 + n^{-1/2} K(v, \cdot)) \\ & - n^{-1/2}\, N\, \tfrac{1}{2}\varphi_\Sigma^{(1)}(v)\, \sigma_{11}^{-1/2} \sum_{j=2}^{p} \sigma_{11}^{(j)} v_j)\big|_{\theta = (\theta_1 + n^{-1/2}t, \theta_2, \ldots, \theta_p)}\, dv + o(n^{-1/2}) \\ & = \Phi(\sigma_{11}^{-1/2} u + n^{-1/2}[G(-\sigma_{11}^{-1/2} u) - N\, \tfrac{1}{2}\sigma_{11}^{-2} \sum_{j=2}^{p} \sigma_{1j}\sigma_{11}^{(j)} u \\ & - t\, \tfrac{1}{2}\sigma_{11}^{-3/2}\sigma_{11}^{(1)} u]) + o(n^{-1/2})\,. \end{aligned} \tag{8.36}$$

Applying (8.36) for

$$u = -N\sigma_{11}^{1/2} - t - n^{-1/2}[\sigma_{11}^{1/2}\, G(N) + N^2\, \tfrac{1}{2}\sigma_{11}^{-1} \sum_{j=2}^{p} \sigma_{1j}\sigma_{11}^{(j)}]$$

and taking into account (8.35), we obtain the assertion. □

Acknowledgment

The author wishes to thank Mr. W. Wefelmeyer for his cooperation during the preparation of the manuscript. He carefully read several versions of the manuscript, elaborated the proofs, and collected the regularity conditions. He also suggested to make the proof of Theorem 6.5 more transparent by stating Theorem 4.8 separately.

Furthermore, the author wishes to thank Mr. K. Michel who did some of the computations (among others, he computed the power function stated in Theorem 4.8). In this connection the author thanks the Deutsche Forschungsgemeinschaft which enabled the cooperation with Mr. K. Michel by a grant.

Finally, the author wishes to thank Mr. K. – L. Bender for helpful discussions.

References

ANDERSON, T. W. (1955). The integral of a symmetric unimodal function over a symmetric convex set and some probability inequalities. *Proc. Amer. Math. Soc.*, **6**, 170—176.

BAHADUR, R. R. (1964). On Fisher's bound for asymptotic variances. *Ann. Math. Statist.*, **35**, 1545—1552.

CHIBISOV, D. M. (1973a). Asymptotic expansions for Neyman's C(α) tests. Proceedings of the Second Japan—USSR Symposium on Probability Theory. (Ed. by G. Maryama, Yu. V. Prokhorov). *Lecture Notes in Mathematics 330,* Springer-Verlag, Berlin, 16—45.

CHIBISOV, D. M. (1973b). Asymptotic expansions for some asymptotically optimal tests. *Proceedings of the Prague Symposium on Asymptotic Statistics.* (Ed. by J. Hájek). Vol. 2, 37—68.

CHIBISOV, D. M. (1973c). An asymptotic expansion for a class of estimators containing maximum likelihood estimators. *Theor. Probability Appl.*, **18**, 295—303.

GNEDENKO, S. V. - KOLMOGOROV, A. N. (1954). "Limit Distributions for Sums of Independent Random Variables". Addison-Wesley, Cambridge, Mass.

LEHMANN, E. (1959). "Testing Statistical Hypotheses". Wiley, New York.

MICHEL, R. (1975). An asymptotic expansion for the distribution of asymptotic maximum likelihood estimators of vector parameters. *J. Multivariate Anal.*, **5**, 67—82.

NAGAEV, S. V. (1965). Some limit theorems for large deviations. *Theor. Probability Appl.*, **10**, 214—235.

NEYMAN, J. (1954). Sur une famille de tests asymptotiques des hypothèses statistiques composées. *Trabajos Estadist.*, **5**, 161—168.

NEYMAN, J. (1959). Optimal asymptotic tests of composite statistical hypotheses. Probability and Statistics. The Harald Cramér Volume. (Ed. by U. Grenander). Almqvist and Wiksell, Stockholm, 213—234.

PFANZAGL, J. (1972). Further results on asymptotic normality I. *Metrika*, **18**, 174—198.

PFANZAGL, J. (1973a). Asymptotic expansions related to minimum contrast estimators. *Ann. Statist.*, **1**, 993—1026. Correction note **2**, 1357—1358.

PFANZAGL, (1973b). Asymptotically optimum estimation and test procedures. *Proceedings of the Prague Symposium on Asymptotic Statistics.* (Ed. by J. Hájek). Vol. 1, 201—272.

PFANZAGL, J. (1975). On asymptotically complete classes. Statistical Inference and Related Topics. *Proceedings of the Summer Research Institute on Statistical Inference for Stochastic Processes.* (Ed. by M. L. Puri). Vol. 2, 1—43.

RAO, R. R. (1962). Relations between weak and uniform convergence of measures with applications. *Ann. Math. Statist.*, **33**, 659—680.

SAZONOV, V. V. (1968). On the multidimensional central limit theorem. *Sankhyā* A, **30**, 181—204.

SCHMETTERER, L. (1974). "Introduction to Mathematical Statistics". Springer, Berlin.

WELCH, B. L. (1965). On comparisons between confidence point procedures in the case of a single parameter. *J. Roy. Statist. Soc.* B, **27**, 1—8.

MATHEMATISCHES INSTITUT DER UNIVERSITÄT ZU KÖLN, KÖLN,
WEST GERMANY

Received March 1976
Revised June 1976

A SIMPLE TEST FOR GOODNESS-OF-FIT BASED ON SPACINGS WITH SOME EFFICIENCY COMPARISONS

by

M. L. PURI (1) *, J. S. RAO (2), AND YOUNGJOO YOON (3)

1. Introduction

Let $X_1, \ldots, X_{n-1}$ be independently and identically distributed random variables with a common distribution (d.f.). The goodness-of-fit problem is to test if this d.f. is equal to a specified one. A simple probability integral transformation on the random variables (r.v.'s) would permit us to equate the specified d.f. to the uniform distribution on $[0, 1]$. Thus from now on, we shall assume that this reduction has been effected and under the hypothesis, the observations have a uniform distribution on $[0, 1]$. The original problem thus is equivalent to one of testing for uniformity viz. whether a given random sample of observations come from a uniform distribution on $[0, 1]$.

Let $X'_1 \leqq X'_2 \leqq \ldots \leqq X'_{n-1}$ be the order statistics. The sample spacings $(T_1, \ldots, T_n)$ are defined by

$$T_i = X'_i - X'_{i-1}, \quad i = 1, \ldots, n$$

where we put $X'_n = 0$, $X'_0 = 1$. Tests for goodness-of-fit (or equivalently uniformity) based on spacings have been proposed by several authors. See for instance Pyke (1965) or Rao and Sethuraman (1975) and the references contained therein. It can be seen (see e.g. Pyke (1965) Section 2.1) that the distribution of $(T_1, \ldots, T_n)$ under the hypothesis of uniformity is Dirichlet $D(1, 1 \ldots 1; 1)$ distribution with any subset $(T_{i_1}, \ldots, T_{i_k})$ of them having $D(1, \ldots, 1; n-k)$ distribution. See Wilks (1962) pp. 177–182 for an elementary discussion on Dirichlet distributions.

In analysing circularly distributed data, testing for uniformity i.e. deciding whether a given set of observations on the circumference of a unit circle indicate a preferred direction, is a very basic problem. This is a necessary preliminary step before estimating or making inferences on the mean direction. Also the goodness-

* Work supported by the Air Force Office of Scientific Research, AFSC, USAF, under Grant No. AFOSR 76-2927. Reproduction in whole or in part permitted for any purpose of the U.S. Government.

of-fit problem on the circle is equivalent to this just as on the line. In the circular case, the spacings may be defined as the arc-lengths between successive observations on the circumference, ignoring the zero-direction. Apart from the minor difference that n observations on the circle lead to n circular spacings while on the line $(n - 1)$ observations make n spacings, the distribution of the spacings in either case is the same (see for instance Rao (1969) pp. 63–67 or Mardia (1972) p. 172). For purposes of inference on the circle, one requires a statistic that is invariant under changes of the origin and a general invariant statistic is of the form $h(T_1, T_2, \ldots, T_n)$ where $h(\cdot)$ is a function that remains invariant under cyclical permutations of the arguments. For instance functions symmetric in all the arguments may be considered though they are not asymptotically efficient. See Sethuraman and Rao (1970). Thus the spacings $\{T_i\}$ play a crucial role in testing goodness-of-fit on the circle whereas for the linear case, one has tests that are not necessarily based on spacings. Therefore all our further discussion on spacings can be related also to the circular case and is indeed more important in that context.

In Section 2, we propose a simple class of test $R_n = R_n(n\delta_n)$ based on spacings and obtain the exact distribution under the hypothesis of uniformity. Section 3 deals with the asymptotic distribution of R_n while sections 4 and 5 respectively discuss the Asymptotic Relative Efficiency (ARE) and Bahadur Efficiency (BE) of R_n relative to U_n, another spacings test discussed by Rao (1969). Since the limiting efficiencies of a number of test-statistics including U_n have already been investigated by Sethurmaman and Rao (1970) and Rao (1972), the results of sections 4 and 5 provide a basis for comparing R_n with any of those tests. Finally in Section 6 we discuss the statistic R_n^*, which has the maximum limiting efficiency in the class of tests $R_n(n\delta_n)$. We also provide a table that can be used to obtain critical values of R_n^* and illustrate, by means of a numerical example, how simple it is to use this R_n^* – test.

2. The statistic R_n and its exact null distribution

Choose and fix a $\delta_n > 0$. We shall call a sample spacing 'small' if it is less than δ_n in length. The test criterion is to reject H_0, the hypothesis of uniformity when we observe too many 'small' spacings, since this clearly indicates clustering of the observations. At this stage we will leave open the choice of δ_n though a suitable value might be to take for instance $\delta_n = 1/n$, the expected length of any spacing under uniformity. Since T_i's are of order $(1/n)$ under H_0, we consider the so-called "normalized" spacings $\{nT_i\}$ and define

$$\begin{aligned} R_n = R_n(n\delta_n) &= \{\text{number of } (nT_i)\text{'s} \leqq n\delta_n\}\,, \\ &= \text{number of spacings } T_i \text{ smaller than } \delta_n\,, \end{aligned}$$

and reject H_0 if R_n is too large. The exact distribution of R_n is given by the following.

Theorem 2.1. *Under the hypothesis of uniformity, the probability function of* R_n *is given by*

$$(2.1) \qquad P(R_n = k) = \binom{n}{k} \sum_{j=0}^{k} (-1)^j \binom{k}{j} \langle 1 - (n-k+j)\,\delta_n \rangle^{n-1}$$

$$\textit{for} \quad k = 0, 1, \ldots, n-1,$$

$$= 0 \quad \textit{otherwise}$$

with the notation $\langle x \rangle = x$ *if* $x > 0$ *and* $= 0$ *if* $x \leqq 0$.

Proof:

Let E_i denote the event that i^{th} spacing T_i exceeds δ_n, $i = 1, \ldots, n$ and let P_m denote the probability that a specified set of m arcs exceed δ_n. Clearly we have to have $m \leqq [1/\delta_n]$, the largest integer contained in $1/\delta_n$. Since the spacings are exchangeable,

$$P_m = P(E_{i_1} \cap \ldots \cap E_{i_m})$$

$$= P(E_1 \cap \ldots \cap E_m)$$

$$= P(T_1 > \delta_n, \ldots, T_m > \delta_n)$$

$$= \int_{\delta_n}^{1} \cdots \int_{\delta_n}^{1 - \sum_{i=1}^{m=2} t_i} \int_{\delta_n}^{1 - \sum_{i=1}^{m=1} t_i} g(t_1, \ldots, t_m)\, \mathrm{d}t_m, \ldots, \mathrm{d}t_1$$

$$= \begin{cases} (1 - m\delta_n)^{n-1} & \text{for} \quad 0 < m \leqq \left[\dfrac{1}{\delta_n}\right], \\ 0 & \text{otherwise} \end{cases}$$

$$= \langle 1 - m\delta_n \rangle^{n-1}$$

with the notation $\langle\ \rangle$ used in (2.1). Now if S_m denotes the probability that any subset m of these n events take place, then because of exchangeability,

$$(2.2) \qquad S_m = \binom{n}{m} Pm$$

$$= \binom{n}{m} \langle 1 - m\delta_n \rangle^{n-1}.$$

Further if Π_m denotes the probability that exactly m of these n events take place, then we have (see e.g. Feller I, p. 106)

$$\Pi_m = \sum_{j=m}^{n} (-1)^{j-m} \binom{j}{m} S_j$$

which on substituting (2.2) gives

$$= \sum_{j=m}^{n} (-1)^{j-m} \binom{n}{j} \binom{j}{m} \langle 1 - j\delta_n \rangle^{n-1}$$

$$= \binom{n}{m} \sum_{j=m}^{n} (-1)^{j-m} \binom{n-m}{n-j} \langle 1 - j\delta_n \rangle^{n-1}$$

using again the notation $\langle \; \rangle$ of (2.1). Finally since $R_n = k$ if and only if exactly $(n-k)$ of the spacings exceed δ_n (hence exactly k arcs are smaller than δ_n), we have

$$P(R_n = k) = \Pi_{n-k}$$

$$= \binom{n}{k} \sum_{j=n-k}^{n} (-1)^{j-(n-k)} \binom{k}{n-j} \langle 1 - j\delta_n \rangle^{n-1}$$

$$= \binom{n}{k} \sum_{j=0}^{k} (-1)^{j} \binom{k}{j} \langle 1 - (n-k+j)\, \delta_n \rangle^{n-1}$$

Q. E. D.

Remark 1

The distribution in (2.1) can also be derived by using the results of Darling (1953) who gives the characteristic function of $N_n(\alpha, \beta)$, the number of spacings with values between α and β. It is given by

$$\mathrm{E}(\mathrm{e}^{i\xi N_n(\alpha,\beta)}) = \frac{(n-1)!}{2\pi i} \int_{c-i\infty}^{c+i\infty} \mathrm{e}^{z} z^{-n} \{1 + (\mathrm{e}^{i\xi} - 1)(\mathrm{e}^{-z\alpha} - \mathrm{e}^{-z\beta})\}^n \, \mathrm{d}z \,.$$

Since our $R_n = N_n(0, \delta_n)$, the characteristic function of R_n is obtained by letting $\alpha = 0$ and $\beta = \delta_n$, i.e.

$$\mathrm{E}(\mathrm{e}^{i\xi R_n}) = \frac{(n-1)!}{2\pi i} \int_{c-i\infty}^{c+i\infty} \mathrm{e}^{z} z^{-n} \{1 + (\mathrm{e}^{i\xi} - 1)(1 - \mathrm{e}^{-z\delta_n})\}^n \, \mathrm{d}z \,.$$

If we expand the factor in braces and select the coefficient of $\mathrm{e}^{i\xi k}$ for any fixed $k = 0, 1, \ldots, n$,

$$P(R_n = k) = \frac{(n-1)!}{2\pi i} \int_{c-i\infty}^{c+i\infty} \mathrm{e}^{z} z^{-n} \{\mathrm{e}^{-(n-k)\delta_n z}(1 - \mathrm{e}^{-\delta_n z})^k\} \, \mathrm{d}z$$

$$= \binom{n}{k} \sum_{j=0}^{k} \binom{k}{j} (-1)^j \frac{(n-1)!}{2\pi i} \int_{c-i\infty}^{c+i\infty} \mathrm{e}^{z(1-(n-k+j)\delta_n)} z^{-n} \, \mathrm{d}z$$

$$= \binom{n}{k} \sum_{j=0}^{k} \binom{k}{j} (-1)^j \langle 1 - (n-k+j)\, \delta_n \rangle^{n-1} \,.$$

The last equality follows from the Residue Theorem.

Remark 2

Another spacings statistic of interest is $U_n = \frac{1}{2}\sum_{i=1}^{n} |T_i - 1/n|$ discussed in detail by Rao (1969) in connection with testing uniformity of circular distributions. Its density function was investigated by Darling (1953), Sherman (1950) and Rao (1969). We show below that this statistic U_n is closely related to $R_n(1)$ with $\delta_n = 1/n$. Let $K = n - R_n(1)$ denote the (random) number of spacings with lengths larger than $1/n$ and

$$S_K = T_{(n-k+1)} + T_{(n-k+2)} + \dots + T_{(n)} = T_{R_n(1)+1} + \dots + T_{(n)}$$

where $T_{(1)} \leqq \dots \leqq T_{(n)}$ are the ordered spacings. Thus S_K denotes the sum of those K largest spacings which exceed $1/n$ in length. Notice that

$$U_n = \tfrac{1}{2}\sum_{i=1}^{n} \left|T_i - \frac{1}{n}\right| = \sum_{\{i:T_i > 1/n\}} \left(T_i - \frac{1}{n}\right) = S_K - \frac{K}{n}. \tag{2.3}$$

Mauldon (1951) derived the distribution of S_k. the sum of the k largest spacings for any fixed k. Treating this as the conditional density of S_K given $K = k$ and using (2.1), we can write the joint density of (S_K, K) and hence obtain the density of U_n through the relation (2.3). The resulting expression for the density of U_n is very complex and attempts to show that this is identical to the density given for instance in Darling (1953), have not been successful.

3. Asymptotic null distribution of R_n

In this section, we establish the asymptotic normality of R_n under the hypothesis of uniformity as well as under a suitable sequence of alternatives. Notice that for computing the Pitman Asymptotic Relative Efficiency (ARE) of R_n, which will be considered in the next section, it is enough to obtain the limiting distributions under a sequence of alternatives which converge to the hypothesis (see for instance Rao and Sethuraman (1975)). Hence we will specify the alternative hypotheses by a sequence of distribution functions $A_n(x)$ depending on n and converging to the uniform distribution, which corresponds to the null hypothesis. Under the alternative hypothesis, we specify the distribution to be

$$A_n(x) = x + L_n(x)/n^{1/4}, \quad 0 \leqq x \leqq 1 \tag{3.1}$$

where $L_n(0) = L_n(1) = 0$. We further assume that $L_n(x)$ is twice differentiable on $[0, 1]$ and there is a function $L(x)$ which is twice continuously differentiable and such that $L(0) = L(1) = 0$, $n^{1/4} \sup_{0 \leqq x \leqq 1} |L_n''(x) - l'(x)| = o(1)$ where $l(x)$ and $l'(x)$ are the first and second derivatives of $L(x)$. This sequence of alternatives is smooth in a certain sense and has been considered before. See for instance Rao and Sethuraman (1975).

We define the empirical distribution function of the "normalized" spacings $\{nT_i, i = 1, \ldots, n\}$ by

$$H_n(x) = \sum_{i=1}^{n} I(nT_i; x)/n \quad \text{for} \quad x \geqq 0 \tag{3.2}$$

where

$$I(z; x) = 1 \quad \text{if} \quad z \leqq x, \qquad 0 \quad \text{otherwise}.$$

Let

$$G_n(x) = 1 - e^{-x} + e^{-x}(x - x^2/2) \cdot \left(\int_1^2 l^2(p)\, dp \right) / \sqrt{n} \quad \text{for} \quad x \geqq 0 \tag{3.3}$$

and

$$\{\zeta_n(x) = \sqrt{(n)}\,(H_n(x) - G_n(x)), \quad x \geqq 0\}.$$

$\zeta_n(\cdot)$ can be considered as a stochastic process with values in $D[0, \infty]$. See Rao and Sethuraman (1975) from which we have the following

Theorem 3.1 (Rao and Sethuraman (1975))

Under the alternatives (3.1), *the sequence of stochastic processes* $\zeta_n(x) = \sqrt{(n)}\,(H_n(x) - G_n(x)), x \geqq 0\}$ *converges weakly to the Gaussian process* $\{\zeta(x), x \geqq 0\}$ *in* $D[0, \infty]$ *with mean function zero and covariance kernel*

$$K(s, t) = e^{-t}(1 - e^{-s} - ste^{-s}) \quad \text{for} \quad 0 \leqq s \leqq t \leqq \infty.$$

Moreover if $g(\cdot)$ is a real-valued measurable function on $D[0, \infty]$ which is a.e. continuous with respect to the probability measure induced by the Gaussian process $\{\zeta(x), x \geqq 0\}$, then the distribution of the real-valued random variable $g(\zeta_n')$ converges weakly to that of $g(\zeta)$ as $n \to \infty$.

At this stage we will assume that δ_n is of the form $\delta_n = \delta/n$ for some $\delta > 0$. Since the individual T_i's are of order $1/n$ in probability under the hypothesis, for asymptotic purposes this would be the correct normalization. When $\delta_n = \delta/n$, we have the following theorem on $R_n = R_n(n\delta_n) = R_n(\delta)$.

Theorem 3.2

Under the sequence of alternatives (3.1), $\sqrt{(n)}\,(R_n(\delta)/n - G_n(\delta))$ *where* $G_n(x)$ *is defined in* (3.3), *has a limiting* $N(0, \sigma^2)$ *distribution with* $\sigma^2 = e^{-\delta}(1 - e^{-\delta} - \delta^2 e^{-\delta})$.

Proof:

$$\text{Note } R_n(\delta) = \text{number of } (nT_i) \leqq (n\delta_n) = \delta$$
$$= n\, H_n(\delta)$$

where $H_n(x)$ is the empirical distribution of the normalised spacings and is defined in (3.2). Thus

$$\sqrt{(n)}\left(\frac{R_n(\delta)}{n} - G_n(\delta)\right) = \sqrt{(n)}\,[H_n(\delta) - G_n(\delta)]$$
$$= \zeta_n(\delta)\,.$$

Therefore the stated result follows from Theorem 3.1. q.e.d.

Corollary 3.3

Under the null hypothesis of uniformity

$$\sqrt{(n)}\left(\frac{(R_n(\delta)}{n} - (1 - e^{-\delta})\right) / \{e^{-\delta}(1 - e^{-\delta} - \delta^2 e^{-\delta})\}^{1/2}$$

has a limiting $N(0, 1)$ *distribution.*

This Corollary 3.3 may also be obtained alternately using Theorem 9.1 of Darling (1953). But unfortunately the expression for the limiting variance given there, is incorrect. We now state the correct version without proof. This result may also be obtained as a corollary from Theorem 3.1 of Rao and Sethuraman (1975).

Theorem 3.4

Denote by $N_n(a/n, b/n)$ *the number of spacings whose length lies between* a/n *and* b/n. *Then the random variable* $N_n(a/n, b/n)$ *is asymptotically normally distributed with an asymptotic mean and a variance given by*

$$\mu_n = n(e^{-a} - e^{-b})$$
$$\sigma_n^2 = n[(e^{-a} - e^{-b}) - (e^{-a} - e^{-b})^2 - (ae^{-a} - be^{-b})^2]\,.$$

4. The ARE of R_n relative to U_n

For a definition of ARE, see Fraser (1957). The ARE of a test relative to another may be defined as the limit of the inverse ratio of sample sizes required to obtain the same limiting power at a sequence of alternatives converging to the null hypothesis.

This limiting power should be a value in between the limiting size α and the maximum power 1, in order that it can give an insight into the power behaviour of the test. If this converges to a number in the interval $(\alpha, 1)$, then a measure of the rate of this convergence, called 'efficacy' can be computed. Under certain standard regularity assumptions (see e.g. Fraser (1957)) which include a condition about the nature of alternatives and the asymptotic normality of the test statistic under the alternatives, which are satisfied here, the 'efficacy' is given by

$$\text{efficacy} = \left(\frac{\mu}{\sigma}\right)^4 \tag{4.1}$$

in this case. Here μ and σ are the mean and variance of the limiting normal distribution under the sequence of alternatives (3.1) when the test-statistic has been normalized to have a limiting normal distribution with mean zero and finite variance under the hypothesis. In such a situation, the ARE of one test with respect to another is simply the ratio of their efficacies.

From Corollary 3.3, $\sqrt{(n)}\,(R_n/n - (1 - e^{-\delta}))$ has a limiting normal distribution with mean zero and variance $e^{-\delta}(1 - e^{-\delta} - \delta^2 e^{-\delta})$ under H_0. On the other hand, from Theorem 3.2, under the sequence of alternatives (3.1) the same statistic has a limiting normal distribution with mean $(\int_0^1 l^2(p)\,dp)\,e^{-\delta}(\delta - \delta^2/2)$ and the same variance. Hence the efficacy of $R_n(\delta)$ is given by

$$\frac{\left(\int_0^1 l^2(p)\,dp\right)^4 \left(\delta - \dfrac{\delta^2}{2}\right)^4}{(e^{\delta} - 1 - \delta^2)^2}. \tag{4.2}$$

Sethuraman and Rao (1970) show that the Pitman efficacy of U_n in this situation is given by

$$\frac{\left(\int_0^1 l^2(p)\,dp\right)^4}{(4(2e - 5))^2}.$$

Hence the ARE of R_n with respect to U_n is given by

$$\frac{16(2e - 5)^2 \left(\delta - \dfrac{\delta^2}{2}\right)^4}{(e^{\delta} - 1 - \delta^2)^2}. \tag{4.3}$$

From the results of Rao and Sethuraman (1970) who compute the efficacies of many other spacings tests, one can compare the ARE of R_n with any of those tests. We will return to the expression (4.2) again in Section 6.

5. Limiting Bahadur Efficiency of R_n relative to U_n:

We refer the reader to Bahadur (1960) for the concepts of Bahadur Approximate slope (BAS) and Bahadur Approximate Efficiency (BAE). We use the same notations as in Bahadur (1960). We consider the class of alternative densities

$$g_k(x) = 1 + k\, l(x)\,, \quad 0 \leqq x \leqq 1 \tag{5.1}$$

where k is a real number and $l(x)$ is any square integrable function on $[0, 1]$ with $\int_0^1 l(x)\,\mathrm{d}x = 0$. For instance in connection with the circle, taking $l(x) = \cos 2\pi x$ yields the so called cardioid curve. Here k is a scale parameter and since uniformity corresponds to $k = 0$, the null hypothesis formulates H_0: $k = 0$. These alternatives are very similar to those formulated earlier in (3.1). We now take as the standard sequence

$$T_n^{(1)} = \left(\frac{R_n(\delta)}{n} - (1 - \mathrm{e}^{-\delta})\right) / \{\mathrm{e}^{-\delta}(1 - \mathrm{e}^{-\delta} - \delta^2 \mathrm{e}^{-\delta})\}^{1/2}\,. \tag{5.2}$$

Since $T_n^{(1)}$ has a $N(0, 1)$ distribution asymptotically from Corollary 3.3, this sequence of test statistics satisfies conditions (1), (2) and (3) on p. 276 of Bahadur (1960) with $a = 1$. To find the probability limit of $T_n^{(1)}/\sqrt{n}$, we state a result from Rao (1969).

Theorem 5.1 (Rao (1969))

Under the alternative distribution $G(x)$ on $[0, 1]$ with continuous density $g(\cdot)$, the statistic $H_n(a)$ defined in (3.2) converges in probability to $1 - \int_0^1 \exp(-a\, g(u)\,.$ $.\,\mathrm{d}G(u)$.

Thus under the alternative (5.1)

$$\frac{R_n}{n} = H_n(\delta) \xrightarrow{\text{Pr}} 1 - \int_0^1 \exp(-\delta\, g_k(u))\, g_k(u)\, du\,, \tag{5.3}$$

$$= 1 - \mathrm{e}^{-\delta} \int_0^1 \mathrm{e}^{-\delta k l(u)} (1 + k\, l(u))\, \mathrm{d}u\,.$$

As in Rao (1972), the comparison of the limiting efficiencies is made easier by considering approximations to the slopes when k is small, since in any case we let $k \to 0$ for obtaining the limiting efficiencies. Thus for k small, by expanding the exponential function in a power series and noting that $\int_0^1 l(x)\,\mathrm{d}x = 0$, the probability limit in (5.3) can be shown to be

$$1 - \mathrm{e}^{-\delta}\left[1 + k^2 \left(\int_0^1 l^2(x)\,\mathrm{d}x\right)\left(\frac{\delta^2}{2} - \delta\right) + o(k^2)\right].$$

Hence the BAS of $T_n^{(1)}$ is given by

$$C_1(k) = \left(\frac{\delta^2}{2} - \delta\right)^2 . k^4 . \left(\int_0^1 l^2(x)\, dx\right)^2 / 4(e^\delta - 1 - \delta^2). \tag{5.4}$$

on the other hand, similar calculations yield the BAS of the standardized U_n to be

$$C_2(k) = k^4 \left(\int_0^1 l^2(x)\, dx\right)^2 / 8(2e - 5).$$

Thus the limiting Bahadur efficiency of R_n relative to U_n is

$$\lim_{k \to 0} \frac{C_1(k)}{C_2(k)} = 4\left(\delta - \frac{\delta^2}{2}\right)^2 (2e - 5)/(e^\delta - 1 - \delta^2). \tag{5.5}$$

This value, it may be noted, is the square root of the ARE derived in (4.3).

6. The statistic R_n^* and a table of significance points

In this section, we consider the class of tests $\{R_n(\delta)\}$ for varying δ and select the one with maximum efficacy. This amounts to finding out the value of δ for which the expression (4.2) (or equivalently (5.4)) is a maximum. The mathematical problem of finding the maximum does not appear simple but using a computer, it may be checked that the maximum efficiency is attained close to a value of $\delta = 0.7379$. For example, it may be seen that the efficiency of $R_n(1)$ relative to $R_n(0.7379)$ is close to 86%. Thus if one were to restrict consideration to this class of tests, then it is clearly best to take $\delta = 0.7379$. But from a practical point of view, we suggest using a more reasonable fraction like $\delta = 0.75$. Since the loss of efficiency in doing this is insignificant, we advocate the use of the statistic

$$R_n^* = R_n(0.75) = \left\{\text{number of } T_i\text{'s} \leqq \frac{3}{4n}\right\} \tag{6.1}$$

as the best among this class. From Theorem 2.1, Corollary 3.3 and equation (4.2), we have the following result regarding the exact, asymptotic distributions of R_n^* as well as its efficacy.

Corollary 6.1

The following results hold for the statistic R_n^* defined in (6.1):

(a) *Exact null distribution:*

$$(6.2) \qquad P(R_n^* = k) = \binom{n}{k} \sum_{j=0}^{k} (-1)^j \binom{k}{j} \left\langle 0.25 + (0.75)\left(\frac{k-j}{n}\right)\right\rangle^{n-1}$$

$$\text{for} \quad k = 0, 1, \ldots, (n-1),$$

$$= 0 \quad \text{otherwise}.$$

with the notation $\langle x \rangle = x$ if $x > 0$ and $= 0$ if $x \leqq 0$.

Table 6.1. Distribution Function of the Statistic R_n^* in the range of 9.90 to 1.00

n	k	$F(k)$	$F(k+1)$	$F(k+2)$	$F(k+3)$	$F(k+4)$	$F(k+5)$
3	1	.6350	1.0000				
4	2	.8418	1.0000				
5	2	.5545	.9392	1.0000			
6	3	.7583	.9780	1.0000			
7	4	.8818	.9923				
8	4	.7030	.9465	.9974			
9	5	.8334	.9772	.9991			
10	5	.6621	.9134	.9907			
15	8	.7280	.9142	.9841	.9985		
20	11	.7752	.9207	.9810	:9971		
25	14	.8112	.9286	.9801	.9960		
30	17	.8399	.9364	.9804	.9954		
35	20	.8632	.9436	.9813	.9951		
40	23	.8825	.9501	.9825	.9950		
45	26	.8986	.9559	.9838	.9950		
50	28	.8281	.9121	.9611	.9852	.9952	
55	31	.8514	.9237	.9657	.9865	.9954	
60	34	.8710	.9336	.9697	.9878	.9957	
65	37	.8878	.9421	.9733	.9890	.9960	
70	39	.8293	.9023	.9494	.9764	.9901	
75	42	.8505	.9147	.9557	.9791	.9911	
80	45	.8689	.9254	.9611	.9816	.9920	
85	48	.8849	.9346	.9659	.9837	.9929	
90	51	.8988	.9427	.9700	.9856	.9936	
95	53	.8538	.9110	.9497	.9736	.9872	.9943
100	56	.8706	.9216	.9558	.9768	.9887	.9949

(b) *Asymptotic Null Distribution:*

$$\sqrt{(n)}\left(\frac{R_n^*}{n} - 0.5276\right)/\{0.3517\} \quad \textit{has a limiting} \quad N(0, 1)$$

distribution.

(c) *Pitman efficiency:*

The Pitman efficacy of R_n^* *against the alternatives* (3.1) *is given by* (0.1570) $(\int_0^1 l^2(p)\,dp)^4$.

Using the exact null distribution of R_n^* given in (6.2), the following table of cumulative probability function $F(k) = P(R_n^* \leq k)$ in the upper tail area has been constructed for sample sizes $n = 3(1)\,10(5)\,100$ if the observed value k of R_n^* is such that $F(k)$ (from Table 6.1) exceeds $(1 - \alpha)$, then we reject the hypothesis of uniformity H_0 at that level α.

It may be remarked here that the data need not be scaled to the interval (0, 1) in order to calculate R_n^*. We now illustrate by means of a numerical example, the extreme simplicity in using the statistic R_n^* for testing uniformity. It may be remarked here that the simplicity in using R_n^* in our view, more than compensates for the lower asymptotic efficiency.

Example

Consider a fire station which received 20 calls on a particular day. We want to know if these calls are randomly distributed over the entire day or if they tend to cluster around some particular time of the day. Suppose that the calls are received at 1 : 100, 4 : 30, 6 : 00, 6 : 10, 7 : 00, 8 : 00, 8 : 30, 8 : 45, 9 : 30, 10 : 05 a.m. and 1 : 00, 2 : 10, 4 : 00, 5 : 50, 7 : 30, 9 : 15, 10 : 00, 10 : 15, 11 : 00, 11 : 30 p.m. Since $\delta_n =$ $= (0.75\ 24/20 = 0.9$ hrs. $= 54$ mts., R_n^* is the number of spacings less than 54 minutes. We see easily that $R_n^* = 10$. This R_n^* value of 10, when $n = 20$, is not significant even at $\alpha = 10\%$ as can be seen from Table 6.1. Hence we have no reason to reject the hypothesis that these calls are randomly distributed throughout the day. We may remark here that for the purpose of this test the data could very well be accumulated over several cycles (days) instead of just one.

References

BAHADUR, R. R. (1960). Stochastic comparison of tests. *Ann. Math. Statist.*, **31**, 276—95.

DARLING, D. A. (1953). On a class of problems related to the random division of an interval. *Ann. Math. Statist.*, **24**, 239—253.

FELLER, W. (1966). "An introduction to the probability theory and its applications". Vol. I, John Wiley, New York.

FRASER, D. A. S. (1957). "Nonparametric methods in statistics". John Wiley, New York.

MARDIA, K. V. (1972). Statistics of directional data, Academic Press.

MAULDON, J. G. (1951). Random division of an interval. *Proc. Camb. Phil. Soc.*, **47**, 331—336.

PYKE, R. (1965). Spacings, *J. Roy. Stat. Soc.* B, **27**, 395—449.

RAO, J. S. (1969). Some contributions to the analysis of circular data. Unpublished Ph. D. thesis, *Indian Statistical Institute*, Calcutta.

RAO, J. S. (1972). Bahadur efficiencies of some tests for uniformity on the circle. *Ann. Math. Statist.*, **43**, 468—479.

RAO, J. S. - SETHURAMAN, J. (1975). Weak convergence of empirical distribution functions of random variables subject to perturbations and scale factors. *Ann. Statist.*, **3**, 299—313.

SETHURAMAN J. - RAO, J. S. (1970). Pitman efficiencies of tests based on spacings: In *Nonparametric Techniques in Statistical Inference*, Ed. M. L. Puri, Cambridge Univ. Press, 405—415.

SHERMAN, B. (1950). A random variable related to the spacing of sample values. *Ann. Math. Statist.*, **21**, 339—51.

WILKS, S. S. (1962). "Mathematical Statistics". John Wiley, New York.

(1) INDIANA UNIVERSITY, BLOOMINGTON, INDIANA, U.S.A.
(2) UNIVERSITY OF CALIFORNIA, SANTA BARBARA, CALIFORNIA, U.S.A.
(3) BLUE CROSS ET BLUE SHIELD OF TEXAS, DALLAS, TEXAS, U.S.A.

Received March 1976

NONPARAMETRIC TESTS FOR INTERCHANGEABILITY UNDER COMPETING RISKS*

by

PRANAB KUMAR SEN

1. Introduction

For a two-component system, let $F(x, y)$ be the joint distribution function (df) of the survival times X and Y of the two components. We desire to test the null hypothesis that X and Y are interchangeable, i.e.,

$$F(x, y) = F(y, x) \quad \text{for all} \quad (x, y) \in E^2 , \tag{1.1}$$

where E^k, $k \geqq 1$, stands for the k-dimensional Euclidean space. Nonparametric tests for (1.1) are due to Sen (1967), Bell and Smith (1969) and others. In competing risks problems, instead of (X, Y), the observable random vector is (Z, Q), where

$$Z = \min(X, Y) \quad \text{and} \quad Q = 1, 0 \quad \text{or} \quad -1 \quad \text{according as} \quad Z = X , \tag{1.2}$$
$$Z = X = Y \quad \text{or} \quad Z = Y .$$

For example, one may be interested in the functioning of the two human kidneys under a toxic dose. Once, one of the two kidneys stops functioning, the other kidney acquires extra assignment and, as a result, its life time may be considerably shorter than its normal case. In this case, in comparing the life times of the two kidneys, say, the left and right ones, it seems that valid statistical information is contained in the life span of the first kidney to fail and its particular position. There are other examples where a two component system may even breakdown completely when at least one of its components stops working. Thus, in the competing risks setups, based on a set of observable random vectors (Z_i, Q_i), $i = 1, \ldots, n$, our problem is to test for (1.1) against suitable alternatives.

Three different types of nonparametric tests are considered: (i) the conventional fixed sample size procedure based on all the n observations through a single statistic, (ii) the first sequential procedure based on the observations when the Z_i are observable sequentially, and (iii) the second sequential procedure suitable under progressive

* Work partially supported by the Air Force office of Scientific Research, A.F.S.C., U.S.A.F., Grant No. AFOSR 74-2746.

censoring. The first sequential procedure is suitable when the observations are not available at the same time, so that if the null hypothesis (1.1) may be rejected based on fewer than n observations, there is a reduction of the total time to perform the test. In the context of life-testing problems, when n independent systems are subject simultaneously to a continuous time-observation process and the (Z_i, Q_i) are observable only at the expiry of the lives of these systems, one may naturally be interested in monitoring the experiment with the objective of rejecting the null hypothesis with the minimum sacrifice of the lives of the units, that is, stopping the experiment at a time point where for the first time the accumulated evidence leads to the rejection of H_0. Unlike the other case, here the ordered random variables corresponding to $Z_1, \ldots, Z_n$ are observed sequentially, and the scheme is known as a *progressively censored scheme*. For simple linear rank statistics, such procedures have been developed by Halperin and Ware (1974) and Chatterjee and Sen (1973), among others. Our second sequential procedure is based on a different type of rank statistics and applies to this situation. Thus, for both the sequential procedures, the stopping times are random variables, and the procedures may lead to reduction of time and cost of experimentation. This will be elaborated more in Section 2.

The best procedures along with the preliminary notions are introduced in Section 2. The theory of these rank tests is based on a stochastic process approach which is motivated in Section 3. The results of Section 3 are then incorporated in Sections 4 and 5 for the study of the properties of the proposed tests. Section 6 is devoted to the choice of optimal scores for the rank statistics used for the tests.

2. Preliminary notions and the proposed test procedures

Under the competing risk setup, the observable random vectors of a sample of size n are (Z_i, Q_i), $i = 1, \ldots, n$. Note that (Z_i, Q_i) corresponds to (X_i, Y_i) by (1.2) and if we assume that $F(x, y)$ has a probability density function (pdf) $f(x, y)$ for all (x, y), then $g(z)$, the pdf of Z_i is given by

$$g(z) = \int_z^{\infty} f(x, z)\,dx + \int_z^{\infty} f(z, y)\,dy \quad \text{for all } z; \tag{2.1}$$

the corresponding distribution function (df) is denoted by $G(z)$. Also, note that, by assumption, ties among $Z_1, \ldots, Z_n$ can be neglected with probability one, and further, $P\{Q_i = 0\} = P\{X_i = Y_i\} = 0$ for all $i \geqq 1$.

For testing H_0 in (1.1) by a parametric method, one needs to know (or assume) the form of the joint df of (Z, Q) so that (i) the alternative hypothesis can be framed in terms of some parameters of this df and (ii) the usual likelihood ratio principle can be employed to construct suitable tests having some desirable (or optimal) properties. Now, specification of the joint df of (Z, Q) in some particular form may naturally localize the scope of the tests to situations where such a df can actually be justified. Moreover, unless this df is of simple form, usually the distribution theory of the associated test statistics is quite complicated for finite sample sizes. Finally, if the true and the assumed forms of the df of (Z, Q) are not the same, the performance of the parametric tests may be quite poor. For these reasons, we propose here some nonparametric tests which remain valid for the entire class of continuous G and are robust in nature. These tests are based on suitable rank statistics, which we introduce first.

Let $c(u) = 1$ or 0 according as u is $\geqq$ or <0 and let $R_{ni} = \sum_{j=1}^{n} c(Z_i - Z_j)$ be the rank of Z_i among $Z_1, \ldots, Z_n$, for $i = 1, \ldots, n$. Then, $\mathbf{R}_n = (R_{n1}, \ldots, R_{nn})$ is some permutation of $(1, \ldots, n)$. For every $n(\geqq 1)$, consider a set of real-valued *scores*

$$a_n(i) = E\,\Phi(U_{ni}) \quad \text{or} \quad \Phi(i/(n+1))\,, \quad i = 1, \ldots, n\,, \tag{2.2}$$

where $U_{n1} < \ldots < U_{nn}$ are the ordered random variables of a sample of size n from the rectangular $[0, 1]$ df and the *score-function* $\Phi(u)$, $0 < u < 1$, is assumed to be square-integrable and non-degenerate, so that

$$0 < A^2 = \int_0^1 \Phi^2(u)\,du < \infty\,. \tag{2.3}$$

Consider first the fixed-sample size test. Define the rank statistics

$$T_n = \sum_{i=1}^{n} Q_i\, a_n(R_{ni})\,, \quad n \geqq 1\,. \tag{2.4}$$

By Lemma 3.1 (to follow), under H_0 in (1.1), $\mathbf{Q}_n = (Q_1, \ldots, Q_n)$ and $\mathbf{R}_n$ are stochastically independent, $E(T_n \mid H_0) = 0$ and $V(T_n \mid H_0) = \sum_{i=1}^{n} a_n^2(i) = nA_n^2$, say. On the other hand, T_n is stochastically shifted to the right or left when H_0 does not hold and one of the variates is stochastically larger than the other. Hence, in the one sided case, for the upper tail version, we consider the following test procedure:

$$\text{accept or reject } H_0 \text{ in (1.1) according as } T_n \text{ is } < \text{ or } \geqq T_{n,\alpha}\,, \tag{2.5}$$

where we select $T_{n,\alpha}$ in such a way that the test has a prescribed level of significance, α: $0 < \alpha < 1$, i.e., $P\{T_n \geqq T_{n,\alpha} \mid H_0\} \leqq \alpha$. The case of the lowertail test follows similarly. For a two-sided version, we accept or reject H_0 according as $|T_n|$ is $<$ or $\geqq T^*_{n,\alpha}$, where $P\{|T_n| \geqq T^*_{n,\alpha} \mid H_0\} \leqq \alpha$.

In Section 4, we shall show that the above test procedure is genuinely distribution-free (i.e., $T_{n,\alpha}$, $T^*_{n,\alpha}$ do not depend on the underlying F when (1.1) holds) and, moreover, as n increases,

$$n^{-1/2}T_{n,\alpha}/A_n \to \tau_\alpha\,, \quad n^{-1/2}T^*_{n,\alpha}/A_n \to \tau_{\alpha/2}\,, \tag{2.6}$$

where τ_α is the upper $100\alpha\%$ point of a standard normal distribution. Other properties of the test will also be studied.

Consider next the first sequential procedure where all the n observations (Z_i, Q_i), $i = 1, \ldots, n$ are not available at the same time. Since, these arrive sequentially, it may be advisable to add a provision to stop at an intermediate stage i.e., when (Z_i, Q_i), $i = 1, \ldots, k$ are observed for some $k \leqq n$, provided the statistical evidence up to that stage advocates the rejection of H_0. Defining the T_n as in (2.4) and conventionally letting $T_0 = 0$, we define

$$M_n^+ = [\max_{0 \leqq k \leqq n} T_k]/(n^{1/2}A_n) \quad \text{and} \quad M_n = [\max_{0 \leqq k \leqq n} |T_k|]/(n^{1/2}A_n) . \tag{2.7}$$

Suppose that it is possible to find two constants $M_{n,\alpha}^+$ and $M_{n,\alpha}$ such that

$$P_0\{M_n^+ \geqq M_{n,\alpha}^+\} \leqq \alpha \quad \text{and} \quad P_0\{M_n \geqq M_{n,\alpha}\} \leqq \alpha , \tag{2.8}$$

where P_0 stands for the probability under H_0 in (1.1). Then, operationally, the test procedure consists sequentially the T_k, $k \geqq 1$, until for the first time for some $k = N(\leqq n)$, $n^{-1/2}A_n^{-1}T_N$ (or $n^{-1/2}A_n^{-1}|T_N|$) exceeds $M_{n,\alpha}^+$ (or $M_{n,\alpha}$), rejecting H_0 at that stage with the termination of the experiment. If no such $N(\leqq n)$ exists, then H_0 is accepted when $(Z_1, Q_1), \ldots, (Z_n, Q_n)$ are observed. We shall see in Section 4 that the test procedure is distribution-free, so that $M_{n,\alpha}^+$ or $M_{n,\alpha}$ does not depend on the underlying F, and further, as n increases,

$$M_{n,\alpha}^+ \to \tau_{\alpha/2} \quad \text{and} \quad M_{n,\alpha} \to W_\alpha , \quad 0 < \alpha < 1 , \tag{2.9}$$

where W_α is the upper $100\alpha\%$ point of the distribution

$$\omega^*(x) = \sum_{k=-\infty}^{\infty} (-1)^k [\omega((2k+1)x) - \omega((2k-1)x)] , \quad x \geqq 0 , \tag{2.10}$$

and $\omega(x)$ is the standard normal df.

Finally, let us consider the progressively censored rank test. In this case, the experimentation starts with the continuous observation on n units and their values are recorded as they are observed sequentially. Thus, here the order statistics $Z_{n,1} < \ldots < Z_{n,n}$ (corresponding to $Z_1, \ldots, Z_n$) are observed in a sequence. We may note that $Z_{n,i} = Z_{S_{ni}}$, $i = 1, \ldots, n$, where $\mathbf{S}_n = (S_{n1}, \ldots, S_{nn})$ (some permutation of $1, \ldots, n$) is the vector of *anti-ranks*. Also, we denote the Q_j corresponding to $Z_{S_{ni}}$ by $Q_{S_{ni}} = Q(n, S_{ni})$, for $i = 1, \ldots, n$. Then, we observe that at the kth stage when $Z_{n,1}, \ldots, Z_{n,k}$ have been observed, we are provided with $Q(n, S_{n1}), \ldots, Q(n, S_{nk})$, for $k = 1, \ldots, n$. We denote by

$$T_{nk} = \sum_{i=1}^{k} Q(n, S_{ni}) \, a_n(i) , \quad k = 1, \ldots, n . \tag{2.11}$$

Note that, by definition,

$$(2.12)\qquad T_{nn} = \sum_{i=1}^{n} Q(n, S_{ni})\, a_n(i) = \sum_{i=1}^{n} Q_i\, a_n(R_{ni}) = T_n\,.$$

Also, conventionally, we let $T_0 = T_{n0} = 0$ for every $n \geqq 1$, and define then

$$(2.13)\qquad D_n^+ = \big[\max_{0\leqq k\leqq n} T_{nk}\big]/(n^{1/2}A_n) \quad\text{and}\quad D_n = \big[\max_{0\leqq k\leqq n} |T_{nk}|\big]/(n^{1/2}A_n)\,.$$

For an one sided test, we use D_n^+ as the test statistic and reject H_0 when

$$(2.14)\qquad D_n^+ \geqq D_{n,\alpha}^+ \quad\text{where}\quad P\{D_n^+ \geqq D_{n,\alpha}^+ \mid H_0\} \leqq \alpha\,,$$

and for a two sided test, we use D_n and reject H_0 when

$$(2.15)\qquad D_n \geqq D_{n,\alpha} \quad\text{where}\quad P\{D_n \geqq D_{n,\alpha} \mid H_0\} \leqq \alpha\,.$$

Operationally, the test procedure consists in continuing the experiment so long as $n^{-1/2}A_n^{-1}T_{nk}$ (or $n^{-1/2}A_n^{-1}|T_{nk}|$), $1 \leqq k \leqq n$, continue to lie below $D_{n,\alpha}^+$ (or $D_{n,\alpha}$), and if $N(\leqq n)$ is the smallest positive integer for which $n^{-1/2}A_n^{-1}T_{nN}$ is $\geqq D_{n,\alpha}^+$ (or $n^{-1/2}A_n^{-1}|T_{nk}|$ is $\geqq D_{n,\alpha}$), the experimentation is terminated along with the rejection of H_0. If no such $N(\leqq n)$ exists, H_0 is accepted. In Section 5, we shall see that the tests based on D_n^+ and D_n are genuinely distribution-free and

$$(2.16)\qquad D_{n,\alpha}^+ \to \tau_{\alpha/2} \quad\text{and}\quad D_{n,\alpha} \to W_\alpha\,, \quad\text{as}\quad n\to\infty\,,\quad \alpha\in(0,1)\,,$$

where τ_α and W_α are defined after (2.6) and (2.9), respectively.

For the study of the various properties of these tests, we require to study first some basic asymptotic results on the two sequences $\{T_n,\ n \geqq 1\}$ and $\{T_{nk},\ 1 \leqq k \leqq n\}$ and these are considered in Section 3.

3. Invariance principles for $\{T_n\}$ and $\{T_{nk}\}$

On defining $\mathbf{R}_n$, $\mathbf{Q}_n$, $Q(n, S_{ni})$, $i = 1, \ldots, n$ and $\mathbf{S}_n$ as in Section 2, we let

$$(3.1)\qquad \mathbf{Q}(\mathbf{S}_n) = (Q(n, S_{n1}), \ldots, Q(n, S_{nn}))\,.$$

Also, let $\mathbf{j}_n = ((-1)^{j_1}, \ldots, (-1)^{j_n})$ where j_i is either 0 or 1, $i = 1, \ldots, n$ and let

$$(3.2)\qquad \mathbf{J}_n = \{\mathbf{j}_n : j_i = 0, 1;\ i = 1, \ldots, n\}\,.$$

Finally, let $\mathscr{S}_n$ be the set of all possible $n!$ permutations of $(1, \ldots, n)$. Then, the following lemma provides the fundamental result on which the invariance principles rest.

Lemma 3.1. *Under H_0 in (1.1), $\mathbf{Q}_n$ and $\mathbf{R}_n$ are stochastically independent and*

$$P\{\mathbf{Q}(\mathbf{S}_n) = \mathbf{j}_n\} = 2^{-n} \quad \textit{for every} \quad \mathbf{S}_n \in \mathcal{S}_n, \quad \mathbf{j}_n \in \mathbf{J}_n . \tag{3.3}$$

Proof. Let $\pi(z) = P\{Z = X \mid Z = z\} = P\{Q = 1 \mid Z = z\}$, so that

$$\pi(z) = \left[\int_z^\infty f(z, y)\, dy\right]/g(z), \quad 0 \leq \pi(z) \leq 1, \quad -\infty < z < \infty, \tag{3.4}$$

where $g(z)$ is defined by (2.1). Note that under H_0 in (1.1), $f(x, y) = f(y, x)$, $\forall(x, y)$, so that $\pi(z) = \frac{1}{2}$, for all z.

Now $Z_1, \ldots, Z_n$ are iidrv, so that $\mathbf{R}_n$ can have all possible $n!$ permutations of $(1, \ldots, n)$ with the common probability $1/n!$. On the other hand, if $\mathbf{r}_n$ is any permutation of $(1, \ldots, n)$, then

$$P\{\mathbf{Q}_n = \mathbf{j}_n, \mathbf{R}_n = \mathbf{r}_n \mid H_0\} = \int \cdots \int_{S(n)} \{\prod_{i=1}^n \{g(z_i)\,[\pi(z_i)]^{1-j_i}\,[1 - \pi(z_i)]^{j_i}\}\, dz_i , \tag{3.5}$$

where the n fold integration extends over the domain $\{-\infty < z_{s_{n1}} < \ldots < z_{s_{nn}} < \infty\}$ and $\mathbf{S}_n$ is the anti-rank vector corresponding to $\mathbf{r}_n$. Since, under H_0, $\pi(z) = \frac{1}{2}$ for all z, (3.5) reduces to

$$2^{-n} \int \cdots \int_{S(n)} g(z_1) \ldots g(z_n)\, dz_1 \ldots dz_n = 2^{-n}(n!)^{-1} = 2^{-n} P\{\mathbf{R}_n = \mathbf{r}_n\} . \tag{3.6}$$

Hence, $P\{\mathbf{Q}_n = \mathbf{j}_n \mid \mathbf{R}_n = \mathbf{r}_n, H_0\} = 2^{-n}$, $\forall \mathbf{r}_n$, and this implies the independence of $\mathbf{R}_n$ and $\mathbf{Q}_n$. Hence,

$$P\{\mathbf{Q}_n = \mathbf{j}_n \mid H_0\} = 2^{-n}, \quad \forall \mathbf{j}_n \in \mathbf{J}_n . \tag{3.7}$$

By virtue of the fact that $\mathbf{S}_n$ is the anti-rank corresponding to some $\mathbf{R}_n$, we have $P\{\mathbf{Q}(\mathbf{S}_n) = \mathbf{j}_n \mid H_0\} = P\{\mathbf{Q}_n = \mathbf{j}_n \mid \mathbf{R}_n, H_0\}$, and hence, (3.3) follows from (3.7) and the independence of $\mathbf{Q}_n$ and $\mathbf{R}_n$. Q.E.D.

Now, having the statistics M_n, M_n^+ and T_n in mind, we intend to provide a basic theorem which yields the large sample distributions of all these statistics.

We denote by $I = \{t : 0 \leq t \leq 1\} = [0, 1]$, the unit interval. Then, for every $n \geq 1$, we define a process $W_n = \{W_n(t) : 0 \leq t \leq 1\}$ by letting

$$W_n\left(\frac{k}{n}\right) = n^{-1/2} T_k / A_n, \quad k = 1, \ldots, n ; \quad W_n(0) = 0, \tag{3.8}$$

and, by linear interpolation, for $k/n \leq t \leq (k + 1)/n$, taking

$$W_n(t) = W_n\left(\frac{k}{n}\right) + (nt - k)\left[W_n\left(\frac{k+1}{n}\right) - W_n\left(\frac{k}{n}\right)\right], \tag{3.9}$$

for $k = 0, 1, \ldots, n-1$. Also, we denote by $W = \{W(t),\ 0 \leqq t \leqq 1\}$ a standard Brownian motion on I. Then, we have the following.

Theorem 3.2. *Suppose that for the score function $\Phi(u)$, $u \in I$ in (2.2), the following conditions hold:*

(a) $\Phi(u) = \Phi_1(u) - \Phi_2(u)$, $0 < u < 1$, *where $\Phi_j(u)$ is non-decreasing and absolutely continuous inside* $(0, 1)$, *and*

*(b) $\int_0^1 |\Phi_j(u)| \{u(1-u)\}^{-1/2}\, du < \infty$, $j = 1, 2$.

Then, under H_0 in (1.1), *W_n defined by* (3.8)–(3.9), *converges in law to a standard Brownian motion W.*

Proof. By virtue of Lemma 3.1, the distributional structure of T_n is quite similar to the classical one sample case where $Z_i = |X_i - Y_i|$ and $Q_i = \operatorname{sgn}(X_i - Y_i)$, $i \geqq 1$. Thus, if the scores in (1.2) are defined by $a_n(i) = E\,\Phi(U_{ni})$, $i = 1, \ldots, n$, it follows by the same arguments as in Section 3 of Sen and Ghosh (1973) that under H_0 in (1.1), $\{T_n\}$ is a martingale sequence. Also, on defining $T_n^* = \sum_{i=1}^{n} Q_i\, \Phi(G(Z_i))$, it follows that T_n and T_n^* are asymptotically equivalent in quadratic mean when H_0 holds. Thus, the convergence of the finite dimensional distributions (f.d.d.) of W_n (to those of W) follows readily by applying the cassical (multivariate version) of the central limit theorem on finitely many T_k^*. Also, by virtue of the martingale property of $\{T_n\}$, the convergence of the f.d.d. of W_n implies (cf. Section 6 of Brown (1971)) that W_n is tight. Hence, the proof is complete for the case of $a_n(i) = E\,\Phi(U_{ni})$, $i = 1, \ldots, n$. Also, under the hypothesis of the theorem,

$$(3.10)\qquad \Big|\sum_{i=1}^{n} Q_i E\,\Phi(U_{nR_{ni}}) - \sum_{i=1}^{k} Q_i\, \Phi((n+1)^{-1} R_{ni})\Big|$$

$$\leqq \sum_{i=1}^{n} \big|E\,\Phi(U_{ni}) - \Phi(i/(n+1))\big| = o(n^{1/2}),$$

where the last step follows from a recent result of Hoeffding (1973). As such, the theorem also holds for the case of $a_n(i) = \Phi(i/(n+1))$, $i = 1, \ldots, n$.

We may note that by virtue of Theorem 3.2, we have for $n \to \infty$,

$$(3.11)\qquad n^{-1/2}\, T_n/A_n = W_n(1) \xrightarrow{\mathscr{D}} W(1), \quad M_n^+ = \sup_{0 \leqq t \leqq 1} W_n(t) \xrightarrow{\mathscr{D}} \sup_{0 \leqq t \leqq 1} W(t)$$

and

$$(3.12)\qquad M_n = \sup_{0 \leqq t \leqq 1} |W_n(t)| \xrightarrow{\mathscr{D}} \sup_{0 \leqq t \leqq 1} |W(t)|,$$

* (b) holds, in particular, if $\int_0^1 |\Phi_j(u)|^r\, du < \infty$ for some $r > 2$; $j = 1, 2$.

where $\xrightarrow{\mathscr{D}}$ stands for the convergence in distribution. The last two equations are of basic importance in our study. We shall comment more on it in Section 4.

For the progressively censored scheme, we have D_n^+ and D_n in mind, and as a result, we like to develop a similar approximation theorem for a suitable stochastic process constructed from the sequence $\{T_{nk}, k = 0, 1, \ldots, n\}$ defined by (2.11)–(2.12). With this objective, for every $n \geqq 1$, we construct a stochastic process $W_n^* = \{W_n^*(t), 0 \leqq t \leqq 1\}$ by letting

$$W_n^*\left(\frac{k}{n}\right) = n^{-1/2}\, T_{nk}/A_n\,, \quad k = 0, 1, \ldots, n\,, \tag{3.13}$$

and, as in (3.9), by linear interpolation, taking

$$W_n^*(t) = W_n^*\left(\frac{k}{n}\right) + (nt - k)\left[W_n^*\left(\frac{k+1}{n}\right) - W_n^*\left(\frac{k}{n}\right)\right], \quad \frac{k}{n} \leqq t \leqq \frac{k+1}{n} \tag{3.14}$$

for $k = 0, 1, \ldots, n-1$. Then, we have the following.

Theorem 3.3. *Under* (1.1), (2.4) *and the condition that**

$$\max_{1 \leqq k \leqq n} \{n^{-1/2} |a_n(k)|\} \to 0 \quad \text{as} \quad n \to \infty\,, \tag{3.15}$$

W_n^* converges in distribution to a standard Brownian motion W.

Proof. Let $\mathbf{S}_n^{(k)} = (S_{n1}, \ldots, S_{nk})$ and $\mathbf{Q}(\mathbf{S}_n^{(k)}) = (Q(n, S_{n1}), \ldots, Q(n, S_{nk}))$, $k = 1, \ldots, n$. Also, we denote by $\xi_{nk} = T_{nk} - T_{nk-1}$, $k = 1, \ldots, n$. Then, by Lemma 3.1, $E[\xi_{nk} \mid H_0] = 0$ for $k = 1, \ldots, n$ and

$$E[\xi_{nk} \mid \mathbf{S}_n^{(k-1)}, \mathbf{Q}(\mathbf{S}_n^{(k-1)}), H_0] = 0\,, \quad 1 \leqq k \leqq n\,, \tag{3.16}$$

$$E[\xi_{nk}^2 \mid \mathbf{S}_n^{(k-1)}, \mathbf{Q}(\mathbf{S}_n^{(k-1)}), H_0] = a_n^2(k)\,, \quad 1 \leqq k \leqq n\,. \tag{3.17}$$

Thus,

$$V_n = \sum_{k=1}^{n} E[\xi_{nk}^2 \mid \mathbf{S}_n^{(k-1)}, \mathbf{Q}(\mathbf{S}_n^{(k-1)}), H_0] = \sum_{k=1}^{n} a_n^2(k) = nA_n^2\,, \tag{3.18}$$

where $A_n^2 \to A^2$ as $n \to \infty$. Therefore, by (3.15) and (3.18), for every $\varepsilon > 0$, as $n \to \infty$,

$$V_n^{-1} \sum_{i=1}^{n} E[\xi_{ni}^2 I(|\xi_{ni}| > \varepsilon V_n^{1/2}) \mid \mathbf{S}_n^{(k-1)}, \mathbf{Q}(\mathbf{S}_n^{(k-1)}), H_0] \to 0\,, \tag{3.19}$$

where $I(A)$ stands for the indicator function of a set A. The convergence of the finite dimensional distributions of W_n^* to those of W follows directly from (3.18), (3.19)

* Note that (3.15) is less restrictive than the conditions of Theorem 3.2.

and Theorem 2.1 of Dvoretzky (1972). Also, by virtue of the martingale property (3.16), the convergence of the f.d.d. of W_n^* (to W) implies that W_n^* is tight. Q.E.D.

Note that by (2.13) and Theorem 3.3, as $n \to \infty$,

(3.20)
$$D_n^+ = \sup_{0 \leq t \leq 1} W_n^*(t) \xrightarrow{\mathscr{D}} \sup_{0 \leq t \leq 1} W(t) \quad \text{and} \quad D_n = \sup_{0 \leq t \leq 1} |W_n^*(t)| \xrightarrow{\mathscr{D}} \sup_{0 \leq t \leq 1} |W(t)| .$$

We shall make use of this result in Section 5.

4. Properties of the tests based on T_n, M_n^+ and M_n

By virtue of Lemma 3.1, under H_0 in (1.1), T_n has the same distribution as of $\tilde{T}_n = \sum_{i=1}^{n} a_n(i)\, U_i$ where the U_i are independent random variables with $P\{U_i = \pm 1\} = \frac{1}{2}$. Since, $\tilde{T}_n$, the classical one-sample rank order statistic is genuinely distribution-free and for various common type of scores, its distribution is fairly extensively tabulated [cf. Owen (1962)], the distribution of T_n for small n can be traced without much pain. For large n, we use (3.11) and this proves (2.6). We shall proceed now to study the Bahadur efficiency of the tests based on T_n. For this reason, we first consider the following.

Lemma 4.1. *Under condition* (a) *of Theorem* 3.2, *as* $n \to \infty$, $n^{-1}T_n \to \mu(\Phi, F)$ *almost surely, where*

$$\mu(\Phi, F) = \int_{-\infty}^{\infty} \{2\pi(z) - 1\}\, \Phi(G(z))\, g(z)\, dz . \tag{4.1}$$

Proof. We make use of the classical Hájek (1968) polynomial approximation of the score function Φ. Thus, for every $\eta > 0$, there exists a decomposition

$$\Phi(u) = \Phi_{(1)}(u) + \Phi_{(2)}(u) - \Phi_{(3)}(u)\,, \quad 0 < u < 1\,, \tag{4.2}$$

where $\Phi_{(1)}$ is a polynomial, $\Phi_{(2)}$ and $\Phi_{(3)}$ are non-decreasing and

$$\int_0^1 \Phi_{(j)}^2(u)\, du < \left(\int_0^1 \Phi^2(u)\, du\right) \eta/2\,, \quad \text{for} \quad j = 2, 3\,. \tag{4.3}$$

By (4.2), we may rewrite $T_n = T_{n,1} + T_{n,2} - T_{n,3}$ where $T_{n,j}$ is defined by (2.2)–(2.4) with Φ being replaced by $\Phi_{(j)}$, $j = 1, 2, 3$. Then, by (4.3) and the Schwarz inequality,

$$(n^{-1}T_{n,j})^2 \leq (n^{-1} \sum_{i=1}^{n} Q_i^2)\, (n^{-1} \sum_{i=1}^{n} a_{n,j}^2(i)) = n^{-1} \sum_{i=1}^{n} a_{n,j}^2(i) < \eta/2\,. \tag{4.4}$$

for $j = 2, 3$,

where $a_{n,j}(i)$ is defined by (2.2) for $\Phi = \Phi_{(j)}$. Since, η is arbitrary, it suffices to show that $n^{-1}T_{n,1} \to \mu(\Phi_{(1)}, F)$ a.s., as $n \to \infty$. For this, we consider the empirical df $G(z) = n^{-1}\sum_{i=1}^{n} c(z - Z_i)$, $n \geqq 1$, $z \in E$. Then, we may rewrite $T_{n,1}$ as

$$n^{-1}T_{n,1} = n^{-1}\sum_{i=1}^{n} Q_i\Phi_{(1)}((n+1)^{-1}R_{ni}) + o(n^{-1/2}) \tag{4.5}$$

$$= n^{-1}\sum_{i=1}^{n} Q_i\,\Phi_{(1)}(G(Z_i)) + n^{-1}\sum_{i=1}^{n} Q_i\left[\Phi_{(1)}\left(\frac{n}{n+1}G_n(Z_i)\right) - \Phi_{(1)}(G(Z_i))\right] + o(n^{-1/2}).$$

Now, $\mathscr{X}_i = Q_i\,\Phi_{(1)}(G(Z_i))$, $i \geqq 1$, are independent and identically distributed with mean $\mu_1(F)$, and hence, by the Khintchine strong law of large numbers,

$$n^{-1}\sum_{i=1}^{n} Q_i\,\Phi_{(1)}(G(Z_i)) \to \mu_1(F) \quad \text{a.s., as} \quad n \to \infty. \tag{4.6}$$

Also, by the Glivenko-Cantelli theorem, $\sup|G_n(z) - G(z)| \to 0$ a.s., as $n \to \infty$, so that on noticing that $|Q_i| \leqq 1$, $\forall i \geqq 1$ and $\Phi_{(1)}$ is a polynomial, we immediately conclude that the second term on the right hand side of (4.5), being bounded by $\max_{1 \leqq i \leqq n}|\Phi_{(1)}(n/(n+1)\,G_n(Z_i)) - \Phi_{(1)}(G(Z_i))|$, converges a.s. to 0 as $n \to \infty$. So, the proof is complete.

By virtue of Lemma 4.1 and Theorem 3.2, the BARE (Bahadur ARE) of $\{T_n\}$, based on the score function Φ, with respect to $\{T_n^*\}$, based on the score function Φ^*, is given by

$$e_1(\Phi, \Phi^*) = \varrho^2(\Psi, \Phi)/\varrho^2(\Psi, \Phi^*), \tag{4.7}$$

where $\Psi(u) = \pi^*(G^{-1}(u)) = 2\pi(G^{-1}(u)) - 1)$, $0 < u < 1$ and

$$\varrho^2(\Phi, \Psi) = \left(\int_0^1 \Phi(u)\,\Psi(u)\,du\right)^2 \Big/ \left(\int_0^1 \Phi^2(u)\,du\right)\left(\int_0^1 \Psi^2(u)\,du\right). \tag{4.8}$$

Thus, form the BARE point of view, the optimal choice of $\Phi(u)$ is $\Psi(u)$, $0 < u < 1$, and, as a result, $e_1(\Phi, \Psi) = \varrho^2(\Phi, \Psi)$ is always bounded by 1.

One may also consider a sequence of Pitman-type alternatives $\{H_n\}$ where H_n: $F(x, y) = F_{(n)}(x, y)$, $G = G_{(n)}$ and $\pi(z) = \pi_{(n)}(z)$ such that

$$\lim_{n\to\infty} G_{(n)}(z) = G(z) \quad \text{exists}, \quad \pi_{(n)}(z) = [1 + n^{-1/2}\gamma(z)]/2, \tag{4.9}$$

and

$$\int_{-\infty}^{\infty} |\gamma(z)|\,|\Phi(G(z))|\,dG(z) < \infty. \tag{4.10}$$

Then, we let $\Psi_0(u) = \gamma(G^{-1}(u))$, $0 < u < 1$, $A_0^2 = \int_0^1 \Psi_0^2(u)\,du$ and

$$(4.11) \qquad \varrho(\Psi_0, \Phi) = \left(\int_0^1 \Psi_0(u)\,\Phi(u)\,du\right)\Big/[AA_0],$$

where A is defined by (2.3). If we proceed as in Chapter 6 of Hájek and Šidák (1967), by using the powerful tool of contiguity, the asymptotic normality of $\{T_n\}$ under $\{H_n\}$ can then be established under standard conditions. In fact, as n increases, under H_n, $n^{-1/2}T_n/A_n$ is asymptotically normal with mean $A_0\varrho(\Psi_0, \Phi)$ and unit variance. Hence, the PARE (Pitman ARE) of $\{T_n\}$ with respect to $\{T_n^*\}$ is equal to

$$(4.12) \qquad e_2(\Phi, \Phi^*) = \varrho^2(\Psi_0, \Phi)/\varrho^2(\Psi_0, \Phi^*),$$

and in the light of the PARE, the optimal score function is $\Phi(u) = \Psi_0(u)$, $0 < u < 1$.

In general, e_1 and e_2, defined by (4.7) and (4.12) are not equal. Whereas (4.7) relates to a fixed alternative situation, (4.12) relates to a sequence of local alternatives. However, for local alternatives, in some situations, these two agree [viz., Section 6.]

Let us now consider the tests based on M_n^+ and M_n. Note that here also the null hypothesis distribution of M_n^+ or M_n is generated by the $2^n n!$ equally likely realizations of $(\mathbf{Q}_n, \mathbf{R}_n)$. [It may be remarked that given $\mathbf{Q}_n$ and $\mathbf{R}_n$, the vector $(T_1, \ldots, T_n)$ assumes a particular value dependent only on the score function and $(\mathbf{Q}_n, \mathbf{R}_n)$.] Thus, here, one can enumerate the distribution of M_n^+ or M_n by direct evaluation of all the $2^n n!$ equally likely realizations of $(\mathbf{Q}_n, \mathbf{R}_n)$; by this constitution, the statistics M_n and M_n^+ are distribution-free under H_0. The process of evaluating the exact null distribution of M_n^+ or M_n becomes prohibitively laborious as n increases. However, for large n, by (3.11), (3.12) and the fact that $\sup\{W(t) : t \in I\}$ and $\sup\{|W(t)| : t \in I\}$ have distributions $2\omega(x) - 1$ and $\omega^*(x)$, $x \geqq 0$, respectively, where ω and ω^* are defined by (2.10) (viz., Billingsley [3], p. 72, p. 79), we have

$$(4.13) \qquad P\{M_n^+ \leqq x \mid H_0\} \to 2\,\omega(x) - 1, \quad x \geqq 0,$$

$$(4.14) \qquad P\{M_n \leqq x \mid H_0\} \to \omega^*(x), \quad x \geqq 0.$$

Note that (2.9) follows then from (2.10), (4.13) and (4.14). Also, note that by Lemma 4.2, $n^{-1/2}M_n^+$ (or $n^{-1/2}M_n$) converges almost surely to $A^{-1}\mu(F, \Phi)$ (or $A^{-1}|\mu(F, \Phi)|$) as $n \to \infty$. Hence, by (2.9), we conclude that the tests based on M_n^+ and M_n are consistent against $\mu(F, \Phi)$ (or $|\mu(F, \Phi)|) > 0$, respectively. Further, note that for every $x > 0$, $2[1 - \omega(x)] \leqq 1 - \omega^*(x) \leqq 4[1 - \omega(x)]$, so that on using Mills ratio for $[1 - \omega(x)]/\omega'(x)$, we conclude that for either (4.13) or (4.14), the right hand side has a logarithm expressible as $-\frac{1}{2}x^2[1 + o(1)]$, as $x \to \infty$. Consequently, by Lemma 4.1 the *efficacy* (in the sense of Bahadur [1]) of either M_n^+ or M_n is

$$(4.16) \qquad \mu^2(F)/A^2 = \varrho^2(\Psi, \Phi)\left(\int_0^1 \Psi^2(u)\,du\right),$$

where $\Psi(u)$ is defined after (4.7). Thus, again the light of the BARE, the relative efficiency of two statistics, based on score functions Φ and Φ^*, is given by (4.7). Also, the choice of optimal score function relates to $\Phi = \Psi$. Further, the BARE of M_n^+ (or M_n) with respect to T_n (based on the same score function $\Phi(u)$, $0 < u < 1$) is equal to one. On the other hand, for the test based on T_n, the sample size n is prefixed, while for the test based on M_n^+ or M_n, the actual sample size (N) is a random variable (bounded from above by n), where $P(N < n) > 0$. Thus, there can be a reduction in the amount of sampling in the sequential procedure without any sacrifice of the BARE. In fact, by Lemma 4.1 and (2.9), it follows that for every $\varepsilon > 0$, as $n \to \infty$,

$$(4.17)\qquad P\{N_n > \varepsilon n \mid \mu(F) > 0\} \leq P\{n^{-1/2} T_{[n\varepsilon]}^{+}/A_n > M_{n,\alpha}^{+} \mid \mu(F) > 0\}$$
$$= P\{[n\varepsilon]^{-1} T_{[n\varepsilon]}/A_n > (n^{1/2}/[n\varepsilon])\, M_{n,\alpha}^{+} \mid \mu(F) > 0\} \to 0\,,$$

and a similar result follows for M_n. Consequently, when H_0 is not true, one may expect a considerable amount of reduction of the ASN of the first sequential procedure, without any loss of the BARE.

5. Properties of the tests based on D_n^+ and D_n

For the study of the second sequential procedure, we note that by Lemma 3.1, under H_0, $\mathbf{Q}(\mathbf{S}_n)$ assumes all possible 2^n realizations $\mathbf{j}_n \in \mathbf{J}_n$, each with the equal probability 2^{-n}. By a look at (2.11) and (2.13), we observe that the set of realizations of $(T_{n1}, \ldots, T_{nn})$, and hence, of D_n^+ or D_n, generated by the set of 2^n equally likely realizations of $\mathbf{Q}(\mathbf{S}_n)$, can be traced, and the exact null distribution can be computed. By virtue of this constitution, the tests based on D_n^+ and D_n are distribution-free.

As in Section 3, on denoting the standard Brownian motion (on $[0, 1]$) by $W = \{W(t),\ t \in [0, 1]\}$, we conclude from (3.20) that as n increases, D_n^+ (or D_n) has a distribution converging to that of $\sup_{0 \leq t \leq 1} W(t)$ (or $\sup_{0 \leq t \leq 1} |W(t)|$). Since, these distributions are given by the right hand sides of (4.13) and (4.14), respectively, we conclude that D_n^+ and M_n^+ (or D_n and M_n) have the same limiting (null) distribution. Hence, (2.16) again follows from (2.19).

With a view to studying the consistency and asymptotic efficiency of tests based on D_n^+ and D_n (for different score functions too), first, we define

$$(5.1)\qquad \tau(x, \Phi) = \int_{-\infty}^{x} \pi^*(z)\, \Phi(G(z))\, g(z)\, \mathrm{d}z\,, \quad x \text{ real}\,;$$

$$(5.2)\qquad \tau_+^0(\Phi) = \sup_x \tau(x, \Phi)\,, \quad \tau^0(\Phi) = \sup_x |\tau(x, \Phi)|\,.$$

The tests based on D_n^+ and D_n are consistent against alternatives for which $\tau_+^0(\Phi) > 0$ and $\tau^0(\Phi) > 0$, respectively. Also, we have the following lemma whose proof follows along the lines of the proof of Lemma 4.1 and is omitted.

Lemma 5.1. *Under the conditions of Theorem* 3.2, *as* $n \to \infty$,

$$n^{-1/2} D_n^+ \to \tau_+^0(\Phi)/A \quad \text{and} \quad n^{-1/2} D_n \to \tau^0(\Phi)/A\,, \quad \text{almost surely}\,. \tag{5.3}$$

Note that by (4.1), (5.1) and (5.2), $\tau_+^0(\Phi) \geqq \mu(F, \Phi)$ and $\tau^0(\Phi) \geqq |\mu(F, \Phi)|$; also, $\tau_+^0(\Phi)$ or $\tau^0(\Phi)$ can be different from 0 even when $\mu(F, \Phi) \neq 0$. Hence, the tests based on D_n^+ and D_n are consistent against a broader class of alternatives than in the case of T_n or $|T_n|$.

Now, the limiting null distributions of D_n^+ and D_n are given by the right hand side of (4.13) and (4.14), proceeding as in the case of M_n^+ and M_n and using Lemma 5.1, it follows that the *efficacy* of D_n^+ (or D_n), in the sense of Bahadur [1], is given by

$$(\tau_+^0(\Phi)/A)^2 \quad [\text{or } (\tau^0(\Phi)/A)^2]\,. \tag{5.4}$$

Let us now denote by

$$\Psi_x(u) = \begin{cases} \pi^*(G^{-1}(u))\,, & \text{if} \quad G^{-1}(u) \leqq x\,, \\ 0\,, & \text{otherwise}\,, \end{cases} \tag{5.5}$$

for every real x, and if we let

$$A_x^2(\Psi) = \int_0^1 \Psi_x^2(u)\, du\,, \tag{5.6}$$

then we have

$$(\tau_+^0(\Phi)/A)^2 = \sup_x \left[A_x^2(\Psi)\, \varrho^2(\Psi_x, \Phi)\right]\,, \tag{5.7}$$

where $\varrho(\Psi_x, \Phi)$ is defined as in (4.8) with Ψ being replaced by Ψ_x. Thus, $\varrho^2(\Psi_x, \Phi) \leqq \leqq 1$, $\forall$ real x. On the other hand, $A_x^2(\Psi)$ is non-decreasing, so that if we let $\Phi(u) = = \Psi_\infty(u) = \pi^*(G^{-1}(u))$, $0 < u < 1$, the right hand side of (5.7) is maximized; for any other $\Phi(u)$ (not proportional to $\Psi_\infty(u)$), the right hand side of (5.7) is bounded from above by $A_\infty^2(\Psi) = \int_0^1 \Psi_\infty^2(u)\, du$, so that

$$\sup_{\{\Phi\}} \left[(\tau_+^0(\Phi)/A)^2\right] = A_\infty^2(\Psi) = \int_{-\infty}^{\infty} [\pi^*(z)]^2\, dG(z)\,. \tag{5.8}$$

Hence, here also, maximizing the BARE leads us to the asymptotically optimal score function $\Psi_\infty(u) = \pi^*(G^{-1}(u))$, $0 < u < 1$. A similar result holds for D_n. In the next section, we shall study the optimal score function, in a little more detail, for some important cases.

6. Asymptotically optimal score functions

In some important special cases, $\pi^*(z)$ can be expressed in suitable simple forms and the optimal score functions $\Psi(u)$ and $\Psi_0(u)$, defined after (4.7) and in (4.10), can also be obtained in some simple forms.

6.1. *Stochastically independent components*

If X and Y are stochastically independent, $F(x, y) = F(x, \infty) F(\infty, y) = F_1(x) F_2(y)$, say. Let f_1 and f_2 be the density functions for F_1 and F_2 respectively. Then,

$$g(z) = f_1(z)(1 - F_2(z)) + f_2(z)(1 - F_1(z)), \tag{6.1}$$

$$\pi(z) = f_1(z)(1 - F_2(z))/g(z)$$

$$\pi^*(z) = 2\,\pi(z) - 1 = \{r_1(z) - r_2(z)\}/\{r_1(z) + r_2(z)\}, \tag{6.2}$$

where the hazard functions r_1 and r_2 are defined by

$$r_i(z) = f_i(z)/[1 - F_i(z)], \quad z \in E, \quad \text{for} \quad i = 1, 2. \tag{6.3}$$

Now, under H_0 in (1.1), $F_1 = F_2$, so that $r_1(z) = r_2(z)$ for all z. We consider two special cases where F_1 and F_2 may differ in locations or scales. First consider the model

$$F_2(x) = F_1(x - \theta), \quad -\infty < x < \infty, \quad \theta \text{ real}. \tag{6.4}$$

Then $r_2(z) = r_1(z - \theta)$, so that by (6.2),

$$\pi^*(z) = [r_1(z) - r_1(z - \theta)]/[r_1(z) + r_1(z - \theta)], \quad z \in E. \tag{6.5}$$

For small θ, (6.5) yields (whenever $r_1(z)$ is differentiable)

$$\pi^*(z) \simeq (\theta/2)\left[\frac{\mathrm{d}}{\mathrm{d}z}\log r_1(z)\right], \quad z \in E. \tag{6.6}$$

Thus, for local translation alternatives, the asymptotically optimal score function is

$$\Psi(u) = \Psi_\infty(u) = [(\mathrm{d}/\mathrm{d}z)\log r_1(z)]_{z = G^{-1}(u)}, \quad 0 < u < 1. \tag{6.7}$$

We may recall that the classical two-sample location problem (viz., Hájek and Šidák [10], p. 66), the locally most powerful rank test corresponds to the score function

$$\tilde{\Psi}(u) = -f_1'(F_1^{-1}(u))/f_1(F_1^{-1}(u)), \quad 0 < u < 1. \tag{6.8}$$

In general, (6.7) and (6.8) are quite different from each other, as may be verified by considering an exponential type of df where $\Psi_\infty(u)$ behaves like $\tilde{\Psi}(u)$ as $u \to 0$ but differently when $u \to 1$.

Let us now consider the scale model where

$$F_2(x) = F_1(x/\theta)\,, \quad \theta > 0 \quad \text{and} \quad H_0 : \theta = 1\,. \tag{6.9}$$

In this case, $r_2(z) = \theta^{-1}\, r_1(z/\theta)$, $z \in E$, so that

$$\pi^*(z) = [r_1(z) - \theta^{-1}\, r_1(z/\theta)]/[r_1(z) + \theta^{-1}\, r_1(z/\theta)]\,, \tag{6.10}$$

and hence for $\theta = 1 + \delta$, δ small, (6.10) tends to

$$(-\delta/2)\,\{1 + z(d/dz)\log r_1(z)\}\,. \tag{6.11}$$

Consequently, for local scale alternatives, the asymptotic optimal score function is

$$\Psi(u) = \Psi_\infty(u) = 1 + [z(d/dz)\log r_1(z)]_{z=G^{-1}(u)}\,, \quad 0 < u < 1\,. \tag{6.12}$$

Here also, the optimal score function is usually different from the case without the competing risk setup.

Now, (6.7) and (6.12) both relate to local shift and scale alternatives. In this sense, they are restricted to the case of θ being close to 0 (or 1). For fixed θ (different from 0 (or 1)), the optimal score functions are usually more complicated in form. If we let $\theta = \theta_n = n^{-1/2}\omega$, ω fixed (or for (6.9), $\theta_n = 1 + n^{-1/2}\omega$), then the optimal score functions according to the PARE criterion agree with (6.7) and (6.12). However, for fixed θ, the two scores are not generally the same.

6.2. *A bivariate exponential model of Marshall and Olkin*

Here, the survival functionn $\bar{F}(x, y) = 1 - F(x, \infty) - F(\infty, y) + F(x, y) = P\{X > x, Y > y\}$ is given by (for $x \geqq 0$, $y \geqq 0$)

$$\bar{F}(x, y) = \exp\{-\lambda_1 x - \lambda_2 y - \lambda_3 \max(x, y)\}\,, \quad \lambda_i > 0\,, \quad i = 1, 2, 3\,, \tag{6.13}$$

and H_0 in (1.1) reduces to H_0: $\lambda_1 = \lambda_2 = \lambda$ (unknown). A little computation leads us to

$$\pi(z) = \lambda_1/(\lambda_1 + \lambda_2) \quad \text{for all} \quad 0 < z < \infty\,, \tag{6.14}$$

so that for $\lambda_1 \neq \lambda_2$, $\pi^*(z) = (\lambda_1 - \lambda_2)/(\lambda_1 + \lambda_2)$ is a constant. This implies that for the model (6.1), the optimal score function is $\Psi(u) =$ constant, $0 < u < 1$, i.e., the sign statistics induced after (2.3), provide the optimal tests in each of the three cases. In this case, $\Psi_0(u)$, defined by (4.10), also remains a constant for all $0 \leqq u \leqq 1$, so that in the light of the PARE also, the sign statistics are the optimal. Thus, for both fixed and local alternatives, sign statistics are optimal. On the other hand, in the usual problem of bivariate interchangeability, without having the competing risks setup, the optimal score is different from above (see for example, Ferguson [8]).

6.3. *Gumbel's bivariate exponential model*

Here, we have

(6.15) $$F(x, y) = 1 - e^{-\lambda_1 x} - e^{-\lambda_2 y} + e^{-(\lambda_1 x + \lambda_2 y + \lambda_3 xy)}, \quad x, y \geqq 0,$$

where $\lambda_3 = \lambda_1 \lambda_2 \theta$ and $\lambda_1 > 0$, $\lambda_2 > 0$, $\theta > 0$. Since, here

(6.16) $$f(x, y) = (\lambda_1 + \lambda_3 y)(\lambda_2 + \lambda_3 x) \exp\{-(\lambda_1 x + \lambda_2 y + \lambda_3 xy)\},$$

it follows by routine steps that

(6.17) $$\pi^*(z) = \frac{\lambda_1 - \lambda_2}{\lambda_1 + \lambda_2 + 2\lambda_3 z} = \frac{(\lambda_1 - \lambda_2)}{\lambda_1 + \lambda_2 + 2\lambda_1 \lambda_2 \theta z}.$$

Also, by (6.16)–(6.17), we have

(6.18) $$1 - G(z) = P\{Z > z\} = P\{X > z,\ Y > z\} = \exp\{-z(\lambda_1 + \lambda_2 + \lambda_3)\}.$$

Hence,

(6.19) $$z = -(\lambda_1 + \lambda_2 + \lambda_3)^{-1} \log[1 - G(z)],$$

so that the optimal score function (in the sense of BARE) is

(6.20) $$\Psi_\infty(u) = \frac{(\lambda_1 - \lambda_2)}{\lambda_1 + \lambda_2 - \{2\lambda_1\lambda_2\theta/(\lambda_1 + \lambda_2 + \lambda_3)\} \log(1 - u)}, \quad 0 < u < 1.$$

For small u (i.e., u close to 0), $\Psi_\infty(u)$ behaves linearly in u while as $u \to 1$, $\Psi_\infty(u) \to 0$. Since λ_1, λ_2 and θ are not known (usually), the exact choice of $\Psi_\infty(u)$ is difficult, but one can advocate the use of a function which, for small u, decreases linearly in u and for $u \to 1$, goes to 0. Here also, we attach more weights to the smaller order statistics, so that for D_n^+ or D_n, a considerable amount of savings of time (and cost) of the experimentation is exptected.

Let us now consider the general bivariate model where X and $Y - \theta$ are interchangeable for some real θ. Thus, here, we have, under alternative,

(6.21) $$F(x, y) = F_0(x, y - \theta),$$

where $F_0(x, y) = F_0(y, x)$ for all (x, y). Then, under (6.28), for small θ, we have

(6.22) $$\pi^*(z) = \theta f_0(z, z) - \int_z^\infty [(\partial/\partial u) f_0(x, u)]_{u=z}\, dx] / \left[2 \int_0^\infty f_0(x, z)\, dx\right] + o(\theta),$$

where f_0 is the density function corresponding to F_0. (6.22) reduces to (6.6) when $f_0(z, z) = f_0^2(z)$, for all z. For specific f_0, such as the bivariate normal density, using the results in Johnson and Kotz (1975), it is possible to evaluate (6.22) and obtain the corresponding $\Psi(u)$. However, in general, this is quite complicated.

In the usual case of testing for bivariate interchangeability (without the competing risks setup), Ferguson [8] has compared the asymptotic efficiency of several rank statistics for some bivariate models with scale alternatives under consideration. In her study, the optimal score functions are complicated but the results throw some light on the performances of some standard procedures for such bivariate distributions. The present author believes that for such distributions, under the competing risks setup, the use of sign statistics ($\Phi(u) = 1$, $0 < u < 1$) or rank sum statistics ($\Phi(u) = u$, $0 < u < 1$) or the score function $\Phi(u) = a/(b - \log(1 - u))$, $0 < u < 1$ should lead to reasonably good tests.

References

[1] BAHADUR, R. R. (1960). Stochastic Comparison of Tests. *Ann. Statist.*, **31**, 276—95.

[2] BELL, C. B. - SMITH, P. J. (1961). "Some Nonparametric Tests for the Multivariate Goodness of Fit, Multisample, Independence and Symmetry Problems." Multivariate Analysis-II (Ed: P. R. Krishnaiah), New York: Academic Press, 3—24.

[3] BILLINGSLEY, P. (1968). "Convergence of Probability Measures". New York: John Wiley.

[4] BRINDLEY, E. C., JR. - THOMPSON, W. A., JR. (1972). Dependence and Aging Aspects of Multivariate Survival. *Journ. Amer. Statist. Assoc.*, **67**, 822—30.

[5] BROWN, B. M. (1971). Martingale Central Limit Theorems. *Ann. Math. Statist.*, **42**, 59—66.

[6] CHATTERJEE, S. K. and SEN, P. K. (1973). Nonparametric Testing under Progressive Censoring". Calcutta Statistical Association Bulletin, 13—50.

[7] DVORETZKY, A. Asymptotic Normality for Sums of Dependent Random Variables. *Proceedings of the Sixth Berkeley Symposium on Mathematical Statistics and Probability* (Ed: L. LeCam et al.), Volume 2, 513—534.

[8] FERGUSON, N. L. (1973). Comparison of Three Nonparametric Tests for Bivariate Interchangeability. *Australian Journal of Statistics*, **15**, 191—209.

[9] HÁJEK, J. (1968). Asymptotic Normality of Simple Linear Rank Statistics under Alternatives. *Ann. Math. Statist.*, **39**, 325—46.

[10] HÁJEK, J. - ŠIDÁK, Z. (1967). "Theory of Rank Tests". New York: Academic Press.

[11] HALPERIN, M. - WARE, J. (1974). Early Decision in a Censored Wilcoxon Two-Sample Test for Accumulating Survival Data. *Journ. Amer. Statist. Assoc.*, **69**, 414—422.

[12] HOEFFDING, W. (1974). On the Centering of a Simple Linear Rank Statistic. *Ann. Statist.*, **1**, 54—66.

[13] JOHNSON, N. J. - KOTZ, S. (1975). A Vector Multivariate Hazard Rate. *Journal of Multivariate Analysis*, 5, 53—66.

[14] LEE, L. - THOMPSON, W. A., JR. (1974). Reliability of Multiple Component Systems, *Technical Report No. 48, Mathematical Sciences*, University of Missouri, Columbia.

[15] OWEN, D. B. (1962). "Handbook of Statistical Tables". Reading, Mass: Addison-Wesley.

[16] PURI, M. L. - SEN, P. K. (1971). "Nonparametric Methods in Multivariate Analysis". New York. John Wiley.

[17] SEN, P. K., Nonparametric Tests for Multivariate Interchangeability. Part I: The Problem of Location and Scale in Bivariate Distributions, *Sankhyā A*, **29**, 351—71.
[18] SEN, P. K. - GHOSH, M. (1973). A law of iterated logarithm for one sample rank order statistics and an application, *Ann. Statist.*, **1**, 568—576.

THE UNIVERSITY OF NORTH CAROLINA, CHAPEL HILL, NORTH CAROLINA, U.S.A.

Received September 1975

CHARACTERIZATION OF DISTRIBUTIONS BY LARGE DEVIATION RATES*

by

J. SETHURAMAN

Let $X_1, X_2, \ldots$ be independent and identically distributed (i.i.d.) random variables with common distribution function (d.f.) F. Let $\bar{X}_n = (X_1 + X_2 + \ldots + X_n)/n$. Suppose that, for some a,

$$n^{-1} \log P_F(\bar{X}_n \geqq a) \to \varrho_F(a) \tag{1}$$

where $-\infty < \varrho_F(a) < 0$. We then say that the exponential rate of convergence to 0 of the probability of a deviation of a of the mean exists, or more simply that the large deviation rate $\varrho_F(a)$ exists for a deviation of a.

Suppose that F is a non-degenerate d.f. with $\int x \, dF(x) = 0$ and cumulant generating function (c.g.f.) $\Phi(t)$ which is finite for some $t > 0$. It is well known that (see Chernoff (1952), Bahadur (1971)) under these conditions, the large deviation rate $\varrho_F(a)$ exists for a in $(0, \delta)$ for some $\delta > 0$ and

$$\varrho_F(a) = \inf_{t>0} \{-ta + \Phi(t)\} . \tag{2}$$

Furthermore, the above infimum is attained at $t = t_a$ where t_a is an interior point of the set on which $\Phi(t)$ is finite, and

$$\Phi'(t_a) = a . \tag{3}$$

Conversely, let F be a d.f. with $\int x \, dF(x) \leqq 0$ and let the large deviation rate $\varrho_F(a)$ exist for some $a > 0$. Then F is non-degenerate and the c.g.f. $\Phi(t)$ of F exists for some $t > 0$ (see Baum, Katz and Read (1962)) and the conclusions stated in the earlier paragraph continue to hold.

We now state a theorem which shows that a d.f. is completely determined if its large deviation rate $\varrho_F(a)$ is known on a interval.

---

* Research sponsored by the United States Army Research Office, Durham, under Grant No. DAHCO4-74-G-1088. Reproduction in whole or in part is permitted for any purpose of the United States Government. AMOS Classification Numbers: 60F10, 62E10.

Theorem. *Let F and G be two d.f.'s satisfying* $\int x\, dF(x) \leqq 0$, $\int x\, dG(x) \leqq 0$. *Let the large deviation rates* $\varrho_F(a)$ *and* $\varrho_G(a)$ *exist for a in* $(\alpha, \alpha + \delta)$ *for some* $\alpha > 0$, $\delta > 0$, *and*

$$(4) \qquad \varrho_F(a) = \varrho_G(a) \quad \text{for} \quad a \quad \text{in} \quad (\alpha, \alpha + \delta)\,.$$

Then $F = G$.

Proof. From the discussion in the paragraphs preceding this theorem, the c.g.f. $\Phi(\Psi)$ of $F(G)$ exists. Also, $F(G)$ is non-degenerate and thus $\Phi'' > 0$ $(\Psi'' > 0)$ and $\Phi'(\Psi')$ is strictly increasing on its domain. Furthermore, for each a in $(\alpha, \alpha + \delta)$ there is a $t_a(t_a^*)$ in the domain of $\Phi'(\Psi')$ such that

$$(5) \qquad \Phi'(t_a) = \Psi'(t_a^*) = a$$

and

$$(6) \qquad \varrho_F(a) = at_a - \Phi(t_a) = \varrho_G(a) = at_a^* - \Psi(t_a^*)\,.$$

Let $z(\theta) = (\Psi')^{-1}(\theta)$ and write x for t_a. Solving for t_a^* from (5) and substituting in (6) yields

$$x\,\Phi'(x) - \Phi(x) = \Phi'(x)\, z(\Phi'(x)) - \Psi(z(\Phi'(x)))$$

for x in some interval. Differentiating with respect to x, we obtain

$$x\,\Phi''(x) = z(\Phi'(x))\,\Phi''(x)\,.$$

Since $\Phi'' > 0$, we have $z(\Phi'(x)) = x$ or $\Phi'(x) = \Psi'(x)$ for x in an interval. Noticing that Φ' and Ψ' are restrictions of analytic functions to an interval on the real line, we conclude that $\Phi' = \Psi'$ on their domain and thus $\Phi = \Psi$ and $F = G$. ||

Remark 1. A direct and simple proof of this theorem can be given if we strengthen the hypothesis (4) to read $\varrho_F(a) = \varrho_G(a)$ for all a in the common domain of ϱ_F and ϱ_G. For this we need only to note that $-\varrho_F(a)\,(-\varrho_G(a))$ is the conjugate convex function of the convex function $\Phi(\Psi)$ and appeal to the one-to-one correspondence between convex functions and their conjugates (see Fenchel (1949)).

Remark 2. Even if $\varrho_F(a) = \varrho_G(a)$ for infinitely many a's, it does not follow that $F = G$ as can be checked with the d.f.'s $F(x) = \int_{-\infty}^{x} n(t)\, dt$, and $G(x) = \int_{-\infty}^{x} n(t)\,.$ $.\,(1 + \sin t)\, dt$ where $n(t) = \exp(-t^2/2)/\sqrt{(2\pi)}$.

The Central limit theorem for i.i.d. random variables with common d.f. F states that the probability of ordinary deviations of the mean converges to a limit which involves only the mean and variance of F. Likewise, results on rates of convergence of probabilities of moderate deviations of the mean (Rubin and Sethuraman (1965) and Michel (1974)) show that these rates depend only on a finite number of moments of F. Thus the limits or rates of probabilities of ordinary and moderate deviations do not determine F. In contrast to this type of robustness, it is interesting to note that the large deviation rate completely determines F.

References

BAHADUR, R. R. (1971). Some Limit Theorems in Statistics. Regional Conference *Series in Applied Mathematics*. S.I.A.M. Philadelphia, PA.

BAUM, L. E. - KATZ, M. - READ, R. R. (1962). Exponential Convergence Rates for the Law of Large Numbers. *Trans. Amer. Math. Soc.*, **102,** 187—199.

CHERNOFF, H. (1952). A Measure of Asymptotic Efficiency of Tests of a Hypothesis Based on the Sum of Observations. *Ann. Math. Statist.*, **23,** 493—507.

FENCHEL, W. (1949). On Conjugate Convex Functions. *Canadian Jour. Math.*, **1,** 73—77.

MICHEL, R. (1974). Results on Probabilities of Moderate Deviations. *Ann. Prob.*, **2,** 349—353.

RUBIN, H. - SETHURAMAN, J. (1965). Probabilities of Moderate Deviations. *Sankhya*, A, **27,** 325—346.

THE UNIVERSITY OF MICHIGAN, ANN ARBOR, MICHIGAN, U.S.A.

Received February 1976

MONTE CARLO COMPARISONS OF SOME RANK TESTS OPTIMAL FOR UNIFORM DISTRIBUTIONS

by

ZBYNĚK ŠIDÁK (1), AND STANISLAV HOJEK (2)

1. Problem

Let us consider two random samples $X_1, \ldots, X_m$ and $Y_1, \ldots, Y_n$ from the densities f_1 and f_2, respectively. (To be in accordance with the used tables, suppose that $m \leqq n$.) We are testing the null hypothesis H_0 that $f_1 \equiv f_2$ but otherwise they may be arbitrary, against the alternative hypothesis of shift in location which is given by $f_1(x) = f(x - \Delta), f_2(x) = f(x)$, where $\Delta > 0$, or $\Delta < 0$ (one-sided alternatives), or $\Delta \neq 0$ (two-sided alternative).

We shall deal with some rank tests based on exceeding observations. To define the necessary test statistics, we first find the quantities A, and B', equal to the number of observations among $X_1, \ldots, X_m$ larger than $\max_{1 \leqq j \leqq n} Y_j$, or smaller than $\min_{1 \leqq j \leqq n} Y_j$, respectively, and the quantities A', and B, equal to the number of observations among $Y_1, \ldots, Y_n$ larger than $\max_{1 \leqq i \leqq m} X_i$, or smaller than $\min_{1 \leqq i \leqq m} X_i$, respectively. (Of course, only one of the quantities A, A' is positive, while the other is zero, and similarly for B, B'.) Now, different functions of A, A', B, B' may be used for testing the hypothesis H_0.

J. Hájek found that some of these functions generate locally most powerful rank tests of H_0 against a one-sided shift in the location parameter Δ, if the underlying density f is uniform over an arbitrary interval (a, b).

In particular, he found that the statistic $A + B$ generates the locally most powerful rank test for the above problem in some neighbourhood of $\Delta = b - a$, i.e. in some interval $b - a - \varepsilon < \Delta < b - a$. (Cf. [2], Problem II.13.) This test has been suggested earlier, on the basis of practical considerations, by Šidák and Vondráček [5] and by Tukey [6]. The two-sided version of this test uses the "symmetrized" statistic

$$H = A + B - A' - B' .$$

The test based on H is suitable for two-sided alternatives $\Delta \neq 0$, and it is also more robust, say against possible differences in variance. It has been introduced and investigated by Haga [1], and therefore in [2] it has been called the Haga test. Here we shall call it briefly the H-test.

J. Hájek further found that the statistic $\min(A, B)$ generates the locally most powerful rank test for the above problem with the uniform density in some neighbourhood of $\Delta = 0$, i.e. in some interval $0 < \Delta < \delta$. (Cf. again [2], Problem II.13.) For similar reasons as above, J. Hájek has suggested in [2], § III.1.2, the use of the "symmetrized" test statistic

$$E = \min(A, B) - \min(A', B'),$$

suitable for testing against two-sided alternatives; the corresponding test has been called by him the E-test.

Thus we see that the tests based on the statistics $A + B$ and $\min(A, B)$ have certain good properties for testing the one-sided shift in location Δ of uniform distributions. In particular, $\min(A, B)$ is better than $A + B$ for Δ close to 0, and $A + B$ is better than $\min(A, B)$ for Δ close to $b - a$.

The aim of the present investigation was to obtain some insight into the problem whether these properties continue to hold also for the two-sided "symmetrized" tests, the H-test and the E-test; in particular, whether E is better than H in some neighbourhood of $\Delta = 0$, whether H is better than E in some neighbourhood of $\Delta = b - a$, and, if so, how wide are these neighbourhoods and how large are the differences in the power functions. (We decided to investigate the "symmetrized" tests because we think that two-sided tests are much more often used in practice, and because the critical values and the precise significance levels of the "symmetrized" H-test and E-test are available in [3] and [4] whereas those of "unsymmetrized" $A + B$ and $\min(A, B)$ seem to be unavailable.)

Furthermore, since the Wilcoxon test is perhaps the most widely used rank test for the two-sample problem we decided to compare the H-test and the E-test also with the Wilcoxon test in the present case of uniform distributions; this latter test will be called briefly the W-test.

2. Monte Carlo computations

In the sequel we work with the uniform density f on $(0, 1)$. The power functions of the H, E, and W-tests were estimated by a Monte Carlo method for the following cases: sample sizes $m = n = 10$; $m = n = 20$; $m = 10$, $n = 20$; significance levels $\alpha = 1\%$ and $\alpha = 5\%$; shifts $\Delta = 0.00; 0.05; 0.10; \ldots$ etc. until such Δ where the power functions were equal, or almost equal, to 1.

To describe our procedure in more detail, for each pair m, n pseudo-random numbers $X_1, \ldots, X_m$, $Y_1, \ldots, Y_n$ from $(0, 1)$ were generated, and then the three tests were performed for the samples $X_1 + \Delta, \ldots, X_m + \Delta$ and $Y_1, \ldots, Y_n$. However, in non-randomized rank tests the significance levels can be fixed precisely at $\alpha = 1\%$ or 5% only exceptionally, which makes a direct comparison of these tests impossible. Therefore, in order to achieve the prescribed α for all three tests, we had to use randomized tests; namely, for each realization (labelled i-th, say) of the two random samples we counted the probability P_i of rejecting H_0 as follows.

E.g., let h be a possible critical value of the statistic H such that $P\{|H| \geqq h\} = \alpha_1/100$, $P\{|H| \geqq h - 1\} = \alpha_2/100$, where $\alpha_1 \leqq \alpha < \alpha_2$. Then, if the actual value of $|H|$ in the i-th realization is $\geqq h$ we set $P_i = 1$; if it is $= h - 1$ we set $P_i = (\alpha - \alpha_1)/(\alpha_2 - \alpha_1)$; and if it is $< h - 1$ we set $P_i = 0$. Similar procedure is applied for E and W. Critical values and corresponding exact significance levels for H were taken from [3], those for E from [4], and those for W from [7]. To give the reader some idea of how close to $\alpha = 1\%$ or 5% are the possible levels α_1 and α_2, they are displayed in Table 1 for all cases studied here. (The critical values are not displayed because we think they are uninteresting for the conclusions.)

We used 1000 realizations of the samples, so that the estimates of the power functions in Tables 2, 3, 4 are, in a formula, equal to $\sum_{i=1}^{1000} P_i/1000$.

All computations were done on the calculator Hewlett-Packard 9830A.

3. Conclusions

The estimated power functions of the H, E, and W-tests are displayed in Tables 2, 3, and 4. Though the present study is somewhat limited, by inspection of these Tables we may perhaps draw the following tentative conclusions:

a) The W-test seems to be never better than the H and E-tests. Therefore the H-test or the E-test should be preferred to the W-test if the underlying distribution is uniform or close to it.

b) Which of the two tests, H or E, is better depends generally strongly on m, n, α, Δ.

c) The E-test seems to be better than the H-test in a comparatively small neighbourhood of $\Delta = 0$, say for $|\Delta| < 0.10, 0.15, 0.20$ or so; for large m, n this neighbourhood seems to be narrower, for small m, n it seems to be wider. Thus the E-test should be preferred to the H-test if it may be expected that $|\Delta|$ is rather small.

d) The H-test seems to be better than the E-test in a comparatively wide neighbourhood of $\Delta = 1$, say for $|\Delta| > 0.30, 0.40$ or so; for large m, n this neighbourhood seems to be wider, for small m, n it seems to be narrower. Thus the H-test should be preferred to the E-test if it may be expected that $|\Delta|$ is large, even if mildly. (Of course, if $|\Delta|$ is rather large, i.e. close to 1, the power functions of both the E-test and the H-test are almost equal to 1, so that possible differences between them are uninteresting from the practical point of view.)

e) The "turning" point of the shift Δ such that for smaller Δ the E-test is better and for larger Δ the H-test is better seems to lie somewhere between 0.00 and 0.40.

f) If we do not know anything about the possible magnitude of $|\Delta|$, it is perhaps more recommedable to use the H-test than the E-test for the following reasons: first, the region where H is better than E is wider; second, possible superiority of H over E tends to be more expressed than possible superiority of E over H.

Table 1. Exact significance levels (in per cents)

	α	H		E		W	
		α_1	α_2	α_1	α_2	α_1	α_2
$m = 10$	1%	.55	1.11	1.00	3.72	.90	1.14
$n = 10$	5%	4.29	8.17	3.72	13.93	4.32	5.24
$m = 20$	1%	.74	1.43	.87	3.39	.94	1.04
$n = 20$	5%	2.72	5.07	3.39	13.17	4.90	5.24
$m = 10$	1%	.64	1.13	.50	2.30	.96	1.10
$n = 20$	5%	3.29	5.45	2.30	10.40	4.90	5.46

Table 2. Power functions for $m = 10$, $n = 10$

Δ	$\alpha = 1\%$			$\alpha = 5\%$		
	H	E	W	H	E	W
.00	.010	.009	.017	.051	.053	.049
.05	.013	.010	.017	.056	.065	.053
.10	.019	.019	.022	.113	.121	.095
.15	.057	.057	.040	.222	.232	.163
.20	.123	.131	.083	.357	.356	.244
.25	.211	.228	.144	.523	.503	.371
.30	.330	.339	.232	.668	.644	.505
.35	.481	.493	.354	.803	.776	.637
.40	.632	.626	.502	.897	.863	.745
.45	.747	.736	.615	.955	.917	.841
.50	.845	.837	.719	.982	.951	.911
.55	.923	.904	.828	.991	.975	.958
.60	.968	.954	.904	.996	.991	.982
.65	.987	.972	.958	.998	.997	.992
.70	.996	.991	.982	1.000	.999	.996
.75	.999	1.000	.993	1.000	1.000	1.000

Table 3. Power functions for $m = 20$, $n = 20$

Δ	$\alpha = 1\%$			$\alpha = 5\%$		
	H	*E*	*W*	*H*	*E*	*W*
.00	.006	.005	.008	.048	.048	.051
.05	.032	.032	.023	.145	.140	.089
.10	.127	.132	.068	.391	.374	.188
.15	.328	.341	.151	.679	.632	.346
.20	.595	.585	.287	.887	.837	.526
.25	.823	.798	.466	.972	.945	.706
.30	.933	.915	.632	.995	.987	.847
.35	.981	.976	.788	1.000	.998	.921
.40	.997	.991	.899	1.000	1.000	.970
.45	.999	.997	.956	1.000	1.000	.989
.50	1.000	.999	.985	1.000	1.000	.997
.55	1.000	1.000	.996	1.000	1.000	.999

Table 4. Power functions for $m = 10$, $n = 20$

Δ	$\alpha = 1\%$			$\alpha = 5\%$		
	H	*E*	*W*	*H*	*E*	*W*
.00	.007	.005	.009	.053	.049	.054
.05	.018	.023	.016	.078	.098	.076
.10	.043	.062	.042	.175	.207	.142
.15	.105	.147	.082	.328	.359	.237
.20	.217	.258	.160	.526	.530	.367
.25	.380	.397	.274	.716	.671	.508
.30	.545	.532	.392	.865	.784	.655
.35	.708	.659	.542	.943	.871	.783
.40	.848	.771	.684	.983	.926	.894
.45	.946	.857	.807	.996	.963	.949
.50	.977	.914	.907	1.000	.982	.972
.55	.994	.952	.954	1.000	.987	.992
.60	.998	.973	.978	1.000	.996	.995
.65	1.000	.985	.992	1.000	.998	1.000

References

[1] HAGA, T. (1959/60). A two-sample rank test on location. *Ann. Inst. Statist. Math.*, **11**, 211—219.

[2] HÁJEK, J. - ŠIDÁK, Z. (1967). "Theory of Rank Tests". Academia, Prague and Academic Press, New York—London.

[3] HOJEK, S. (1978). Tables for the two-sample Haga test of location. *Aplikace matematiky*, **23**, 237—247.

[4] ŠIDÁK, Z. (1977). Tables for the two-sample location E-test based on exceeding observations. *Aplikace matematiky*, **22**, 166—175.

[5] ŠIDÁK, Z. - VONDRÁČEK, J. (1957). A simple non-parametric test of difference in location of two populations. (In Czech.) *Aplikace matematiky*, **2**, 215—221.

[6] TUKEY, J. W. (1959). A quick, compact, two sample test to Duckworth's specifications. *Technometrics*, **1**, 31—48.

[7] WILCOXON, F. - KATTI, S. K. - WILCOX, A. (1963). Critical values and probability levels for the Wilcoxon rank sum test and the Wilcoxon signed rank test. *American Cyanamid Company and The Florida State University*, August.

(1), (2) MATHEMATICAL INSTITUTE OF CZECHOSLOVAK ACADEMY OF SCIENCES, PRAGUE, CZECHOSLOVAKIA

Received January 1976

SIMPLICIAL MEASURES*

by

JOSEF ŠTĚPÁN

1. Introduction

A probability measure μ on a compact convex set X is called simplicial if it is an extreme point in the set of all probability measures which have the same barycenter as μ. It has been proved (E. M. Alfsen (1971), Ch. F. Skau (1969)) that if $X \subset R^n$ then μ is simplicial if and only if it is supported by a set of (at most $n + 1$) affinely independent points, i.e., if it is a boundary measure on some regular simplex in X.

The basic tool for the study of the concept of a simplicial measure in infinite-dimensional case is a theorem, due to Douglas (1964), which states that a measure μ is simplicial if and only if the space $A(X)$ of continuous affine functions on X is dense in $L_1(\mu)$. We will restrict ourselves to metrizable compacts X and show that the convex support $S_\mu = \overline{\text{co}}\,(\text{support}\ \mu)$ of any simplicial measure μ can be expressed as $\overline{\bigcup_1^\infty K_n}$, where the K_n's are regular simplexes such that K_n is a face in K_{n+1} and μ is supported by the set of extreme points of $\bigcup_1^\infty K_n$. Obviously, this result generalizes that of Alfsen (1971) from finite to infinite dimensional case. In section 3 we introduce the concept of a μ-simplex and employing this concept we obtain (in section 4) a necessary and sufficient condition for a probability measure to be simplicial.

Our reasoning depends heavily on the metrizability of underlying compact convex set X, since separability of $C(X)$ plays a fundamental role in all the constructions.

2. Notations

Let X be a *metrizable compact convex set* in a locally convex Hausdorff topological vector space E (over R), and let $M(X)$, $M^+(X)$, $M^1(X)$ denote the spaces of finite

---

* AMS 1970 subject classification. Primary 60B05; Secondary 52A05. Key words and phrases. Compact convex set, simplex, regular simplex, simplicial measure, μ — simplex.

Borel measures, non-negative finite Borel measures and Borel probability measures on X, respectively.

For any Borel set $B \subset X$ put

$$M(B) = \{\nu \in M(X) : |\nu| (B^c) = 0\},$$

$$M^+(B) = M^+(X) \cap M(B),$$

$$M^1(B) = M^1(X) \cap M(B),$$

where B^c denotes the complement to X.

For any element ν of $M(X)$ we shall denote by $\nu \mid B$ the restriction of ν to B, i.e., for any Borel set $C \subset X$ we put

$$(\nu \mid B)(C) = \nu(BC);$$

obviously, $\nu \mid B \in M(B)$.

As usual, $C(X)$ denotes the space of bounded continuous real valued functions on X. For $B \subset X$ a convex set we shall agree to write $A(B)$ for the space of real valued continuous affine functions on B. Further, write[1]

$$A(X)^\perp = \{\nu \in M(X) : \nu(a) = 0 \quad \text{for all} \quad a \in A(X)\}.$$

If $\nu \in M^1(X)$, we denote by $b\nu$ the unique point in X for which

$$a(b\nu) = \nu(a), \quad \text{all} \quad a \in A(X),$$

i.e., $b\nu$ is the barycenter of ν.

Note that $b : M^1(X) \to X$ is affine and continuous provided the space $M^1(X)$ is considered in the weak* topology.

Consider a compact convex set $K \subset X$ and write

$$\nu(f) = \nu(f \,.\, I_K),$$

where $f \in C(K)$, $\nu \in M^1(K)$ and I_K denotes the indicator of K; clearly $b\nu \in K$.

For later reference note that

$$\text{(1)} \qquad \nu(a) = a(b\nu), \quad \text{all} \quad \nu \in M^1(K), \quad a \in A(K).$$

Indeed, by 23.1.6, Semadeni (1971), there is an $\tilde{a} \in A(X)$ such that $\|a - \tilde{a}\|_K < \varepsilon$[2]. Hence

$$|\nu(a) - \nu(\tilde{a})| < \varepsilon, \quad |a(b\nu) - \tilde{a}(b\nu)| < \varepsilon,$$

---

1 Throughout this note we write $\nu(f) = \int_X f \, d\nu$.

2 We write $\|f\|_B = \sup\{|f(x)|, x \in B\}$ for $B \subset X$ and $f : B \to R$.

so that

$$|\nu(a) - a(b\nu)| < 2\varepsilon,$$

which proves (1).

At the end of this section we will summarize briefly the parts of Choquet's theory which will be needed later. For detailed references see R. R. Phelps (1966), Z. Semadeni (1971), E. M. Alfsen (1971).

Denote by ex K the set of extreme points of a compact convex set $K \subset X$. The metrizability of X insures that ex K is a Borel set in X.

We shall say that $\nu \in M(X)$ is a *boundary measure* on K if $\nu \in M(\text{ex}\, K)$. It is known that ν is boundary on K if and only if $|\nu|$ is a maximal element in $M^+(K)$ with respect to the ordering of Choquet.

By Choquet's Theorem we have

$$b : M^1(\text{ex}\, K) \to K \tag{2}$$

is a surjection and K is said to be *a simplex* if the map (2) is a bijection.

3. Simplicial measures and μ-simplexes

Put

$$M_x^1(X) = \{\nu \in M^1(X) : b\nu = x\}, \quad x \in X$$

and note that $M_x^1(X)$ being a nonempty compact convex set has an extreme point.

We define $\mu \in M^1(X)$ to be a simplicial measure if $\mu \in \text{ex}\, M_{b\mu}^1(X)$.

The following theorem, which is due to Douglas (1964), gives a useful characterization of simplicial measures.

Theorem 1. *Let $\mu \in M^1(X)$. Then the following conditions are equivalent:*

(i) μ is simplicial;

(ii) $A(X)$ is dense in $L_1(\mu)$.

For the sake of completeness we present a short proof of the theorem.

Proof. (Ch. F. Skau (1969)). If $A(X)$ is not dense in $L_1(\mu)$ we can find a function $f \in L_\infty(\mu)$, $f \neq 0$, such that

$$\mu(a \,.\, f) = 0, \quad \text{all} \quad a \in A(X). \tag{3}$$

We may assume that

$$0 < \text{sup vrai}\, |f| < 1 \tag{4}$$

and to define the measure $\nu \in M(X)$ by $d\nu = f \,.\, d\mu$. By (4) we get that the measures

$$\mu_1 = \mu + \nu\,, \quad \mu_2 = \mu - \nu$$

have the following properties: $\mu_1 \neq \mu_2$, $\mu_i > 0$, $i = 1, 2$. Without loss of generality we may assume that $\mu_i \in M^1(X)$ and it follows from (3) that $b\mu_i = b\mu$, $i = 1, 2$. Since $\mu = \frac{1}{2}(\mu_1 + \mu_2)$, μ is not simplicial.

If μ is not simplicial there is a $\mu_1 \in M^1_{b\mu}(X)$ such that $\mu_1 \neq \mu$ and $\mu_1 \leqq 2\mu$. Hence

$$0 \neq 1 - f \in L_\infty(\mu)\,, \quad \text{where} \quad f = \frac{d\mu_1}{d\mu}$$

and $\mu[(1 - f) \,.\, a] = 0$ for all $a \in A(X)$, i.e., $A(X)$ is not dense in $L_1(\mu)$. □

Now, consider a $\mu \in M^1(X)$ and put

$$H(\mu) = \left\{\nu \in M(X) : \nu \ll \mu \quad \text{and} \quad \frac{d\nu}{d\mu} \in L_\infty(\mu)\right\}.$$

Since we have

(5) $$\nu(|a - f|) = \mu\left(|a - f|\,\frac{d\nu}{d\mu}\right) \leqq \sup \text{vrai}\left|\frac{d\nu}{d\mu}\right| . \,\mu(|a - f|)\,,$$

$$\text{all} \quad a \in A(X)\,, \quad f \in L_1(\mu)\,,$$

we conclude from Theorem 1 that the simpliciality of μ implies the simpliciality of $\nu \in M_1(X) \cap H(\mu)$.

Now, we can introduce the concept of μ-simplex which will be our basic tool when describing the structure of the support of a simplicial measure.

Let $\mu \in M^1(X)$. A compact convex set $K \subset X$ is said to be a *μ-simplex* if for

(6) $$\lambda \in M(X) \quad \text{satisfying} \quad \lambda \mid K \in M(\text{ex}\, K)\,, \quad \lambda \mid K^c \in H(\mu)\,,$$

$$\lambda \in A(X)^\perp \quad \text{implies that} \quad \lambda \mid K = 0\,.$$

It will be convenient to introduce the notation

$$B_1(\mu, C) = b(M^1(C) \cap H(\mu))\,, \quad C \subset X \quad \text{a Borel set}\,, \quad \mu \in M^1(X)\,,$$

and we shall agree to write B_1 for $B_1(\mu, X)$ when there is no danger of confusion as to the measure μ under consideration. Note that $B_1(\mu, C)$ is a convex set.

The following theorem gives a necessary and sufficient condition for a simplex $K \subset X$ to be a μ-simplex. Recall that two convex subsets K, B of X are said to be *affinely independent* if every point y of their convex hull can be expressed by a unique convex combination:

$$y = \alpha k + (1 - \alpha)\, x\,, \quad k \in K\,, \quad x \in B\,, \quad \alpha \in [0, 1]\,.$$

Theorem 2. *Let $\mu \in M^1(X)$. A compact set $K \subset X$ is μ-simplex if and only if the following two conditions hold:*

(*i*) *K is a simplex*

(*ii*) *the convex sets K and $B_1(\mu, K^c)$ are affinely independent.*

It is worth noting that condition (ii) implies that

$$\operatorname{aff} K \cap \operatorname{aff} B_1(\mu, K^c) = \emptyset$$

and that K is a μ-face in the following sense: if $\lambda \in M^1(X) \cap H(\mu)$ and $b\lambda \in K$ then $\lambda \in M(K)$.

Proof of Theorem 2

First, assume that K is a μ-simplex. Take $\lambda_1, \lambda_2 \in M^1(\operatorname{ex} K)$ such that $b\lambda_1 = b\lambda_2$. Obviously, $\lambda = \lambda_1 - \lambda_2$ satisfies condition (6) and $\lambda \in A(X)^{\perp}$. Hence $\lambda = \lambda \mid K = 0$, i.e., $\lambda_1 = \lambda_2$, and therefore K is a simplex.

Further, let

$$\alpha k_1 + (1 - \alpha)\, x_1 = \beta k_2 + (1 - \beta)\, x_2 \quad k_i \in K\,, \quad x_i \in B_1(\mu, K^c)\,,$$

$$i = 1, 2\,; \quad \alpha, \beta \geqq 0\,.$$

By Choquet's theorem and by the definition of $B_1(\mu, K^c)$ we may write

$$k_i = b\nu_i \quad \text{for some} \quad \nu_i \in M^1(\operatorname{ex} K) \quad \text{and} \quad x_i = b\eta_i$$

$$\text{for some} \quad \eta_i \in M^1(K^c) \cap H(\mu)\,, \quad i = 1, 2\,.$$

Clearly, $\lambda = \alpha\nu_1 + (1 - \alpha)\,\eta_1 - \beta\nu_2 - (1 - \beta)\,\eta_2$ satisfies (6) and therefore $\alpha\nu_1 - \beta\nu_2 = \lambda \mid K = 0$. Hence $\alpha = \beta$, $k_1 = k_2$, $x_1 = x_2$ and condition (ii) is verified.

Now assume that (i) and (ii) hold and take $\lambda \in A(X)^{\perp}$ satisfying condition (6). As $\lambda^+(1) = \lambda^-(1)$ we may assume without loss of generality that $\lambda^+, \lambda^- \in M^1(X)$. Further assume, also without loss of generality, that

$$\lambda^-(K) \in (0, 1)\,, \quad \lambda^+(K) \in (0, 1)\,.$$

We may write

$$\lambda^+ = \lambda^+(K)\,\frac{\lambda^+ \mid K}{\lambda^+(K)} + \lambda^+(K^c)\,\frac{\lambda^+ \mid K^c}{\lambda^+(K^c)}\,,$$

$$\lambda^- = \lambda^-(K)\,\frac{\lambda^- \mid K}{\lambda^-(K)} + \lambda^-(K^c)\,\frac{\lambda^- \mid K^c}{\lambda^-(K^c)}\,.$$

As $b\lambda^+ = b\lambda^-$ we arrive at

$$\lambda^+(K)\, k^+ + (1 - \lambda^+(K))\, x^+ = \lambda^-(K)\, k^- + (1 - \lambda^-(K))\, x^-\,,$$

where

$$k^+ = b\left(\frac{\lambda^+ \mid K}{\lambda^+(K)}\right), \quad k^- = b\left(\frac{\lambda^- \mid K}{\lambda^-(K)}\right),$$

$$x^+ = b\left(\frac{\lambda^+ \mid K^c}{\lambda^+(K^c)}\right), \quad x^- = b\left(\frac{\lambda^- \mid K^c}{\lambda^-(K^c)}\right).$$

Clearly, $k^+, k^- \in K$ and $x^+, x^- \in B_1(\mu, K^c)$. Hence, by (ii) we have $k^+ = k^-$ and $\lambda^+(K) = \lambda^-(K)$. It follows from (i) and from the fact that $\lambda^+ \mid K$ and $\lambda^- \mid K$ are boundary measures on K that $\lambda^+ \mid K = \lambda^- \mid K$, i.e., $\lambda \mid K = 0$. Thus, K is a μ-simplex and the proof is finished.

We conclude this section by the following simple observation:

(7) *Let $\mu \in M^1(X)$ be a simplicial measure and v_1, v_2 elements of $M^1(X) \cap H(\mu)$ such that $bv_1 = bv_2$. Then $v_1 = v_2$.*

Actually, by (5) we have

$$v_i(|a_n - f|) \to 0, \quad n \to \infty, \quad i = 1, 2 \quad \text{whenever}$$

$$\mu(|a_n - f|) \to 0, \quad n \to \infty, \quad a_n \in A(X), \quad f \in C(X).$$

Since μ is a simplicial measure we get by Theorem 1 that $v_1(f) = v_2(f)$ for all $f \in C(X)$. Hence (7) holds.

It is worth noting that our observation (7) leads immediately to a simple proof of the following:

Theorem (Ch. F. Skau (1969)). *If $\mu \in M^1(X)$ is a simplicial measure, then μ may be approximated in the weak* topology by a discrete measure $v = \sum_{i=1}^{n} \alpha_i \varepsilon_{x_i}$ such that $\{x_1, \ldots, x_n\} \subset X$ are affinely independent and $b\mu = bv$.*

Proof. Take $f \in C(X)$, $\varepsilon > 0$ and cover X by convex open sets C_i, i.e., $X = \bigcup_{i=1}^{n} C_i$, such that the oscillation of the function f is smaller than ε on each of the C_i's. Putting

$$B_i = C_i - \bigcup_{j<i} C_j, \quad i = 1, 2, \ldots, n, \quad J = \{i : \mu(B_i) > 0\},$$

$$\alpha_i = \mu(B_i), \quad x_i = b\left(\frac{\mu \mid B_i}{\alpha_i}\right), \quad v = \sum_{i \in J} \alpha_i \varepsilon_{x_i}, \quad i \in J,$$

we can see that $|v(f) - \mu(f)| < \varepsilon$.

To conclude the proof assume that $\{x_j, j \in J\}$ are affinely dependent, i.e. we may write

$$\sum_{j \in J} \beta_j x_j = 0 \quad \text{where} \quad \sum_{j \in J} \beta_j = 0 \quad \text{and not all} \quad \beta_j = 0.$$

Obviously,

$$\sum_{j\in J^+} \delta_j x_j = \sum_{j\in J^-} \delta_j x_j\,, \quad \text{where} \quad \delta_j = \frac{\beta_j}{\sum_{J\in J^-} \beta_j} \quad \text{for} \quad j \in J^+\,,$$

$$\delta_j = \frac{\beta_j}{\sum_{j\in J^-} \beta_j} \quad \text{for} \quad j \in J^- \quad \text{and} \quad J^+ = \{j \in J : \beta_j > 0\}\,, \qquad J^- = \{j \in J : \beta_j < 0\}\,.$$

Denote

$$\mu_1 = \sum_{J^+} \delta_j \frac{\mu \mid B_j}{\mu(B_j)}\,, \quad \mu_2 = \sum_{J^-} \delta_j \frac{\mu \mid B_j}{\mu(B_j)}$$

and observe that μ_1, μ_2 are singular elements of $M^1(X) \cap H(\mu)$ with $b\mu_1 \equiv b\mu_2$. Hence, by (7) we have $\mu_1 = \mu_2$ which is a contradiction with the singularity of the probability measures μ_1, μ_2.

4. A characterization of simplicial measures

In this section we shall show how the concept of μ-simplex may be used to characterize a simplicial measure in a little more geometrical way than Theorem 1 does.

First, we establish a definition.

We call $\mu \in M^1(X)$ to be *S-regular* if for every $\varepsilon > 0$ there exist a μ-simplex K such that $\mu(K) > 1 - \varepsilon$ and $\mu \mid K$ is a boundary measure on K.

Now, we are in a position to formulate our main result.

Proposition. *Let $\mu \in M^1(X)$. Then μ is a simplicial measure if and only if μ is S-regular.*

As to the *sufficiency part* of the proof we do not meet any difficulties.

Assume that μ is a S-regular measure and let

$$\mu = \alpha\mu_1 + (1 - \alpha)\,\mu_2\,, \quad \text{where} \quad b\mu_1 = b\mu_2 = b\mu\,,$$

$$\mu_1\, \mu_2 \in M^1(X)\,, \quad \alpha \in (0, 1)\,.$$

Take an arbitrary $\varepsilon > 0$ and a μ-simplex K such that $\mu(K) > 1 - \varepsilon$, $\mu_1(K) > 1 - \varepsilon$ and such that $\mu \mid K$ is a boundary measure on K. It is easy to see that $\mu - \mu_1 \in M(X)$ satisfies condition (5). Moreover, $(\mu - \mu_1)(a) = 0$ for all $a \in A(X)$ and hence by the definition of a μ-simplex we conclude that $(\mu - \mu_1) \mid K = 0$. Letting $\varepsilon \to 0$ we arrive at $\mu = \mu_1$. Thus, μ is a simplicial measure.

A little more complicated is the way leading to the verification of the contrary implication. It will be convenient to introduce some more notation. Take $\mu \in M^1(X)$

a simplicial measure and recall that $B_1 = b(M^1(X) \cap H(\mu))$. For every $f \in C(X)$ define an affine map

$$A_f : B_1 \to R \quad \text{by} \quad A_f(x) = \nu_x(f)\,, \quad x \in B_1\,,$$

where $b\nu_x = x$, $\nu_x \in M^1(X) \cap H(\mu)$. Note that this definition is legitimate by virtue of (7).

Furthermore, choose a subset, say φ, of $C(X)$ which is *countable, uniformly dense in* $C(X)$ and which *contains a strictly convex continuous function.* The existence of such a subset follows from the metrizability and compactness of X. For the proof of the existence of a strictly convex continuous function on a metrizable compact convex set see, for example, E. M. Alfsen (1971), (Theorem I.4.3.).

Later we will be able to approximate the total mass of a *simplicial measure* μ by its values on compact convex sets K which satisfy the following[1]:

Condition A. *For every* $f \in \varphi$ *there is an affine map* $a : \mathrm{co}\,(K \cup B_1) \to R$ *such that*

(8) $$a \mid K \in A(K)\,,$$

(9) $$a = f \quad \text{on} \quad \mathrm{ex}\,K\,,$$

(10) $$a = A_f \quad \text{on} \quad B_1$$

holds.

To obtain the desired result (the implication "⇒" in Proposition) we need three rather technical lemmas. These will be used in the next section, too.

Recall that a simplex $K \subset X$ is said to be a *regular simplex* if ex K is closed.

Lemma 1. *Let* μ *be a simplicial measure,* K *a compact convex set satisfying Condition A. Then* K *is a regular simplex.*

Proof. First, we show that K is a simplex. To this end take $\lambda_1, \lambda_2 \in M^1(\mathrm{ex}\,K)$ with $b\lambda_1 = b\lambda_2$ and $f \in \varphi \mid K = \{h \mid K,\ h \in \varphi\}$. By (8) and (9) there is an $a \in A(K)$ such that $a = f$ on ex K and we may write $\lambda_1(f) = \lambda_1(a) = \lambda_2(a) = \lambda_2(f)$. As $\varphi \mid K$ is dense in $C(K)$ we arrive at the conclusion $\lambda_1 = \lambda_2$. Hence, K is a simplex.

To see that ex K is closed we choose the $g \in \varphi \mid K$ which is strictly convex on K. Again by (8) and (9) we have an $a \in A(K)$ such that $a = g$ on ex K. Obviously, $g \leqq a$ on K and it follows from the strict convexity of g that $\{x \in K : g(x) = a(x)\} = \mathrm{ex}\,K$. Thus, ex K is closed and K is a regular simplex.

[1] We write $f \mid C$ for the restriction of $f : X \to R$ to $C \subset X$ and co C for the convex hull of the C.

Lemma 2. *Let μ be a simplicial measure and $K \subset X$ a compact convex set satisfying Condition A. Then K is a μ-simplex and $\mu \mid K$ is a boundary measure on K.*

Proof. First, we prove that K is a μ-simplex. According to Theorem 2 and Lemma 1 it is sufficient to verify that convex sets K and $B_1(\mu, K^c)$ are affinely independent. Let

$$\alpha_1 k_1 + (1 - \alpha_1) x_1 = \alpha_2 k_2 + (1 - \alpha_2) x_2 , \tag{11}$$

$$k_i \in K , \quad x_i \in B_1(\mu, K^c) , \quad \alpha_i \in (0, 1) , \quad i = 1, 2 ,$$

and write

$$k_i = b\lambda_i , \quad x_i = b\nu_i \quad \text{where} \quad \lambda_i \in M^1(\text{ex}\, K) ,$$

$$\nu_i \in M^1(K^c) \cap H(\mu) , \quad i = 1, 2 .$$

Put $\delta_i = \alpha_i\lambda_i + (1 - \alpha_i) \nu_i$ and note that $b\delta_1 = b\delta_2$. Now, take an $f \in \varphi$ and choose to f, $a : \text{co}\,(K \cup B_1) \to R$, an affine map such that (8), (9) and (10) hold. Keeping in mind that λ_i's are boundary measures on K and that $B(\mu, K^c) \subset B_1$ we deduce that for $i = 1, 2$

$$\delta_i(f) = \alpha_i \lambda_i(a \mid K) + (1 - \alpha_i) A_f(x_i) = a(\alpha_i k_i + (1 - \alpha_i) x_i)$$

$$= a(b\delta_i)$$

holds. Hence $\delta_1(f) = \delta_2(f)$ for all $f \in \varphi$. Since the φ is dense in $C(X)$ we see that $\delta_1 = \delta_2$. Hence, $\alpha_1 = \alpha_2$, $\lambda_1 = \lambda_2$, $\nu_1 = \nu_2$, and decomposition (11) is unique. Thus, K and $B_1(\mu, K^c)$ are two affinely independent sets.

To prove that $\mu \mid K$ is a boundary measure on K we assume, without loss of generality, that $\mu(K) > 0$. Put $\nu = \mu \mid K/\mu(K)$ and take g the strictly convex element of φ. We choose to g the affine map $s : \text{co}\,(K \cup B_1) \to R$ satisfying (8), (9) and (10). As $b(\nu) \in B_1$ and $\nu \in M^1(X) \cap H(\mu)$ it follows from (10) that

$$a(b\nu) = A_g(b\nu) = \frac{1}{\mu(K)} \cdot \int_K g \, d\mu . \tag{12}$$

Moreover, since $b\nu \in K$ and $\nu \in M^1(K)$ we have by (8)

$$a(b\nu) = (a \mid K)(b\nu) = \frac{1}{\mu(K)} \int_K a \, d\mu . \tag{13}$$

Combining (12) and (13) we obtain that $\int_K a \, d\mu = \int_K g \, d\mu$ and the strict convexity of g implies that $a > g$ on K-ex K.

Now, if $\mu \mid K$ happens not to be a boundary measure on K then $\mu(K\text{-ex}\, K) > 0$ and we may write

$$\int_K a \, d\mu > \int_{[K - \text{ex} K]} g \, d\mu + \int_{\text{ex} K} g \, d\mu = \int_K g \, d\mu ,$$

which is a contradiction. Hence, $\mu \mid K$ is a boundary measure on K and the proof of Lemma 2 is completed.

Next, we formulate a lemma which provides our basic construction.

Lemma 3. *Let μ be a simplicial measure. Then there is a sequence of compact convex sets K_n satisfying Condition A such that*

$$\operatorname{ex} K_n \subset \operatorname{ex} K_{n+1}, \quad n \in N \tag{14}$$

$$\mu(K_n) > 1 - 1/n, \quad n \in N \tag{15}$$

hold.

Proof. As the set φ is countable we can write $\varphi = \{f^1, f^2, \ldots\}$. By Riesz theorem and Theorem 1 we may find $a_n^i \in A(X)$ such that

$$a_n^i \to f^i \quad \text{a.s.} \quad [\mu], \quad n \to \infty, \quad i \in N \tag{16}$$

and

$$\mu(|a_n^i - f^i|) \to 0, \qquad n \to \infty, \quad i \in N. \tag{17}$$

We state that

(18) for every $\varepsilon > 0$ there is a $D \subset X$ closed such that

$$\mu(D) > 1 - \varepsilon \quad \text{and} \quad \|a_n^i - f^i\|_D \to 0 \quad \text{as} \quad n \to \infty, \quad i \in N.$$

Indeed, in view of (16), Egorov Theorem implies the existence of Borel sets C_i with $\mu(C_i) > 1 - \varepsilon/2^i$ such that $\|a_n^i - f^i\|_{C_i} \to 0$ as $n \to \infty$ for $i \in N$. Hence, putting $C = \bigcap_1^\infty C_i$ we have $\mu(C) > 1 - \varepsilon$ and $\|a_n^i - f^i\|_C \to 0$ for $n \to \infty$ and $i \in N$. In virtue of continuity of a_n^i and f^i we may put $D = \bar{C}$ to get (18).

A simple argument permits to change (18) to the following observation:

(19) *There exists a non-decreasing sequence $\{D_k\}$ of closed sets with the property $\mu(D_k) > 1 - 1/k$ such that $\|a_n^i - f^i\|_{D_k} \to 0$, $n \to \infty$, $i \in N$, $k \in N$.*

Put $K_k = \overline{\text{co}}\, D_k$ for $k \in N$. Obviously, the closed convex set K_k satisfies condition (15). To see that these sets satisfy Condition A and (14) fix $k \in N$ and put $K_k = K$, $D_k = D$. Since $a_n^i \in A(X)$ we get from (19)

$$\|a_n^i - a_m^i\|_K \to 0, \quad n, m \to \infty, \quad i \in N. \tag{20}$$

Hence, the limit $\lim_{n \to \infty} a_n^i(x) \in R$ exists for $x \in K$ and $i \in N$. Moreover, if $x \in B_1$ and $\nu \in M^1(X) \cap H(\mu)$ are such that $b\nu = x$ it follows from (5) and (17) that

$$\lim_{n \to \infty} a_n^i(x) = \nu(f^i) = A_{f^i}(x), \quad i \in N. \tag{21}$$

We claim that $a^i : \text{co}\,(K \cup B_1) \to R$, the affine function defined by

$$a^i(x) = \lim_{n\to\infty} a_n^i(x)\,, \quad x \in \text{co}\,(K \cup B_1)\,, \quad i \in N\,,$$

satisfies (8), (9) and (10) with respect to $f^i \in \varphi$. Actually, (21) implies (10), by (20) we have (8) and as $\text{ex}\,K \subset D$ (by Krein-Milman Theorem), we arrive through (19) to (9).

To show that $\text{ex}\,K_k \subset \text{ex}\,K_{k+1}$ we need in view of (19) to verify that $D_k = \text{ex}\,K_k$. To this end assume that f^1 is the strictly convex element of φ. The strict convexity together with (19) yields the non-trivial part of the equality:

$$\text{ex}\,K_k = [a^1 \,|\, K_k = f^1 \,|\, K_k] \supset D_k\,.$$

This completes proof of Lemma 3.

Now, we may finish the *proof of the proposition*. Suppose that $\mu \in M^1(X)$ is simplicial measure and take $\varepsilon > 0$. By Lemma 3 we have a compact convex set K with $\mu(K) > 1 - \varepsilon$ satisfying Condition A. It follows from Lemma 2 that K is a μ-simplex and $\mu \,|\, K$ is a boundary measure on K. Hence, the measure μ is S-regular and the proof is completed.

5. Corollaries

In this section we employ the preceding lemmas once more to obtain a detailed description of the support of a simplicial measure μ. We denote[1]

$$\text{sp}\,\mu = \bigcap\{F \subset X \text{ closed},\ \mu(F) = 1\} \quad \text{and} \quad S_\mu = \overline{\text{co}}\,\text{sp}\,\mu\,.$$

The theorem below provides a description of the convex support S_μ of a simplicial measure in the infinite-dimensional case.

Theorem 3. *Let μ be a simplicial measure. Then there are regular simplexes $K_n \subset S_\mu$ such that*

(22) $$\mu(K_n) > 1 - \frac{1}{n}\,, \quad \mu \,|\, K_n \in M(\text{ex}\,K_n)\,,$$

$$\text{ex}\,K_n \subset \text{ex}\,K_{n+1}\,, \quad n \in N\,,$$

and such that $\overline{\bigcup_n K_n} = S_\mu$.

Proof. We may assume that $X = S_\mu$. Lemma 3 provides a sequence of compact

1 We write $\overline{\text{co}}\,C$ for the closed convex hull of a $C \subset X$.

convex sets $K_n \subset S_\mu$, satisfying (22) and Condition A. These sets are regular simplexes by Lemma 1. According to (22) we get

$$(23) \qquad \mu(\bigcup_n \text{ex } K_n) = 1$$

and hence, $\text{sp } \mu \subset \overline{\bigcup K_n}$. Since $\bigcup_1^\infty K_n$ is convex, we may write

$$\overline{\bigcup_n K_n} = \overline{\text{co}} \bigcup_n K_n \supset \overline{\text{co}} \left[\overline{\bigcup_n \text{ex } K_n}\right] \supset S_\mu .$$

Thus, $\overline{\bigcup_n K_n} = S_\mu$ and the proof is completed.

We do not know whether the conditions employed in Theorem 3 are sufficient for a measure μ to be simplicial or not. However, note that the set $\bigcup_{n=1}^\infty \text{ex } K_n$ introduced in the preceding theorem is a set of affinely independent points. Thus, it follows from (23) that *simplicial measure is supported by a set of affinely independent points.* Now, if the linear span of S_μ, say $\mathscr{L}(S_\mu)$, happens to be finite-dimensional then such a set is finite. As the contrary implication is trivial we get the following characterization of simplicial measure in finite-dimensional case (Ch. F. Skau (1969), E. M. Alfsen (1971)[1]):

Let $\mu \in M^1(X)$ and $\dim \mathscr{L}(S_\mu) < \infty$. Then μ is simplicial if and only if μ is supported by a finite set of affinely independent points.

Equivalently we may assert:

Theorem 4. *Let $\mu \in M^1(X)$ and* $\dim \mathscr{L}(S_\mu) < \infty$. *Then μ is simplicial if and only if S_μ is a regular simplex and μ is a boundary measure on S_μ.*

As to the infinite-dimensional case, the result contained in Theorem 4 provides a motivation to the study of the set

$$S_{1r}(X) = \{\mu \in M^1(X) : S_\mu \text{ is a regular simplex and } \mu \text{ is a boundary measure on } S_\mu\} .$$

Further denote by $S_1(X)$ the set of all simplicial measures on X. Clearly, $S_{1r}(X) \subset S_1(X)$ and $S_{1r}(X) = S_1(X)$ when $\mathscr{L}(X)$ has a finite dimension.

We close this section with an easy consequence of Theorem 3.

Theorem 5. *$S_{1r}(X)$ is dense in $S_1(X)$ with respect to the uniform convergence on Borel subsets of X.*

---

[1] Proposition I.6.10., p. 60 and Proposition I.6.11., p. 61.

Proof. Put

$$\|\lambda\| = \sup\{|\lambda(B)|,\ B\text{ - Borel set}\} \quad \text{for} \quad \lambda \in M(X).$$

Take $\mu \in S_1(X)$ and arbitrary $\varepsilon > 0$. By Theorem 3 there exists a regular simplex K such that $\mu(K) > 1 - \varepsilon$, $\mu \mid K \in M(\operatorname{ex} K)$. If $\nu = \mu \mid K/\mu(K)$ then $\operatorname{sp} \nu \subset \operatorname{ex} K$ and hence, $\nu \in S_{1r}(X)$. Moreover, it is easy to see that

$$\|\nu - \mu\| = \frac{1}{\mu(K)} (\|\mu \mid K - \mu\| + \|\mu - \mu(K) \cdot \mu\|) = \frac{2\varepsilon}{1 - \varepsilon}.$$

Letting $\varepsilon \to 0$ we obtain the assertion.

References

ALFSEN, E. M. (1971). "Compact Convex Sets and Boundary Integrals". Springer-Verlag, Berlin.

DOUGLAS, R. G. (1964). On extremal measures and subspace density. *Michigan Math. J.* **11**, 243—246.

PHELPS, R. R. (1966). "Lectures on Choquet's Theorem". Van Nostrand, Princeton.

SEMADENI, Z. (1971). "Banach Spaces of Continuous Functions". Polish Scientific Publishers, Warszava.

SKAU, C. F. (1969). Existence of simplicial boundary measures on compact convex sets. *Åarhus Univ. Various Publ. Series*, no. 5.

CHARLES UNIVERSITY, PRAGUE, CZECHOSLOVAKIA

Received December 1975

ON THE CRAMÉR-FRÉCHET-RAO INEQUALITY IN THE NON-REGULAR CASE

by

I. VINCZE

Introduction

Many papers pay attention to the basic and important inequality derived independently by M. Fréchet [5], H. Cramér [3] and C. R. Rao [9]. In the textbooks of mathematical statistics mainly the regular case is considered, i.e. the case when the supports of the underlying density functions in the sample space coincide. But important investigations were done for the non regular case, too; we mention here the initiative papers only, by D. G. Chapman and H. Robbins [2], by J. M. Hammersley [6], by J. Kiefer [7], by D. A. Fraser and I. Guttman [4]. (A more exhaustive collection of the literature and further references are given by Th. Polfeldt [8].)

Although many of these papers refer to the possibility of further generalisations, they investigate the case of a real or perhaps the case of a vector parameter having real components – using sometimes further restrictions. The aim of the present paper is to give a brief account of the main results for the non regular case pointing out that almost no assumption is needed concerning the structure of the parameter space; (see also Barankin [1]), accordingly certain points of our treatment slightly differ from the earlier investigations.

§ 1. The case of an arbitrary parameter set

1.1. As sample space the Euclidean n-space E_n will be taken, the σ-algebra of its Borel sets will be denoted by $\mathscr{B}_n$. The random variable $X = (X_1, X_2, \ldots, X_n)$ – $X \in E_n$ – is distributed according to a law having density function $f(x; \vartheta) = f(x_1, x_2, \ldots, x_n; \vartheta)$ with respect to a σ-finite measure μ on $\mathscr{B}_n$, where ϑ is an element of a parameter set Θ. For $f(x; \vartheta)$ will be assumed that it is completely determined by ϑ. For the set of the underlying densities the following notation will be used:

$$f = \{f(x; \vartheta);\ \vartheta \in \Theta\} \tag{1.1.1}$$

and for the support of the distributions

$$A_\vartheta = \{x : f(x; \vartheta) > 0\} .$$

Let Θ be an arbitrary set and let us assume that the real valued function $g(\vartheta)$ has the unbiased estimator $t(x)$

$$E_\vartheta(t(x)) = g(\vartheta) , \quad \vartheta \in \Theta .$$

Given two elements ϑ, ϑ' of Θ the following set of densities will be constructed

$$(1.1.2) \qquad \mathscr{f}_\alpha = \{f_\alpha(x; \vartheta, \vartheta') = (1 - \alpha) f(x; \vartheta) + \alpha f(x; \vartheta'),\ 0 < \alpha < 1\} ,$$

which depends on the real parameter α. Evidently $\mathscr{f}_\alpha$ is a homogeneous set of densities with respect to μ on the set $A_{\vartheta,\vartheta'} = A_\vartheta \cup A_{\vartheta'}$, i.e., the set $\{x : f_\alpha(x; \vartheta, \vartheta') > 0\}$ does not depend on α for $0 < \alpha < 1$. – We turn to our first statement.

Theorem 1.1.1. *Using the notation and assumptions given above, the statistic*

$$(1.1.3) \qquad \hat{\alpha} = \frac{t(x) - g(\vartheta)}{g(\vartheta') - g(\vartheta)}$$

is an unbiased estimator for the parameter α and the following relation holds

$$(1.1.4) \qquad D_\alpha^2(\hat{\alpha}) \geqq \frac{1}{\displaystyle\int_{A_{\vartheta,\vartheta'}} \frac{[f(x; \vartheta') - f(x; \vartheta)]^2}{f_\alpha(x; \vartheta, \vartheta')} \, d\mu(x)} = \frac{1}{\mathscr{I}_\alpha(\vartheta, \vartheta')} .$$

Proof: We do not refer to the C–F–R theorem, which is valid in this case, as we do not need the assumption of differentiability under the integral sign.

The statement for the unbiasedness of $\hat{\alpha}$ is trivial. Let $0 < \alpha$, $\alpha + \Delta < 1$, then

$$\alpha + \Delta = \int_{A_{\vartheta,\vartheta'}} \hat{\alpha} f_{\alpha+\Delta}(x; \vartheta, \vartheta') \, d\mu(x)$$

and

$$\alpha = \int_{A_{\vartheta,\vartheta'}} \hat{\alpha} f_\alpha(x; \vartheta, \vartheta') \, d\mu(x) ,$$

further

$$\alpha \int_{A_{\vartheta,\vartheta'}} [f_{\alpha+\Delta}(x; \vartheta, \vartheta') - f_\alpha(x; \vartheta, \vartheta')] \, d\mu(x) = 0$$

and

$$f_{\alpha+\Delta}(x; \vartheta, \vartheta') - f_\alpha(x; \vartheta, \vartheta') = \Delta(f(x; \vartheta') - f(x; \vartheta)) .$$

From these relations follows

$$1 = \int_{A_{\vartheta,\vartheta'}} (\hat{\alpha} - \alpha)(f(x;\vartheta') - f(x;\vartheta))\, d\mu(x)$$

$$= \int_{A_{\vartheta,\vartheta'}} (\hat{\alpha} - \alpha)\sqrt{[f_\alpha(x;\vartheta,\vartheta')]}\, \frac{f(\alpha;\vartheta') - f(x;\vartheta)}{f_\alpha(x;\vartheta,\vartheta')} \sqrt{[f_\alpha(x;\vartheta,\vartheta')]}\, d\mu(x)\,.$$

Now the Bunjakowski-Cauchy-Schwarz inequality completes the proof.

1.2. We turn now to some consequences of Theorem 1.1.1. A direct calculation yields the relation

$$D^2_\alpha(\hat{\alpha}) = \frac{(1-\alpha)\, D^2_\vartheta(t(x)) + \alpha\, D^2_{\vartheta'}(t(x))}{(g(\vartheta') - g(\vartheta))^2} + \alpha(1-\alpha)\,.$$

Hence for the weighted variance, it follows (see Kiefer [7], Polfeldt [8]).

(1.2.1)
$$(1-\alpha)\, D^2_\vartheta(t(x)) + \alpha\, D^2_{\vartheta'}(t(x)) \geqq (g(\vartheta') - g(\vartheta))^2 \left\{\frac{1}{\mathscr{I}_\alpha(\vartheta,\vartheta')} - \alpha(1-\alpha)\right\}.$$

We obtain now the

Corollary 1.2.1. *If the variance of the unbiased estimator $t(x)$ does not depend on ϑ, then we have*

(1.2.2)
$$D^2(t(x)) \geqq \sup_{(\alpha)} \sup_{(\vartheta')} [g(\vartheta') - g(\vartheta)]^2 \left\{\frac{1}{\mathscr{I}_\alpha(\vartheta,\vartheta')} - \alpha(1-\alpha)\right\}.$$

If on the other hand in 1.2.1 α tends to zero, we obtain the second Kiefer bound (see l.c. p. 628, (4)):

Corollary 1.2.2.

(1.2.3)
$$D^2_\vartheta(t(x)) \geqq \sup_{(\vartheta')} \frac{(g(\vartheta') - g(\vartheta))^2}{\displaystyle\int_{A_{\vartheta,\vartheta'}} \frac{[f(x;\vartheta') - f(x;\vartheta)]^2}{f(x;\vartheta)}\, d\mu(x)}\,.$$

As was pointed out by Kiefer, the last bound is not always the best one, which happens in the case of uniformly distributed variables in $(0, \vartheta)$. The same is true for

(1.2.2). But in both cases the right order of magnitude can be attained which is $1/n^2$ in contrary to the $1/n$ valid for the regular case. Some questions concerning the uniform distribution will be considered in Section 2.

1.3. Let now $\mathscr{A}_\Theta$ be a σ-algebra of subsets of Θ and let $\lambda(\cdot)$ and $\lambda'(\cdot)$ be two probability measures on $\mathscr{A}_\Theta$. We shall assume that the densities $f(x; \vartheta)$ in $\boldsymbol{f}$ for fixed $x \in \mathbf{E}_n$ are $\mathscr{A}_\Theta$ measurable, and further, that they are integrable with respect to $\lambda(\cdot)$ and $\lambda'(\cdot)$.

The mixed densities

$$f_\lambda(x) = \int_\Theta f(x; \vartheta)\,\mathrm{d}\lambda(\vartheta)\,, \quad f_{\lambda'}(x) = \int_\Theta f(x, \vartheta)\,\mathrm{d}\lambda'(\vartheta)\,,$$

and their weighted sum

$$f_\alpha(x; \lambda, \lambda') = (1 - \alpha) f_\lambda(x) + \alpha f_{\lambda'}(x)\,, \quad 0 < \alpha < 1 \tag{1.3.1}$$

will be considered; the latter has the support $A_{\lambda,\lambda'} = A_\lambda \cup A_{\lambda'}$, where

$$A_\lambda = \{x : f_\lambda(x) > 0\}\,, \quad A_{\lambda'} = \{x : f_{\lambda'}(x) > 0\}\,.$$

When $t(x)$ is unbiased estimator for the $\mathscr{A}_\Theta$ measurable function $g(\vartheta)$ and

$$g_\lambda = \int_\Theta g(\vartheta)\,\mathrm{d}\lambda(\vartheta)\,, \quad g_{\lambda'} = \int_\Theta g(\vartheta)\,\mathrm{d}\lambda'(\vartheta)\,,$$

then we can apply the procedure given in 1.1 for the parameter values g_λ and $g_{\lambda'}$; the statistic

$$\hat{\alpha} = \frac{t(x) - g_\lambda}{g_{\lambda'} - g_\lambda}$$

is an unbiased estimator for α in the one-parameter set of densities (1.3.1). In this way we may obtain the corresponding bound for the mixed variance

$$D_\alpha^2(\hat{\alpha}) \geqq \frac{1}{\displaystyle\int_{A_{\lambda,\lambda'}} \frac{[f_{\lambda'}(x) - f_\lambda(x)]^2}{f_\alpha(x; \lambda, \lambda')}\,\mathrm{d}\mu(x)} = \frac{1}{\mathscr{I}_\alpha(\lambda, \lambda')}\,.$$

From this relation we obtain

$$(1 - \alpha)\, D_\lambda^2(t(x)) + \alpha\, D_{\lambda'}^2(t(x)) \geqq (g_{\lambda'} - g_\lambda)^2 \left\{\frac{1}{\mathscr{I}_\alpha(\lambda, \lambda')} - \alpha(1 - \alpha)\right\}.$$

If $\lambda(\cdot)$ gives probability 1 to a fixed parameter value ϑ, then we have

$$(1-\alpha)\, D^2_\vartheta(t(x)) + \alpha\, D^2_{\lambda'}(t(x)) \geqq (g_{\lambda'} - g(\vartheta))^2 \left\{\frac{1}{\mathscr{I}_\alpha(\vartheta, \lambda')} - \alpha(1-\alpha)\right\}$$

with

$$\mathscr{I}_\alpha(\vartheta, \lambda') = \int_{A_\vartheta \cup A_{\lambda'}} \frac{[f_{\lambda'}(x) - f(x;\vartheta)]^2}{f_\alpha(x;\vartheta,\lambda')}\, d\mu(x).$$

If we let α tend to zero, we obtain the first bound given by Kiefer (see l.c.p. 628, formula (3)):

$$D^2_\vartheta(t(x)) \geqq \sup_{(\lambda')} \frac{(g_{\lambda'} - g(\vartheta))^2}{\displaystyle\int_{A_\vartheta \cup A_{\lambda'}} \frac{[f_{\lambda'}(x) - f(x;\vartheta)]^2}{f(x;\vartheta)}\, d\mu(x)}. \tag{1.3.2}$$

1.4. In this section a "discrete" parameter set will be considered but in the regular case only, i.e. when the set A_ϑ does not depend on ϑ for $\vartheta \in \Theta$. The parameter set will be called *discrete* if the following condition (1.4.2) is fulfilled: Denoting by $I(\vartheta, \vartheta')$ the Fisher information belonging to a single element X_i of the sample $(X_1, X_2, \ldots, X_n)$ having independent variables, i.e.

$$I(\vartheta, \vartheta') = \int_{E_1} \frac{[f_1(x_1, \vartheta')]^2}{f_1(x_1, \vartheta)}\, d\mu_1(x_1), \tag{1.4.1}$$

where $f_1(x_1; \vartheta)$ is the density function of X_i, $i = 1, 2, \ldots, n$, then there exists a $\delta > 0$, for which

$$I(\vartheta, \vartheta') > 1 + \delta \tag{1.4.2}$$

holds, for any pair of (different) elements (ϑ, ϑ') in Θ.

In this case the denominator of the relation (1.2.3) has the form

$$\int_{E_n} \frac{[f(x;\vartheta') - f(x;\vartheta)]^2}{f(x;\vartheta)}\, d\mu(x) = \int_{E_n} \frac{[f(x;\vartheta')]^2}{f(x;\vartheta)}\, d\mu(x) - 1 = I(\vartheta, \vartheta') - 1.$$

Hence we can state under our assumptions

$$D^2_\vartheta(t(x)) \geqq \sup_{(\vartheta')} \frac{[g(\vartheta') - g(\vartheta)]^2}{[I(\vartheta, \vartheta')]^n - 1} \tag{1.4.3}$$

having a bound which – due to (1.4.2) – tends to zero exponentially. This case was considered by Hammersley (see l.c.).

§ 2. The case of uniform distribution

2.1. For independent random variables distributed uniformly in the interval $(0, \vartheta)$, $0 < \vartheta < \infty$, Kiefer showed that the bound (1.2.3) is not attainable. On the other hand he has given a measure $\lambda'(\cdot)$ for which in the relation (1.3.2) the equality sign holds. In both cases ϑ is estimated by means of the largest sample element.

We shall consider there the uniform distribution with given length of the range: Let ϑ be real $(-\infty < \vartheta < \infty)$ and let $X = (X_1, X_2, \ldots, X_n)$ be distributed according to the uniform distribution in the n-dimensional unit-cube:

$$A_\vartheta = \{(x_1, x_2, \ldots, x_n) : \vartheta - \tfrac{1}{2} \leqq x_i \leqq \vartheta + \tfrac{1}{2}, \quad i = 1, 2, \ldots, n\},$$

$$f(x; \vartheta) = \begin{cases} 1, & \text{if } \ x \in A_\vartheta, \\ 0 & \text{otherwise}. \end{cases}$$

For $\vartheta < \vartheta' < \vartheta + \frac{1}{2}$ we have

$$f_\alpha(x; \vartheta, \vartheta') = \begin{cases} 1 - \alpha, & \text{if } \ x \in A_\vartheta \cap \bar{A}_{\vartheta'}, \\ 1, & \text{if } \ x \in A_\vartheta \cap A_{\vartheta'}, \\ \alpha, & \text{if } \ x \in \bar{A}_\vartheta \cap A_{\vartheta'}, \end{cases}$$

further

$$|f(x; \vartheta') - f(x; \vartheta)| = \begin{cases} 1, & \text{if } \ x \in A_\vartheta \cap \bar{A}_{\vartheta'}, \\ 0, & \text{if } \ x \in A_\vartheta \cap A_{\vartheta'}, \\ 1, & \text{if } \ x \in \bar{A}_\vartheta \cap A_{\vartheta'}. \end{cases}$$

For the Lebesgue-measure we have – using the notation $\delta = \vartheta' - \vartheta$ –

$$L(A_\vartheta \cap A_{\vartheta'}) = (1 - \delta)^n,$$

$$L(A_\vartheta \cap \bar{A}_{\vartheta'}) = 1 - (1 - \delta)^n,$$

$$L(\bar{A}_\vartheta \cap A_{\vartheta'}) = 1 - (1 - \delta)^n.$$

Hence we have

$$\mathscr{I}_\alpha(\vartheta, \vartheta') = \int_{A_\vartheta \cap \bar{A}_{\vartheta'}} \frac{1}{1 - \alpha}\, dx + \int_{\bar{A}_\vartheta \cap A_{\vartheta'}} \frac{1}{\alpha}\, dx = \frac{1}{\alpha(1 - \alpha)} [1 - (1 - \delta)^n].$$

Applying (1.2.2) to the case when the variance of the unbiased estimator $t(X)$ does not depend on ϑ and taking into account that $\max\limits_{(\alpha)} \alpha(1 - \alpha) = \frac{1}{4}$ – we have

$$D^2(t(x)) \geqq \tfrac{1}{4} \sup_{0 < \delta < \frac{1}{2}} \left(\frac{1}{1 - (1 - \delta)^n} - 1 \right) \delta^2 . \tag{2.1.1}$$

The choice $\delta^2 = [(n+1)(n+2)]^{-1}$ yields the relation

$$D^2(t(x)) \geqq \frac{1}{4} \frac{1}{(n+1)(n+2)} \frac{(1-[(n+1)(n+2)]^{-1/2})^n}{1-(1-[(n+1)(n+2)]^{-1/2})^n}.$$

As the last quotient decreases when n increases, its limiting value $(e-1)^{-1}$ can be replaced for it and so we obtain

$$D^2(t(x)) > \frac{1}{6.88} \frac{1}{(n+1)(n+2)},$$

which still does not agree with the variance of the mid-range:

$$D^2(\tfrac{1}{2}(\max_{(i)} X_i + \min_{(i)} X_i)) = \frac{1}{2(n+1)(n+2)},$$

however the right order of magnitude is attained.

2.2. Dropping the assumption that the variance does not depend on the parameter ϑ, the situation is quite different. The relation (1.2.3) gives now the variance bound zero, the denominator being $+\infty$ $(f(x;\vartheta') - f(x;\vartheta) = 1$ and $f(x;\vartheta) = 0$ on the set $\bar{A}_\vartheta \cap A_{\vartheta'}$ of positive measure). This bound is attainable, at least for certain values of the parameter, as the case of the following, known estimator shows:

$$t(x) = [X_1] + \tfrac{1}{2}, \tag{2.2.1}$$

where X_1 is the first (or any other) sample element in the random order, $[a]$ means integer part of a.

The estimator (2.2.1) is unbiased, but – of course – not consistent. The unbiasedness can be obtained in the following way: Let $a - \frac{1}{2} \leqq \vartheta < a + \frac{1}{2}$ with a integer, then the possible values of $t(x)$ are $a - \frac{1}{2}$ and $a + \frac{1}{2}$; consequently

$$E(t(x)) = (a - \tfrac{1}{2})(a + \tfrac{1}{2} - \vartheta) + (a + \tfrac{1}{2})(\vartheta - a + \tfrac{1}{2}) = \vartheta.$$

A direct calculation yields the formula for the variance

$$D^2(t(x)) = \tfrac{1}{4} - (\vartheta - [\vartheta + \tfrac{1}{2}])^2,$$

which is zero whenever $\vartheta - \frac{1}{2}$ is an integer and really our estimation in these cases is exact.

§ 3. The case of a vector function of the parameter

3.1. In this paragraph the procedure given in section 1.1 will be extended considering the covariance matrix of the unbiased estimator

$$\mathbf{t}(x) = (t_1(x), t_2(x), \ldots, t_p(x))' \tag{3.1.1}$$

of the vector function

(3.1.2) $$\mathbf{g}(\vartheta) = (g_1(\vartheta), g_2(\vartheta), \ldots, g_p(\vartheta))',$$

i.e.

(3.1.3) $$E_\vartheta(\mathbf{t}(x)) = \mathbf{g}(\vartheta), \quad \vartheta \in \Theta.$$

Here and in the sequel the sign $(\ldots)'$ denotes the transpose, i.e. the column vector. For our purposes we choose $(p+1)$ different parameter-value from Θ which will be held fixed using the vector notation

$$\boldsymbol{\vartheta} = (\vartheta_0, \vartheta_1, \ldots, \vartheta_p).$$

From the underlying set $\boldsymbol{f}$ in (1.1.1) the following set of densities with respect to the measure μ will be formed:

(3.1.4) $$f_{\boldsymbol{\alpha}}(\boldsymbol{\vartheta}) = \sum_{j=0}^{p} \alpha_j f(x; \vartheta_j), \quad \vartheta_j \in \Theta, \quad f(x; \vartheta) \in \boldsymbol{f},$$

depending on the vector parameter

(3.1.5) $$\boldsymbol{\alpha} = (\alpha_0, \alpha_1, \ldots, \alpha_p)', \quad \sum_0^p \alpha_j = 1, \quad \alpha_j > 0, \quad j = 0, 1, 2, \ldots, p.$$

The set in (3.1.4) is a homogeneous set on the support $A_{\boldsymbol{\vartheta}} = \bigcup_0^p A_{\vartheta_j}$ in E_n.

The following further notations will be introduced the meaning and role of which being obvious:

$$\bar{g}_i = \sum_{j=0}^{p} \alpha_j g_i(\vartheta_j), \quad i = 1, 2, \ldots, p.$$

Let be, further

(3.1.6) $$\bar{\mathbf{g}} = (\bar{g}_1, \bar{g}_2, \ldots, \bar{g}_p)' = \sum_0^p \alpha_j \mathbf{g}(\vartheta_j).$$

For the $p \times p$ matrix

(3.1.7) $$\mathbf{G} = (\mathbf{g}(\vartheta_1) - \mathbf{g}(\vartheta_0), \mathbf{g}(\vartheta_2) - \mathbf{g}(\vartheta_0), \ldots, \mathbf{g}(\vartheta_p) - \mathbf{g}(\vartheta_0))$$

we shall assume that it is nonsingular.

The covariance of $t_i(x)$ and $t_k(x)$ will be denoted by $U_{ik}(\vartheta)$, $\vartheta \in \Theta$ and the $p \times p$ covariance matrix of $\mathbf{t}(x)$

(3.1.8) $$\mathbf{U}_{\mathbf{t}(x)}(\vartheta) = (U_{ik}(\vartheta))$$

will be considered, more precisely its mixed form

$$\mathbf{U}^{(\boldsymbol{\alpha})}_{\mathbf{t}(x)}(\boldsymbol{\vartheta}) = \sum_0^p \alpha_j \mathbf{U}_{\mathbf{t}(x)}(\vartheta_j).$$

Let us introduce now the vector variable

(3.1.9) $$\hat{\boldsymbol{\alpha}} = (\hat{\alpha}_1, \hat{\alpha}_2, \ldots, \hat{\alpha}_p)',$$

defined uniquely by the system of equations

$$(3.1.10) \qquad t_i(x) - g_i(\vartheta_0) = \sum_{j=0}^{p} \hat{\alpha}_j(g_i(\vartheta_j) - g_i(\vartheta_0)), \quad i = 1, 2, \ldots, p.$$

This system has the matrix form

$$G\hat{\boldsymbol{\alpha}} = \boldsymbol{t}(x) - \boldsymbol{g}(\vartheta_0)$$

from which – due to the regularity of G – the relation

$$(3.1.11) \qquad \hat{\boldsymbol{\alpha}} = G^{-1}\,\boldsymbol{t}(x) - \boldsymbol{g}(\vartheta_0)$$

follows.

Finally, the $p \times p$ information matrix – assumed that it is nonsingular – will be introduced

$$\mathcal{I}_{\boldsymbol{\alpha}} = \mathcal{I}_{\boldsymbol{\alpha}}(\boldsymbol{\vartheta}) = (I_{ik})$$

with elements

$$I_{ik} = \int_{A_{\vartheta}} \frac{[f(x;\vartheta_i) - f(x;\vartheta_0)]\,[f(x;\vartheta_k) - f(x;\vartheta_0)]}{f_{\boldsymbol{\alpha}}(x;\vartheta)}\, \mathrm{d}\mu(x).$$

Now we formulate our

Theorem 3.1.1. *In the frame of the set of densities in* (3.1.4) *the vector variable* $\hat{\boldsymbol{\alpha}}$ *is an unbiased estimator of the parameter vector* α *and the covariance matrix* $U_{\hat{\boldsymbol{\alpha}}}(\boldsymbol{\alpha})$ *of* $\hat{\alpha}$ *satisfies the following statement*:

$$U_{\hat{\boldsymbol{\alpha}}}(\boldsymbol{\alpha}) - \mathcal{I}_{\alpha}^{-1}$$

is positive semidefinite.

Being this theorem the known generalization of the C–F–R theorem, in a simple case, we omit the proof. We remark only that carrying out the proof in the usual way from the regularity conditions the homogenity is needed only, but the differentiability under the integral sign not, as happened in the case of our Theorem 3.1.1.

3.1.2. The question is now the corresponding statement for the (mixed) covariance matrix of $\boldsymbol{t}(x)$. For this case, the following theorem is valid.

Theorem 3.1.2. *Under the above assumptions and notation, the matrix*

$$\sum_{j=0}^{p} \alpha_j\, U_{\boldsymbol{t}(x)}(\vartheta_j) + \sum_{j=0}^{p} \alpha_j(\boldsymbol{g}(\vartheta_j) - \bar{\boldsymbol{g}})\,(\boldsymbol{g}(\vartheta_j) - \bar{\boldsymbol{g}})' - G\mathcal{I}_{\boldsymbol{\alpha}}^{-1}G'$$

is positive semidefinite for any unbiased estimator $\boldsymbol{t}(x)$ *of the vector-parameter* $\boldsymbol{g}(\boldsymbol{\vartheta})$.

For the proof we have to expand $U_{\hat{\alpha}}(\alpha)$, which can be done in the following way: According to (3.1.11)

$$U_{\hat{\alpha}}(\alpha) = G^{-1}\, U_{t(x)}^{(\alpha)}(\vartheta)\, G'^{-1}$$

furthermore

$$E_{\alpha}(t(x)) = \sum_{0}^{p} \alpha_j\, g(\vartheta_j) = \bar{g}\,,$$

consequently

$$U_{t(x)}^{(\alpha)}(\vartheta) = E_{\alpha}[(t(x) - \bar{g})\,(t(x) - \bar{g})'] = \sum_{0}^{p} \alpha_j\, E_{\vartheta_j}(t(x) - \bar{g})\,(t(x) - \bar{g})']$$

$$= \sum_{0}^{p} \alpha_j\, E_{\vartheta_j}[(t(x) - g(\vartheta_j))\,(t(x) - g(\vartheta_j))'] + \sum_{0}^{p} \alpha_j (g(\vartheta_j) - \bar{g})\,(g(\vartheta_j) - \bar{g})']$$

$$= \sum_{0}^{p} \alpha_j\, U_{t(x)}(\vartheta_j) + \sum_{0}^{p} \alpha_j (g(\vartheta_j) - \bar{g})\,(g(\vartheta_j) - \bar{g})'\,,$$

which has the same form as in our theorem.

The author is indebted to E. Csáki, T. Móri and G. Tusnády for their valuable remarks.

References

[1] BARANKIN, E. W. (1949). Locally best unbiased estimator. *Ann. Math. Statist.*, **20**, 477—501.

[2] CHAPMAN, D. C. - ROBBINS, H. (1951). Minimum variance estimation without regularity assumptions. *Ann. Math. Statist.*, **22**, 581—586.

[3] CRAMÉR, H. (1946). Mathematical Methods of Statistics, *Princeton University Press*. Princeton.

[4] FRASER, D. A. S. - GUTTMAN, I. (1952). Bhattacharyya-bounds without regularity assumptions. *Ann. Math. Statist.*, **23**, 629—632.

[5] FRÉCHET, M. (1943). Sur l'extension de certaines évaluations statistiques au cas de petits échantillons. *Rev. Inst. Internat. Statist.*, **11**, 182—205.

[6] HAMMERSLEY, J. M. (1950). On estimating restricted parameters. *J. Royal Statist. Soc.*, B., **12**, 192—240.

[7] KIEFER, J. (1952). On minimum variance estimators. *Ann. Math. Statist.*, **23**, 627—629.

[8] POLFELDT, TH. (1970). Asymptotic Results in Non-Regular Estimation. Skandinavisk Aktuarietidskrift. 1—2 Supplement.

[9] RAO, C. R. (1945). Information and accuracy attainable in the estimation of statistical parameters. *Bull. Calcutta Math. Soc.*, **37**, 81—91.

MATHEMATICAL INSTITUTE OF HUNGARIAN ACADEMY OF SCIENCES,
BUDAPEST, HUNGARY

Received December 1975
Revised May 1977

ASYMPTOTIC DISTRIBUTION OF SIMPLE ESTIMATE FOR REJECTIVE, SAMPFORD AND SUCCESSIVE SAMPLING

by

JAN ÁMOS VÍŠEK

0. Introduction.

In this paper we shall prove asymptotic normality of the simple estimate for rejective, Sampford and successive sampling. Asymptotic normality of the simple estimate in the case of rejective sampling has been already proved by Hájek (1964) and only a simpler proof is presented here. Let $\mathcal{N}$ be the set of all nonnegative integers. Let S be a population, s a sample and N and n be the size of population S and sample s respectively. (For all definitions see J. Hájek (1964), (1973).)

1. General considerations

First, we derive two theorems that will be employed in the sequel.

Theorem 1. *Let X be a random variable, $X \in \mathcal{N}$, let Y be a discrete random variable. Let $r \in \mathcal{N}$, $\psi(v \mid X = r)$ be the characteristic function of Y under the condition $\{X = r\}$ and $\varphi(u, v)$ be the characteristic function of (X, Y). Then*

$$\psi(v \mid X = r) = \frac{\left(\frac{1}{2\pi}\right) \int_{-\pi}^{\pi} e^{-iur} \varphi(u, v)\, du}{P(X = r)} .$$

Proof. Put

$$B = \{k : P(Y = k) > 0\} .$$

We have

$$\psi(v \mid X = r) = \sum_{k \in B} e^{ivk} P(Y = k \mid X = r)$$

$$= \frac{1}{P(X = r)} \sum_{k \in B} e^{ivk} P(Y = k, X = r) .$$

On the other hand

$$\frac{1}{2\pi}\int_{-\pi}^{\pi} e^{-iur}\,\varphi(u, v)\,du = \frac{1}{2\pi}\sum_{k\in B}\sum_{l\in\mathcal{N}}\int_{-\pi}^{\pi} e^{i\{u(e-r)+vk\}}\,P(X = l,\ Y = k)\,du$$

$$= \sum_{k\in B} e^{ivk}\,P(X = r,\ Y = k)\,,$$

as

$$\frac{1}{2\pi}\int_{-\pi}^{\pi} e^{iu(l-r)}\,du = \begin{cases}1 & \text{for } l = r\,,\\ 0 & \text{for } l \neq r\,,\end{cases} \qquad l, r \in \mathcal{N}\,.$$

Theorem 2. *Let $P_N(s)$ and $Q_N(s)$ be probability functions defined on the set of all $s \subset S$ and let*

$$\frac{Q_N(s)}{P_N(s)} \xrightarrow[N\to\infty]{P_N} 1\,.$$

(P_N is probability measure generated by $P_N(s)$ on the algebra $\mathcal{A}$ of all subsets of the set of all samples $s \subset S$; analogously Q_N.) Let $X_N(s)$ be a random variable and $F_N(x)$ and $G_N(x)$ be its distribution function with respect to P_N and Q_N respectively. Then

$$F_N(x) - G_N(x) \xrightarrow[N\to\infty]{} 0$$

uniformly in x.

Proof. Let us define for $N \in \mathcal{N}$ and $\alpha > 0$

$$A_N(\alpha) = \left\{s : s \subset S; \left|\frac{Q_N(s)}{P_N(s)} - 1\right| \geqq \alpha\right\},$$

then

$$\underset{\alpha>0}{\forall}\quad \underset{N_0(\alpha)\in\mathcal{N}}{\exists}\quad \underset{N\geqq N_0(\alpha),\,N\in\mathcal{N}}{\forall}$$

we have

$$\sum_{s\in A_N(\alpha)} P_N(s) \leqq \alpha\,.$$

From

$$\left|\frac{Q_N(s)}{P_N(s)} - 1\right| \leqq \alpha \quad \text{for} \quad s \in A_N^c(\alpha) \quad \text{and} \quad N \geqq N_0(\alpha)\,, \quad N \in \mathcal{N}$$

we easily find

$$0 \leqq \sum_{s\in A_N(\alpha)} Q_N(s) \leqq 1 - (1 - \alpha)^2\,.$$

Let $B_N(x) = \{s : s \subset S; X_N(s) < x\}$. We recall that $F_N(x) = \sum_{s \in B_N(x)} P_N(s)$ and $G_N(x) = \sum_{s \in B_N(x)} Q_N(s)$; hence

$$(1 - \alpha)\{F_N(x) - \alpha\} \leqq G_N(x) \leqq 1 - (1 - \alpha)^2 + (1 + \alpha) F_N(x) .$$

As the distribution function is bounded, the theorem is proved.

2. Rejective sampling

Let π_i, p_i be the probabilities of inclusion of unit i with respect to rejective and Poisson sampling respectively. Let us have the whole class of Poisson samplings for all $N = 1, 2, \ldots$ and $p_1, p_2, \ldots$, $0 < p_i < 1$. Let every unit have a physical measurable property y_i. Let

$$\hat{Z} = \sum_{i=1}^{N} (y_i/p_i) I_i , \quad \hat{Y} = \sum_{i=1}^{N} (y_i/\pi_i) I_i , \quad Y = \sum_{i=1}^{N} y_i$$

and $K = \sum_{i=1}^{N} I_i$ i.e. $\hat{Y}$ and $\hat{Z}$ are the *simple estimate* of Y with respect to rejective and Poisson sampling respectively, and K is the size of the sample. Let all populations be normed by conditions

$$\sum_{i=1}^{N} y_i(1 - p_i) = 0 , \tag{1}$$

$$\sum_{i=1}^{N} y_i^2 \left(\frac{1}{p_i} - 1\right) = 0 .$$

Define for a population and for arbitrary $\varepsilon > 0$

$$A_\varepsilon = \{i : |y_i| > \varepsilon p_i\} ,$$

$$b(\varepsilon) = \sum_{i \in A_\varepsilon} y_i^2 \left(\frac{1}{p_i} - 1\right)$$

and

$$e = \inf\{\varepsilon : b(\varepsilon) < \varepsilon\} .$$

Let us construct the class of rejective samplings from the class of Poisson samplings given the condition $\{K = EK\}$.

Theorem 3. *Let $e \to 0$. Then the random variable $\hat{Z} - Y$ has asymptotically normal distribution with $\mu = 0$ and $\sigma^2 = 1$ in the class of rejective samplings.*

Proof. We shall prove this theorem in two steps.

1st step. $(K - n, \hat{Z} - Y)$ is asymptotically normal with zero expectation and covariance matrix

$$\begin{pmatrix} d, & 0 \\ 0, & 1 \end{pmatrix} \tag{2}$$

where $d = \sum_{i=1}^{N} p_i(1 - p_i)$. To show this, we make use of a limit theorem for random vectors, saying:

$(K - n, \hat{Z} - Y)$ is asymptotically normal with zero expectation and covariance matrix (2) if $\lambda_1(K - n) + \lambda_2(\hat{Z} - Y)$ is asymptotically normal for all λ_1, λ_2 which do not vanish simultaneously, with zero expectation and covariance $\lambda_1^2 d + \lambda_2^2$. As inequality

$$d \geqq \sum_{i \notin A_\varepsilon} p_i(1 - p_i) \geqq \frac{1}{\varepsilon^2} \sum_{i \notin A_\varepsilon} y_i^2 \left(\frac{1}{p_i} - 1\right) = \frac{1}{\varepsilon^2}(1 - b(\varepsilon))$$

implies $d \to \infty$, it is easy to verify that Feller-Lindeberg condition is fulfilled for $\lambda_1(K - n) + \lambda_2(\hat{Z} - Y)$ and for arbitrary λ_1, λ_2 not vanishing simultaneously.

2nd step. Let $\psi(u)$ be the characteristic function of $\hat{Z} - Y$ conditioned by $K - n = 0$, $\varphi(u, v)$ the characteristic function of the vector $(K - n, \hat{Z} - Y)$. Making use of Theorem 1, we have

$$\psi(u) = \frac{1}{P(K - n = 0)} \frac{1}{2\pi} \int_{-\pi}^{\pi} e^{-iu0} \varphi(u, v) \, du .$$

In Hájek (1972), it is proved that

$$P(K - n = 0) \sim (2\pi \, . \, d)^{-1/2}$$

("$\sim$" denotes asymptotic equality) so that

$$\psi(u) \sim \left(\frac{d}{2\pi}\right)^{1/2} \int_{-\pi}^{\pi} \varphi(u, v) \, du .$$

For arbitrary $\delta > 0$, fixed v and $u \in [-M/\sqrt{d}, M/\sqrt{d}]$, where $M > 0$, we can find such d_0, that

$$|\varphi(u, v) - \exp\{-\tfrac{1}{2}(u^2 d + v^2)\}| < \delta$$

for all $d \geqq d_0$, which gives

$$\left| \left(\frac{d}{2\pi}\right)^{1/2} \int_{-M/d^{1/2}}^{M/d^{1/2}} [\varphi(u, v) - \exp\{-\tfrac{1}{2}(u^2 d + v^2)\}] \, du \right| \leqq \left(\frac{2}{\pi}\right)^{1/2} M\delta .$$

By means of the transformation $ud^{1/2} = t$, we conclude (put M equal to the α-quantile of the normal distribution, $K = \exp\{-\frac{1}{2}v^2\}$)

$$\left|\frac{1}{(2\pi)^{1/2}}\int_{-M}^{M} \exp\{-\tfrac{1}{2}(t^2 + v^2)\}\, dt - \exp\{-\tfrac{1}{2}v^2\}\right| \leqq 2\alpha K$$

and finally

$$\left|\left(\frac{d}{2\pi}\right)^{1/2}\int_{-M/d^{1/2}}^{M/d^{1/2}} \varphi(u, v)\, du - \exp\{-\tfrac{1}{2}v^2\}\right| \leqq 2\alpha K + \left(\frac{2}{\pi}\right)^{1/2} M\delta\,. \tag{3}$$

Let $W = \{u : M/d^{1/2} \leqq |u| \leqq \pi\}$. From the independence of indicators of inclusion of units in Poisson sampling it results

$$\varphi(u, v) = \prod_{i=1}^{N} [(1 - p_i)\, e^{-ip_i\{u+v(y_i/p_i)\}} + p_i e^{i(1-p_i)\{u+v(y_i/p_i)\}}] \tag{4}$$

For $|a| \leqq \frac{3}{2}\pi$ we have

$$|(1 - p_i)\, e^{-ip_i a} + p_i e^{i(1-p_i)a}| \leqq e^{-p_i(1-p_i)4a^2/9\pi^2}\,. \tag{5}$$

Let $\varepsilon < \pi/2|v|$. Using (5) we estimate the expresion

$$(1 - p_i)\, e^{-ip_i\{u+v(y_i/p_i)\}} + p_i e^{i(1-p_i)\{u+v(y_i/p_i)\}} \tag{6}$$

for $i \notin A_\varepsilon$. For $i \in A_\varepsilon$, (6) is bounded by one and thus

$$\left(\frac{d}{2\pi}\right)^{1/2}\int_W \varphi(u, v)\, du \leqq \left(\frac{d}{2\pi}\right)^{1/2}\int_W \left\{\prod_{i\notin A_\varepsilon} \exp\left[-p_i(1 - p_i)\frac{4[u + v(y_i/p_i)]^2}{9\pi^2}\right]\right\} du\,.$$

Hence we easily get

$$\left(\frac{d}{2\pi}\right)^{1/2}\int_W \varphi(u, v)\, du \leqq \frac{1}{(2\pi)^{1/2}} \exp\left\{\frac{4}{9\pi^2}\, b\left(\frac{\varepsilon}{|v|}\right)\left[\frac{\pi^2}{\varepsilon^2} + 2\frac{\pi|v|}{\varepsilon} + v^2\right] - \frac{v^2}{\pi^2}\right\}$$

$$\cdot \int_{(M/\pi) \leqq t \leqq d} \exp\{-t^2\}\, dt$$

which completes the proof.

J. Hájek (1972) has shown that the equality

$$\pi_i = p_i\left[1 - \frac{(\bar{p} - p_i)(1 - p_i)}{d} + o(d^{-1})\right]$$

holds if $d \to \infty$ with $\bar{p} = [\sum_{i=1}^{N} p_i^2(1 - p_i)]/d$. As the condition $d \to \infty$ is symmetric for sampling with probabilities of inclusion p_i $(i = 1, 2, \ldots, N)$ and $1 - p_i$ $(i = 1, 2, \ldots, N)$, we can verify the validity of the equality

$$\pi_i = p_i \left[1 - \frac{(\bar{p} - p_i)(1 - p_i)}{d} \right] + p_i(1 - p_i)\, o(d^{-1}) . \tag{7}$$

Now we shall prove a lemma that enables us to prove the asymptotic normality of $\hat{Y}$.

Lemma 1. Let $X = \sum_{i=1}^{N} y_i(1/p_i - 1/\pi_i) I_i$, $R(s)$ *be the probability function describing a rejective sampling with probabilities of inclusion* π_i *generated by Poisson sampling with probabilities of inclusion* p_i *by the condition* $\{K = n\}$ *where* n *is the size of rejective sampling and let* $e \to 0$. *Then*

$$X \xrightarrow{R} 0 .$$

Proof. From (7) it follows

$$X = \sum_{i=1}^{N} \frac{y_i}{p_i} I_i(s) \left[\frac{(p_i - \bar{p})(1 - p_i)}{d} + (1 - p_i)\, o(d^{-1}) \right] .$$

Let

$$X' = \sum_{i=1}^{N} \frac{|y_i|}{p_i} I_i(s) \left| \frac{(p_i - \bar{p})(1 - p_i)}{d} + (1 - p_i)\, o(d^{-1}) \right|$$

and P be the probability measure generated by Poisson sampling (see Theorem 2) above. Then $X' > |X|$ and so it suffices to show

$$X' \xrightarrow{R} 0 . \tag{8}$$

We use the lemma 4.1 from the paper of J. Hájek (1964), where we only verify that an event $\{X' \geqq \varepsilon\}$ is hereditary with respect to above defined algebra $\mathscr{A}$. From this lemma it follows that for proving (8) it suffices to show that

$$X' \xrightarrow{P} 0 . \tag{9}$$

As the inequality

$$E_P X' = \sum_{i \in A_\varepsilon} |y_i| \left| \frac{(p_i - \bar{p})(1 - p_i)}{d} + (1 - p_i)\, o(d^{-1}) \right|$$

$$+ \sum_{i \notin A_\varepsilon} |y_i| \left| \frac{(p_i - \bar{p})(1 - p_i)}{d} + (1 - p_i)\, o(d^{-1}) \right|$$

$$\leqq \left(\frac{1}{\varepsilon} + \varepsilon d \right) \left(\frac{1}{d} + o(d^{-1}) \right)$$

holds, we have $E_P X' \to 0$ and (9) is equivalent to

$$X' - E_P X' \xrightarrow{P} 0 .$$

As also

$$\operatorname{var}_P X' = \sum_{i=1}^{N} p_i(1 - p_i) \frac{y_i^2}{p_i^2} \left(\frac{(p_i - \bar{\bar{p}})(1 - p_i)}{d} + (1 - p_i)\, o(d^{-1})^2 \right)$$
$$\leqq \left(\frac{1}{d} + o(d^{-1}) \right)^2 \sum_{i=1}^{N} y_i^2 \left(\frac{1}{p_i} - 1 \right)$$

holds, the lemma is proved.

Theorem 4. *Let $\hat{Y}$ be a simple estimate and let all populations be normed by conditions (1). Then, if $e \to 0$, $\hat{Y} - Y$ has an asymptotic normal distribution with $\mu = 0$ and $\sigma^2 = 1$ in the class of all rejective samplings.*

Proof follows from Theorem 3 and Lemma 1.

Remark. Let us denote

$$A'_\varepsilon = \{ i : |y_i| > \varepsilon \pi_i \} ,$$

$$b'(\varepsilon) = \sum_{i \in A_\varepsilon'} y_i^2 \left(\frac{1}{\pi_i} - 1 \right)$$

and

$$e' = \inf \{ \varepsilon : b'(\varepsilon) \leqq \varepsilon \} .$$

If the conditions (1) are fulfilled for some class of Poisson samplings and $e \to 0$, then for the class of rejective samplings, that is in the above mentioned relation with this class of Poisson samplings, it holds:

$$\sum_{i=1}^{N} y_i(1 - \pi_i) \xrightarrow[N \to \infty]{} 0 ,$$

$$\sum_{i=1}^{N} y_i^2 \left(\frac{1}{\pi_i} - 1 \right) \xrightarrow[N \to \infty]{} 1$$

and

$$e' \to 0 ,$$

as can be easily seen.

3. Sampford sampling

Sampford sampling is a modified rejective sampling and the modification is made to achive precisely the a priori given probabilities of inclusion. It is usually easy to give the probabilities of inclusion so, that the estimate has good properties, unfortunately we do not know the precise expresions for relationships between the probabilities of inclusion π_i and the probabilities of drawing α_i in the case of rejective sampling. We have an asymptotic formula for these relationship, but computation may be lengthy and expensive. From it follows that relation between rejective and Sampford sampling with the same probabilities of inclusion is very interesting. To study this situation we prove a theorem for a little different situation at first.

Theorem 5. *Let $Q(s)$ be the probability function describing Sampford sampling with probabilities of inclusion π_i; let $P(s)$ describe Poisson sampling with the same probabilities of inclusion π_i, let $R(s)$ describe rejective sampling generated by Poisson sampling $P(s)$ conditioned by the requirement that the sample size is n. Then*

$$\frac{Q(s)}{R(s)} \xrightarrow[d \to \infty]{R} 1 \tag{10}$$

Remark. As $R(s) = Q(s) = 0$ for all samples with size different from n, $\frac{Q(s)}{R(s)}$ has sense with probability one.

Proof. Let us denote as $\varkappa_i$ the probabilities of inclusion in the rejective sampling $R(s)$. In Hájek (1972), Chapter ii, Section 8, the equality is presented:

$$\frac{Q(s)}{R(s)} = \frac{\sum_{i=1}^{N} \pi_i(1 - I_i(s))}{\sum_{i=1}^{N} \pi_i(1 - \varkappa_i)} .$$

Let us define a random variable

$$X(s) = \frac{\sum_{i=1}^{N} \pi_i(1 - I_i(s))}{\sum_{i=1}^{N} \pi_i(1 - \varkappa_i)}$$

for all $s \subset S$. Let P be the probability measure generated by Poisson sampling mentioned in this theorem and let us use the inequality:

$$R\left(\left\{s : \left|\frac{Q(s)}{R(s)} - 1\right| \geqq \varepsilon\right\}\right) \leqq \frac{1}{\varepsilon^2} E_R \left(\frac{Q(s)}{R(s)} - E_P X(s)\right)^2 .$$

Then we have

$$R\left(\left\{s : \left|\frac{Q(s)}{R(s)} - 1\right| \geqq \varepsilon\right\}\right) \leqq \sum_{\substack{s \subset S \\ \text{size } s = n}} \frac{1}{\varepsilon^2} (X(s) - E_P X(s))^2 \, c \prod_{i \in s} \pi_i$$

$$\cdot \prod_{j \notin s} (1 - \pi_j) \leqq c \frac{1}{\varepsilon^2} \operatorname{var}_P X(s) .$$

From the definition of rejective sampling (as conditional Poisson sampling) we know that c^{-1} is equal to the probability that the size of a sample, drawn by Poisson sampling, equals n. This probability is asymptotically equal to $(2\pi d)^{-1/2}$ (see J. Hájek (1972), Theorem 7.A.) hence

$$c \operatorname{var}_P X(s) \xrightarrow[N \to \infty]{} 0 .$$

The proof is finished.

Let us study now the case when the probabilities of inclusion are the same. Let us have the class $\mathscr{A}$ of all rejective samplings and to every element of class $\mathscr{A}$ let us define Sampford and Poisson sampling with the same probabilities of inclusion π_i as this rejective sampling has. Let all populations be normed by the conditions (1) with respect to the probabilities π_i and let $e \to 0$. Then $\hat{Z}$ (defined in the paragraph 2) has in respective class of rejective samplings (respective to the class of Poisson samplings just mentioned) asymptotically normal distribution with expectation Y and $\sigma^2 = 1$. As the Theorem 5 shows Sampford sampling defined above is with just mentioned rejective sampling asymptotically equivalent and so $\hat{Z}$ ($\hat{Z}$ is the simple estimate with respect to this Sampford sampling) has asymptotically normal distribution. Now we find such class of Poisson samplings, with the probabilities of inclusion p_i, respective to rejective sampling having the probabilities of inclusion π_i. As the Remark 1 shows the conditions (1) are also asymptotically fulfilled with respect to p_i and also $e' \to 0$. Finally $\hat{Y}$ has asymptotically normal distribution in the class of rejective samplings with probabilities of inclusion π_i, too (see Theorem 4); so we have proved the theorem:

Theorem 6. *Let $\mathscr{A}$ be the class of all rejective samplings and $\mathscr{B}$ the class of all Sampford samplings. Let us take two sequences of samplings so that the first sequence is from the class $\mathscr{A}$, the second one is from $\mathscr{B}$ so, that the k-th element in the first sequence and the k-th element in the second one have the same probabilities of inclusion. Then in every couple of such sequences, if $e \to 0$, the simple estimate has the same asymptotic distribution with respect to the first sequence as with respect to the second one. This distribution is normal with expectation Y and $\sigma^2 = 1$.*

4. Successive sampling

Under conditions

(11) $$N \to \infty ,$$

(12) $$\frac{\max\limits_{1 \leqq i \leqq N} \alpha_i}{\min\limits_{1 \leqq i \leqq N} \alpha_i} < K \quad \text{for all} \quad N$$

(where α_i are the probabilities of drawing),

(13) $$0 < a < \frac{n}{N} < b < 1$$

and

(14) $$\max_{1 \leqq i \leqq N} |y_i - \bar{y}| \left[\frac{1}{N-1} \sum_{i=1}^{N} (y_i - \bar{y})^2 \right]^{-1/2} < K'$$

for all N and $\bar{y} = (1/N) \sum_{i=1}^{N} y_i$. B. Rosén (1972) has proved normality of simple estimate for the class of successive samplings. Successive sampling has been sugested to remove the disadvantage of rejective sampling, that lies in the frequent rejection of a part of sample in the case, when a replication occurs. In the case, when the probabilities of drawing are the same for rejective and successive sampling and the conditions (1) are fulfilled, the conditions (11), (12), (13) and (14) are not generally fulfilled. This situation is not satisfactory. Our goal is to find a condition under which the simple estimate has the same asymptotic distribution for both rejective and successive sampling.

Theorem 7. *Let us have the class of rejective samplings and the class of successive ones such, that for every element in the first class there is an element of second one with the same probabilities of drawing. Let $R(s)$ and $S(s)$ be the probability functions describing the rejective sampling and the successive one, respectively and let* (12) *be fulfilled. Then, under*

(15) $$\frac{n^{5/2}}{N} \xrightarrow[N \to \infty]{} 0 ,$$

(16) $$\frac{S(s)}{R(s)} \xrightarrow[N \to \infty]{} 1 .$$

Proof. From (12) we deduce

$$\frac{1}{NK} \leqq \alpha_i \leqq \frac{K}{N} .$$

We recall that

$$R(s) = \begin{cases} \dfrac{n! \prod_{i \in s} \alpha_i}{P(A(n))} & \text{for} \quad K(s) = n\,, \\ 0 & \text{otherwise}\,, \end{cases} \tag{17}$$

where $A(n)$ denotes the event, when two units are not the same among the n units in n independent draughts.

$$S(s) = \begin{cases} \prod_{i \in s} \alpha_i \, J(s) & \text{for} \quad K(s) = n\,, \\ 0 & \text{otherwise}\,, \end{cases} \tag{18}$$

where

$$J(s) = \sum_{(r_1, \ldots, r_n)} [(1 - \alpha_{r_1})(1 - \alpha_{r_2} - \alpha_{r_1}) \ldots (1 - \alpha_{r_1} - \alpha_{r_2} - \ldots - \alpha_{r_{n-1}})]^{-1}$$

where $\sum_{(r_1, \ldots, r_n)}$ denotes sum over all permutations of all elements contained in the sample s. Let us first estimate

$$m(r) = (1 - \alpha_{r_1})(1 - \alpha^r{}_1 - \alpha_{r_2}) \ldots (1 - \alpha_{r_1} - \alpha_{r_2} - \ldots - \alpha_{r_{n-1}})\,.$$

We have

$$\left(1 - \frac{K}{N}\right)\left(1 - \frac{2K}{N}\right)\left(1 - \frac{3K}{N}\right) \ldots \left(1 - \frac{(n-1)\,K}{N}\right) \leqq m(r) \leqq 1\,,$$

$$m(r) \geqq 1 - \left[\frac{K}{N} + \frac{2K}{N} + \frac{3K}{N} + \ldots + \frac{(n-1)\,K}{N}\right] + \frac{K^2}{N^2} \sum_{i=1}^{n-1} i \sum_{j=i+1}^{n-1} j$$

$$- \frac{K^3}{N^3} \sum_{i=1}^{n-1} i \sum_{j=i+1}^{n-1} j \sum_{k=j+1}^{n-1} k + \ldots\,. \tag{19}$$

Let us compare two elements for fixed $i, j, \ldots, l$ of this sum such, that the first is from some odd member of (19) and second is from the following even one.

$$\frac{K^{2k}}{N^{2k}} ij \ldots l - \frac{K^{2k+1}}{N^{2k+1}} ij \ldots l \sum_{m=l+1}^{n-1} m \geqq \frac{K^{2k}}{N^{2k}} ij \ldots l \, \frac{N - K\, n(n-1)/2}{N} \geqq 0\,.$$

From it follows

$$n! \leqq J(s) \leqq n! \left[1 - \frac{K}{N} \frac{n(n-1)}{2}\right]^{-1}$$

and the Theorem is proved.

Lemma 2. *From* (11), (12), *and* (15) *follows*

$$|\pi_i(R) - \pi_i(S)| \leqq \pi_i(S)\, z(N) \tag{20}$$

where $\pi_i(R)$ *and* $\pi_i(S)$ *are the probabilities of inclusion in rejective sampling and successive one, respectively and* $z(N) \to 0$ *when* $N \to \infty$.

Proof. From (17) and (18) follows that with probability 1

$$\frac{R(s)}{S(s)} = \frac{n!}{P(A(n))\, J(s)} \tag{21}$$

and above mentioned estimate of $J(s)$ gives

$$R(s)\, P(A(n)) \leqq S(s) \leqq R(s)\, P(A(n)) \left[1 - \frac{K\, n(n-1)}{2N}\right]^{-1}.$$

Let us take sum over $s \subset S$, and we obtain

$$P(A(n)) \leqq 1 \leqq P(A(n)) \left[1 - \frac{K\, n(n-1)}{2N}\right]^{-1}.$$

From (15) now we can find that

$$\lim_{n \to \infty} P(A(n)) = 1\,.$$

It follows from (21) that

$$S(s) \left[\frac{1 - K \dfrac{n(n-1)}{2N}}{P(A(n))} - 1\right] \leqq R(s) - S(s) \leqq S(s) \left[\frac{1}{P(A(n))} - 1\right].$$

Let us take a sum over $s \ni i$ and the assertion is proved.

Theorem 8. *Let* (11), (12) *and* (15) *hold. Let every population is normed by conditions* (1) *with respect to* $\pi_i(R)$ *and* $e \to 0$. *Then the simple estimate* $\hat{Y}$ *has in the class of successive samplings asymptotically normal distribution with expectation* Y *and* $\sigma^2 = 1$.

Proof. Let us denote

$$X' = \sum_{i=1}^{N} y_i\, I_i(s) \left(\frac{1}{\pi_i(R)} - \frac{1}{\pi_i(S)}\right).$$

$$|EX'| \leqq \sum_{i=1}^{N} |y_i|\, \frac{\pi_i(R) - \pi_i(S)}{\pi_i(S)} \leqq K_1 \sqrt{(n)}\, K_2\, \frac{n^2}{N},$$

where K_1 and K_2 do not depend on n and N.

Analogously,

$$\operatorname{Var} X' = \sum_{i=1}^{N} \pi_i(R)\,(1 - \pi_i(R))\, y_i^2 \left(\frac{1}{\pi_i(R)} - \frac{1}{\pi_i(S)}\right)^2$$

$$\leqq \sum_{i=1}^{N} \frac{y_i^2}{\pi_i(R)} \left(\frac{\pi_i(S) - \pi_i(R)}{\pi_i(S)}\right)^2$$

$$\leqq \left(1 + \frac{K_3\, n}{N}\right) K_4\, Z(N)$$

where K_3 and K_4 again do not depend on n and N.

So we can conclude that

$$X'(s) \xrightarrow{R} 0$$

when $N \to \infty$. So the Theorem is proved.

Acknowledgment

I am deeply grateful to Professor Jaroslav Hájek for suggesting this problem to me when I was his postgraduate student.

References

HÁJEK, J. (1964). Asymptotic Theory of Rejective Sampling with Varying Probabilities from a Finite Population. *Ann. Math. Statist.*, **35**, 1491—1523.

HÁJEK, J. (1972). Theory of Sampling from Finite Populations. (In Czech, mimeographed texts edited for seminar that J. Hájek guided in year 1972. Text is a part of the forthcoming monograph).

HÁJEK, J. (1973). Asymptotic Theories of Sampling with Varying Probabilities without Replacement. *Proceedings of the Prague Symposium on Asymptotic Statistics*, 3—6 September 1973, Volume I, 127—138.

RAO, C. R. (1965). "Linear statistical inference and its applications", Wiley, New York.

ROSÉN, B. (1972). Asymptotic Theory for Successive Sampling with Varying Probabilities without Replacement I, II. *Ann. Math. Statist.*, **43**, 373—397 and 748—776.

INSTITUTE OF INFORMATION THEORY AND AUTOMATION OF CZECHOSLOVAK ACADEMY OF SCIENCES, PRAGUE, CZECHOSLOVAKIA

Received December 1975

DISCRETE COMMUNICATION CHANNELS DECOMPOSABLE INTO FINITE-MEMORY COMPONENTS

by

K. WINKELBAUER

As a contribution to asymptotic methods of statistics, this paper is devoted to the study of asymptotic behaviour of the maximum length of n-dimensional error-correcting codes having their maximum probability of error less than a prescribed level.

The investigation is performed for the case of discrete communication channels which are decomposable into any finite or infinite number of component channels having each finite memory. The contents of the paper represents a natural extension of the author's results published in [12] (cf. the list of references at the end), where the asymptotic estimate of the maximum length was given for the case of channels decomposable into memoryless components.

Contents

1. Coding theorem and its strong converse

As mentioned in the introductory remark, the whole paper is closely connected with [12] so that some of the notations and terminology used in the sequel coincide with those given there. On the other hand, the problem considered here calls for a little more general approach, and that is why most of the symbolism and nomenclature is taken from the author's paper [9].

1.1. Prerequisities. Let us begin with some notational conventions. If a mapping u of a set X into a set Y, i.e. $u \in Y^X$, is considered as a family of elements in Y with parameters x running over X, then the mapping u will be written as $\{u(x), x \in X\}$ or $\{u_x\}_{x \in X}$.

The letter n will be exclusively employed to designate a natural number (i.e. $n = 1, 2, \ldots$); for example, we shall write

$$\sup_n x_n \quad \text{instead of} \quad \sup_{n=1,2,\ldots} x_n$$

for the supremum of a sequence $\{x_n\}$ of real numbers, and analogously in similar cases.

Throughout the entire paper the letter I means the set of all integers and R the set of all (finite) real numbers. Let us agree that the range of variables i and j will always be the set I; it plays the role of standard index set whereas i and j may be interpreted as time indices.

If V is a random variable on a probability space $(X, \mathbf{X}, \pi)$, then the (lower) Θ-quantile of V will be denoted by $q(\Theta, \pi; V)$; it is the infimum

$$q(\Theta, \pi; V) = \inf \{t \in R : \pi\{V \leqq t\} \geqq \Theta\}, \quad 0 < \Theta \leqq 1. \tag{1.1.1}$$

Notice that the essential supremum of V with respect to measure π is expressed by

$$\operatorname*{ess.\,sup}_{x \in X[\pi]} V(x) = q(1, \pi; V),$$

and that

$$\begin{aligned} q(\Theta, \pi; kV) &= k\, q(\Theta, \pi; V) \quad \text{for} \quad k \geqq 0, \\ q(\Theta, \pi; V_1) &\leqq q(\Theta, \pi; V_2) \quad \text{for} \quad V_1 \leqq V_2, \end{aligned} \tag{1.1.2}$$

where V_1 and V_2 are random variables.

If M is a finite non-empty set, then a probability M-vector is defined as a family $p = \{p(a)\}_{a \in M}$ of non-negative numbers which add to one. The set of all probability M-vectors will be denoted in accordance with [12] by $P(M)$; in symbols:

$$P(M) = \{p \in R^M : \sum_{a \in M} p(a) = 1, \quad p(a) \geqq 0 \ (a \in M)\}. \tag{1.1.3}$$

By $\mathbf{F}_M$ we shall designate the σ-algebra of subsets in the space M^I generated by the class of all sets of the form

$$[z]_i = \{\zeta \in M^I : \{\zeta_{i+j}\}_{0 \leqq j < n} = z\}, \quad z \in M^n;$$

the convention

$$M^n = M^{\{0,1,\ldots,n-1\}}$$

means that the components of vectors in M^n will be indexed in what follows from 0 to $n-1$. Let us set, for a measure μ on $\mathbf{F}_M$,

$$\mu[z] = \mu[z]_0 \quad \text{where} \quad \mu[z]_i = \mu([z]_i), \quad z \in M^n; \tag{1.1.4}$$
$$\mu[E] = \sum_{z \in E} \mu[z], \quad E \subset M^n.$$

A measure μ on $\mathbf{F}_M$ will be called n-invariant if it is a probability measure satisfying the condition that

$$\mu[z]_{in} = \mu[z] \quad \text{for} \quad z \in M^k \quad (k = 1, 2, \ldots).$$

A standard reasoning leads to the conclusion that the notion of n-invariance coincides with that given in [9], p. 231. An n-invariant measure is said to be decomposable if there are mutually distinct n-invariant measures μ' and μ'' such that $\mu = \lambda\mu' + (1-\lambda)\mu''$ for some λ, $0 < \lambda < 1$; μ is said indecomposable if it is not decomposable.

An indecomposable n-invariant measure on $\mathbf{F}_M$ will be called (as in [9]) n-ergodic: the class of all n-ergodic; the class of all n-ergodic measures on $\mathbf{F}_M$ will be denoted in what follows by $\mathcal{M}_{n-\mathrm{erg}}(M)$. A measure which is 1-ergodic, is called ergodic; the class of all ergodic measures on $\mathbf{F}_M$ is designated, more simply, by $\mathcal{M}_{\mathrm{erg}}(M)$; in symbols:

$$\mathcal{M}_{\mathrm{erg}}(M) = \mathcal{M}_{1-\mathrm{erg}}(M). \tag{1.1.5}$$

If A and B are finite non-empty sets, then $\mathbf{F}_{AB}$ will stand for $\mathbf{F}_{A \times B}$ and xy will mean the element in $(A \times B)^I$ for which

$$(xy)_i = (x_i, y_i) \quad \text{where} \quad x \in A^I, \quad y \in B^I.$$

The canonical mapping $(x, y) \to xy$ establishes a one-to-one correspondence between the measurable spaces $(A^I \times B^I, \mathbf{F}_A \times \mathbf{F}_B)$ and $((A \times B)^I, \mathbf{F}_{AB})$; here $\mathbf{F}_A \times \mathbf{F}_B$ means the σ-algebra generated by the class of measurable rectangles. We shall make use of a special notation for some cylinder sets in $\mathbf{F}_{AB}$ which is given by

$$[x, y]_{i,j} = \{x'y' : (x', y') \in [x]_i \times [y]_j\} \tag{1.1.6}$$

for $x \in A^n$, $y \in B^{m+n}$, where m is a non-negative integer.

We shall associate with a probability measure ω on $\mathbf{F}_{AB}$ its marginal measures ω' on $\mathbf{F}_A$ and ω'' on $\mathbf{F}_B$ uniquely determined by the conditions that

$$\omega'[x]_i = \omega\{x'y' : x' \in [x]_i\}, \quad x \in A^n, \tag{1.1.7}$$
$$\omega''[y]_i = \omega\{x'y' : y' \in [y]_i\}, \quad y \in B^n,$$

and together with a non-negative integer m the quantities (rates)

$$(1.1.8) \qquad R_{n,m}(\omega) = \sum_{x \in A^n, y \in B^{m+n}} \omega[x, y]_{m,0} \log \frac{\omega[x, y]_{m,0}}{\omega'[x]_m \, \omega''[y]_0} .$$

If, for some positive integer k, ω is k-invariant, then the sequence $\{(m+n)^{-1} \cdot \mathcal{R}_{n,m}(\omega)\}$ is convergent with a limit independent of m, called the information rate of ω and denoted here by $I(\omega)$; in symbols:

$$(1.1.9) \qquad I(\omega) = \lim_n \frac{1}{m+n} \mathcal{R}_{n,m}(\omega) = \lim_n \frac{1}{n} \mathcal{R}_n(\omega) ,$$

where we have set

$$(1.1.10) \qquad \mathcal{R}_n(\omega) = \mathcal{R}_{n,0}(\omega) .$$

A point z in the space $(A \times B)^I$ is, by definition, regular iff there is an ergodic measure on $\mathbf{F}_{AB}$, say ω_z, uniquely associated with the point z by the property that

$$(1.1.11) \qquad \omega_z[v]_i = \lim_n \frac{1}{n} \sum_{j=0}^{n-1} \chi[v]_{i+j}(z) , \quad v \in (A \times B)^n ;$$

here χ_E means the characteristic function of $E \subset (A \times B)^I$. The set of all regular points in the space $(A \times B)^I$ will be denoted by R_{AB}. The function V_{AB} defined by

$$(1.1.12) \qquad V_{AB}(z) = I(\omega_z) \quad \text{for} \quad z \in R_{AB} , \quad V_{AB}(z) = 0 \quad \text{for} \quad z \notin R_{AB}$$

is measurable on $(A \times B)^I, \mathbf{F}_{AB})$ as immediately follows from (1.1.9), (1.1.8), (1.1.11), and from the fact that $R_{AB} \in \mathbf{F}_{AB}$; the measurability of R_{AB} was proved in [8], Lemma 2, p. 138 (the above definition of regularity coincides with that given in Sect. 2, loc. cit.).

1.2. Channels. If there are finite non-empty sets A and B, and if $v = \{v_\eta, \eta \in B^I\}$ is a family of probability measures v_η on $\mathbf{F}_A$ with parameters η running over the index set B^I which satisfies the following stationarity-and-finite-past-history condition (cf. convention (1.1.4))

$$(1.2.1) \quad v_\zeta[x]_i = v_\eta[x] \quad \text{for} \quad \zeta \in [y]_{i-m} , \quad \eta \in [y]_{-m} , \quad x \in A^n , \quad y \in B^{m+n}$$

for some non-negative integer m, then the family v will be called a (discrete communication) channel whereas A will be referred to as the output alphabet and B the input alphabet of the channel v. The least non-negative integer m for which condition (1.1.2) is satisfied, will be denoted by $m(v)$; it may be called the duration of past history (or the duration of input memory). In what follows we shall set

$$\nu[E \mid y] = \sum_{x \in E} \nu[x \mid y], \quad E \subset A^n, \tag{1.2.2}$$

$$\nu[x \mid y] = \nu_\eta[x], \quad x \in A^n, \quad y \in B^{m+n},$$

$$\text{where} \quad \eta \in [y]_{-m}, \quad m \geqq m(\nu).$$

If convenient, the channel ν will be referred to as a channel with alphabets A and B whereas elements in A^I and in A^n, respectively, may be referred to as output sequences, and elements in B^I and in B^n, respectively, as input sequences.

We shall make the convention that letters A and B will always mean in what follows the alphabets of channels being under consideration, if not stated otherwise.

Remark. Channels satisfying condition (1.2.1) were studied by the author in [7] and called there stationary channels with finite past history.

If ν is a channel, and if m is a non-negative integer such that $m \geqq m(\nu)$, then an n-dimensional multiple code of past history m (for the channel ν) is defined as a family $Q = \{Q(y), y \in Y\}$ of subsets $Q(y)$ of the space A^n of n-dimensional output sequences with parameters y running over an index set Y of $(m + n)$-dimensional input sequences (i.e. $Y \subset B^{m+n}$), and the (maximum probability of) error of the multiple code Q, denoted by $e(Q, \nu)$ is defined by the relation (for $Y \neq \emptyset$)

$$e(Q, \nu) = \max_{y \in Y} (1 - \nu[Q(y) \mid y]); \tag{1.2.3}$$

as to the sense of the latter definition cf. (1.2.2). The length of the multiple code Q is, by definition, the number of input sequences lying in the index set Y; in symbols:

$$l(Q) = \operatorname{card}(Y), \tag{1.2.4}$$

where card (Y) designates the cardinal number of Y.

Let σ be a positive integer. We shall say that the multiple code Q is of multiplicity $\leqq \sigma$, if every n-dimensional output sequence belongs at most to σ sets of the family Q; more precisely, if

$$\operatorname{card}\{y \in Y : x \in Q(y)\} \leqq \sigma \quad \text{for every} \quad x \in A^n. \tag{1.2.5}$$

If Q is of multiplicity $\leqq 1$, then the family Q is disjoint and will be called a (simple) code or, in more detail, an error-correcting code. This notion coincides with that defined by Wolfowitz in [14].

Let us point out that multiple codes will play in our considerations only an auxiliary role. In the centre of our interest will be the codes that are simple.

An n-dimensional code and an n-dimensional multiple code for the channel ν are defined as being of past history $m = m(\nu)$. If $0 < \varepsilon < 1$, and if Q is an n-dimensional code for ν, such that $e(Q, \nu) < \varepsilon$, then Q is said to be an ε-code. The maximum length of n-dimension al ε-codes for channel ν will be denoted in what follows by $S_n(\varepsilon, \nu)$; in symbols:

$$(1.2.6) \qquad S_n(\varepsilon, \nu) = \max \{l(Q) : Q \text{ is an } n\text{-dimensional } \varepsilon\text{-code for } \nu\} .$$

If ν is a channel, then a probability measure on $\mathbf{F}_B$ (where B is the input alphabet of ν) is said to be an input of the channel ν. If μ is an input of ν, then we shall denote by $\nu\mu$ the measure on $\mathbf{F}_{AB}$ uniquely determined by the property that (cf. (1.1.6))

$$(1.2.7) \quad \nu\mu[x, y]_{i+m,i} = \nu[x \mid y] \, \mu[y]_i , \quad x \in A^n , \quad y \in B^{m+n} , \quad m \geqq m(\nu) .$$

It follows from the measurability of the function V_{AB} defined by (1.1.12) that V_{AB} is a random variable on the space $((A \times B)^I, \mathbf{F}_{AB}, \nu\mu)$, and that, in accordance with (1.1.1), $q(\Theta, \nu\mu; V_{AB})$ is its Θ-quantile. Let us set (cf. (1.1.5))

$$(1.2.8) \qquad c(\Theta, \nu\mu) = q(\Theta, \nu\mu; V_{AB}) \quad \text{for} \quad \mu \in \mathcal{M}_{\mathrm{erg}}(B) ,$$
$$c(\Theta, \nu) = \sup \{c(\Theta, \nu\mu) : \mu \in \mathcal{M}_{\mathrm{erg}}(B)\} ; \quad 0 < \Theta \leqq 1 .$$

The latter quantity represents the least upper bound of Θ-quantiles of V_{AB} associated with measures $\nu\mu$, where parameter μ runs over the class of ergodic inputs.

The capacity function c_ν of the channel ν is defined by the relation $c_\nu(\Theta) = c(\Theta, \nu)$ for $0 < \Theta \leqq 1$; it is monotonically increasing and left-continuous as immediately follows from (1.2.8). The set $\mathcal{D}_\nu$ of discontinuity points of the function c_ν in the open interval (0, 1), which is at most denumerable, is expressed by

$$(1.2.9) \qquad \mathcal{D}_\nu = \{\Theta \in R : 0 < \Theta < 1, \ c_\nu(\Theta) < c_\nu(\Theta + 0)\} .$$

If $0 < \varepsilon < 1$, and if $\varepsilon \notin \mathcal{D}_\nu$ (i.e., if ε is a continuity point of the capacity function c_ν), then the number $c(\varepsilon, \nu)$, i.e. $c_\nu(\varepsilon)$, will be called the ε-capacity of the channel ν.

1.3. The theorem. Let us first remind that a channel is said to have finite memory if it satisfies (together with condition (1.2.1)) the following requirement of stochastic independence

$$(1.3.1) \qquad \nu_\eta([x]_i \cap [v]_j) = \nu_\eta[x]_i \, \nu_\eta[v]_j$$
$$\text{for} \quad x \in A^n , \quad v \in A^{n'} , \quad \eta \in B^I , \quad i + n \leqq j - k$$

for some non-negative integer k. Let us denote the least k for which the condition is fulfilled by $k(\nu)$ (the duration of output memory), and put

$$(1.3.2) \qquad s(\nu) = \max (m(\nu), k(\nu)) .$$

The non-negative integer $s(\nu)$ associated with the finite-memory channel ν is called the duration of its memory.

The (finite-memory) channel ν is, by definition, memoryless if $s(\nu) = 0$; it is uniquely determined by the matrix $\{\nu[a \mid b],\ a \in A,\ b \in B\}$ because

$$\nu[x \mid y] = \prod_{i=0}^{n-1} \nu[x_i \mid y_i]\,, \quad x \in A^n\,, \quad y \in B^n\,; \tag{1.3.3}$$

cf. also (1.2.2), (1.3.1), and take into account that the probability measure ν_η is uniquely determined by its values at cylinder sets $[x]_i$ according to a well-known theorem of Kolmogorov.

A channel ν is said to be decomposable into finite-memory components if there are a probability space $(\Lambda, \mathbf{A}, \xi)$ and a family $\{\nu^\alpha, \alpha \in \Lambda\}$ of finite-memory channels ν^α having each the same alphabets as the channel ν, with parameters α running over the space Λ so that the following conditions are satisfied:

1. $(\nu^\alpha)_\eta\,[x]_i$ as a function of parameter α is measurable on $(\Lambda, \mathbf{A})$ for every $x \in A^n$ and $\eta \in B^I$;

2. the channel ν is composed of the finite-memory channels ν^α by the relation

$$\nu_\eta[x]_i = \int (\nu^\alpha)_\eta\,[x]_i\, d\xi(\alpha)\,, \quad x \in A^n\,, \quad \eta \in B^I\,. \tag{1.3.4}$$

For convenience, we shall say that the channel ν is decomposable into finite-memory components $\nu^\alpha(\alpha \in \Lambda)$ with respect to a probability field $(\mathbf{A}, \xi)$ in the index set Λ, and write symbolically

$$\nu = (\mathbf{A}) \int \nu^\alpha\, d\xi(\alpha)\,, \quad \nu^\alpha \text{ f.mem.}\,, \quad \alpha \in \Lambda\,, \tag{1.3.5}$$

or, more briefly (compare with (1.27) in [9], p. 116),

$$\nu = \int \nu^\alpha\, d\xi(\alpha)\,. \tag{1.3.6}$$

In the sequel a family $\{\nu^\alpha, \alpha \in \Lambda\}$ of channels satisfying Condition 1 stated above will be called measurable on $(\Lambda, \mathbf{A})$. With such a family and a probability measure ξ on $\mathbf{A}$ we may associate the family ν by definition (1.3.4), which constitutes a channel in our sense in case that, for example, $m(\nu^\alpha) \leqq m$ for all $\alpha \in \Lambda$. The latter condition is automatically fulfilled if the family $\{\nu^\alpha\}$ is finite, i.e. if the index set Λ has a finite number of elements. In this case we may write relation (1.3.4) in the symbolic form

$$\nu = \sum_{\alpha \in \Lambda} \xi_\alpha \nu^\alpha \tag{1.3.7}$$

provided that $\sum_{\alpha} \xi_\alpha = 1$, $\xi_\alpha \geqq 0$ $(\alpha \in \Lambda)$, and say that the channel ν is decomposable into a finite number of components; without any loss of generality ν^α in (1.3.7) may be supposed to be mutually distinct with $\xi_\alpha > 0$.

Another special case where relation (1.3.4) or its symbolic version (1.3.6) defines a channel, is described by the condition that $s(\nu^\alpha) = 0$ for all $\alpha \in \Lambda$; in the latter case we shall say that the channel ν is decomposable into memoryless components.

The question what is the asymptotic behaviour of the maximum length of n-dimensional ε-codes for a channel decomposable into finite-memory components is answered in the Corollary to the following theorem that represents the main result of this paper.

Before stating the theorem, let us associate with any channel ν the positive integer

$$d_\nu = \max(\operatorname{card}(A), \operatorname{card}(B)), \tag{1.3.8}$$

where A and B are the alphabets of ν.

Theorem 1. *Let ν be a channel decomposable into finite-memory components, and let ε be a continuity point of the capacity function c_ν. Then for $\lambda > 0$ there exists an index n_0 such that, for $n \geqq n_0$ and $m \geqq m(\nu)$ where $m(\nu)$ is the duration of past history, it holds the*

I. *Coding theorem: There is an n-dimensional code Q of past history m with the error $e(Q, \nu) < \varepsilon$ and the length*

$$l(Q) > \exp_2(n[c_\nu(\varepsilon) - \lambda]);$$

II. *Strong converse: If Q is an n-dimensional code of past history m and of the error $e(Q, \nu) < \varepsilon$, then its length satisfies the inequality*

$$l(Q) < (d_\nu)^{m - m(\nu)} \exp_2(n[c_\nu(\varepsilon) + \lambda]),$$

where d_ν is the number of elements in the alphabet of ν that is of higher cardinality.

The proof of Theorem 1 is given in Sec. 5 and is based on a series of lemmas stated in the subsequent sections.

As an immediate consequence of Theorem 1 and of definition (1.2.6) we obtain the following corollary concerning the asymptotic behaviour the of maximum length $S_n(\varepsilon, \nu)$ of n-dimensional ε-codes for channel ν.

Corollary. *If ν is a channel decomposable into finite-memory components, then (cf. (1.2.6), (1.2.9))*

$$\lim_n \frac{1}{n} \log_2 S_n(\varepsilon, \nu) = c_\nu(\varepsilon) \quad \textit{for} \quad \varepsilon \notin \mathcal{D}_\nu, \quad 0 < \varepsilon < 1.$$

In words, the sequence $\{n^{-1} \log_2 S_n(\varepsilon, \nu)\}$ converges to the ε-capacity of the channel ν (for $\varepsilon \notin \mathscr{D}_\nu$, $0 < \varepsilon < 1$).

Remark. The statement of Theorem 1 was proved for the special case of channels decomposable into memoryless components by the author in [12] (cf. also Theorem 4 in Section 2). The proof of Theorem 1 is substantially based upon the results stated in [12] (see Section 3).

Let us point out that Theorem 1 remains valid for multiple codes with the upper bound for $l(Q)$ given by

$$l(Q) < \sigma(d_\nu)^{m-m(\nu)} \exp_2 (n[c_\nu(\varepsilon) + \lambda]) ,$$

where Q is a multiple code of multiplicity $\leqq \sigma$.

2. Theorems on the capacity function

In this section we shall assume that we are given a channel ν decomposable into finite-memory components ν^α $(\alpha \in \Lambda)$ with respect to a probability field $(\mathbf{A}, \xi)$ in the index set Λ; it is the assumption that

$$\nu = (\mathbf{A}) \int \nu^\alpha \, d\xi(\alpha) , \quad \nu^\alpha \text{ f.mem.} , \quad \alpha \in \Lambda ; \tag{2.1}$$

cf. (1.3.5). If μ is any k-invariant input $(k = 1, 2, \ldots)$, then the function V_μ defined by

$$V_\mu(\alpha) = I(\nu^\alpha \mu) , \quad \alpha \in \Lambda , \tag{2.2}$$

is a random variable on the probability space $(\Lambda, \mathbf{A}, \xi)$. The latter fact is an immediate consequence of definitions (1.1.9), (1.2.7), and of the measurability of the family $\{\nu^\alpha\}$. The Θ-quantile of V_μ we shall denote by

$$q(\Theta, \nu\mu) = q(\Theta, \xi; V_\mu) , \quad 0 < \Theta \leqq 1 . \tag{2.3}$$

Inputs of the given channel which will play the most important role in our study, will be those called here n-independent. A probability measure μ on $\mathbf{F}_B$ is said to be n-independent if (cf. (1.1.4))

$$\mu[z]_{kn} = \prod_{t=0}^{k-1} \mu[\bar{z}_t] , \quad \bar{z}_t = \{z_j, \, tn \leqq j < (t+1)\, n\} \tag{2.4}$$
$$\text{for} \quad z \in B^{kn} , \quad t = 0, 1, \ldots, k-1 \quad (k = 1, 2, \ldots) .$$

The class of all n-independent inputs will be denoted by $\mathscr{M}_n(B)$; since any n-independent input is n-ergodic, it holds that

$$\mathscr{M}_n(B) = \{\mu \in \mathscr{M}_{n-\mathrm{erg}}(B) \colon \mu \text{ is } n\text{-independent}\} . \tag{2.5}$$

We shall see in Section 3 that in the proof of Theorem 1 an important role will be played by the suprema taken over $\mathcal{M}_n(B)$ of the quantiles

$$r_{n,m}(\Theta, \nu\mu) = q(\Theta, \xi; \{\mathcal{R}_{n,m}(\nu^\alpha\mu), \alpha \in \Lambda\}) ; \tag{2.6}$$

cf. (2.1), (1.2.7), (1.1.1), and (1.1.8); $m = 0, 1, \ldots; 0 < \Theta \leqq 1$.

In the following theorem we have set

$$\begin{aligned} c_n(\Theta, \nu) &= \sup \{q(\Theta, \nu\mu) : \mu \in \mathcal{M}_n(B)\} , \\ \bar{c}_n(\Theta, \nu) &= \sup \{q(\Theta, \nu\mu) : \mu \in \mathcal{M}_{n-\mathrm{erg}}(B)\} , \\ r_{n,m}(\Theta, \nu) &= \sup \{r_{n,m}(\Theta, \nu\mu) : \mu \in \mathcal{M}_n(B)\} , \end{aligned} \tag{2.7}$$

where $0 < \Theta \leqq 1$, $m = 0, 1, \ldots$ (cf. (2.3), (2.6), and (2.5)). It is immediately seen that

$$c_n(\Theta, \nu) \leqq \bar{c}_n(\Theta, \nu) , \quad c_n(\Theta, \nu) \leqq c_{kn}(\Theta, \nu) , \quad k = 1, 2, \ldots ; \tag{2.8}$$

the latter inequality holds because $\mathcal{M}_n(B) \subset \mathcal{M}_{kn}(B)$. Another notation of which we shall make use in the theorem is given by

$$\mathbf{A}_\Theta = \{\mathcal{A} \in \mathbf{A} : \xi(\mathcal{A}) \geqq \Theta\} , \quad 0 < \Theta \leqq 1 . \tag{2.9}$$

Theorem 2. *If ν is a channel decomposable into finite-memory components ν^α $(\alpha \in \Lambda)$ such that condition (2.1) is satisfied, then its capacity function c_ν assumes at a point Θ in the semiclosed interval $(0, 1]$ the value which is expressed by*

I. $c(\Theta, \nu) = \bar{c}_n(\Theta, \nu)$, *esp.* $c(\Theta, \nu) = \bar{c}_1(\Theta, \nu)$;

II. $c(\Theta, \nu) = \sup_n c_n(\Theta, \nu)$ *whereas* $c_n(\Theta, \nu) = \sup_{\mu \in \mathcal{M}_n(B)} \inf_{\mathcal{A} \in \mathbf{A}_\Theta} \operatorname{ess.\,sup}_{\alpha \in \mathcal{A}[\xi]} I(\nu^\alpha\mu)$;

III. $c(\Theta, \nu) = \sup_n (1/(m + n))\, r_{n,m}(\Theta, \nu) = \lim_n (1/(m + n))\, r_{n,m}(\Theta, \nu)$,

$m = 0, 1, \ldots$

whereas $r_{n,m}(\Theta, \nu) \leqq (m + n)\, c_{m+n}(\Theta, \nu)$.

Remark 1. If V is a random variable on the probability space $(\Lambda, \mathbf{A}, \xi)$, then its Θ-quantile is expressed by

$$q(\Theta, \xi; V) = \inf_{\mathcal{A} \in \mathbf{A}_\Theta} \operatorname{ess.\,sup}_{\alpha \in \mathcal{A}[\xi]} V(\alpha) , \quad 0 < \Theta \leqq 1 ,$$

as immediately follows from definition (1.1.1) and from the equality

$$\{t \in R : \xi\{V \leqq t\} \geqq \Theta\} = \bigcup_{\mathcal{A} \in \mathbf{A}_\Theta} \{t \in R : \operatorname{ess.\,sup}_{\alpha \in \mathcal{A}[\xi]} V(\alpha) \leqq t\} .$$

From here we obtain the following expression for the capacity function

$$c(\Theta, \nu) = \bar{\bar{c}}_1(\Theta, \nu) = \sup_{\mu \in \mathcal{M}_{\mathrm{erg}}(B)} \inf_{\mathscr{A} \in \mathbf{A}_\Theta} \operatorname{ess.sup}_{\alpha \in \mathscr{A}[\xi]} I(\nu^\alpha \mu)$$

and a similar expression for $r_{n,m}(\Theta, \nu)$. Consequently, also the formula for $c_n(\Theta, \nu)$ given in Theorem 2 is only a consequence of definition (2.7).

We shall postpone the proof of Theorem 2 to Section 5, but we shall show here the validity of the following corollary to Theorem 2.

Corollary. *If ν is a channel decomposable into a finite number of finite-memory components ν^α ($\alpha \in \Lambda$; i.e. Λ finite) such that (cf. (1.3.7))*

$$\nu = \sum \xi_\alpha \nu^\alpha, \quad \xi_\alpha > 0 \ (\alpha \in \Lambda),$$

then every discontinuity point of the capacity function c_ν is of the form $\sum\{\xi_\alpha : \alpha \in \mathscr{A}\}$ for some $\mathscr{A} \subset \Lambda$; in symbols (cf. (1.2.9)):

$$\mathscr{D}_\nu \subset \{\sum_{\alpha \in \mathscr{A}} \xi_\alpha : \mathscr{A} \subset \Lambda\}$$

so that $\mathscr{D}_\nu$ is a finite set. Moreover, the ε-capacity $c(\varepsilon, \nu)$ is constant in every open interval $(\varepsilon_1, \varepsilon_2)$ not containing any discontinuity points from $\mathscr{D}_\nu$ and so is the function $c_n(\varepsilon, \nu)$ whereas

$$c_n(\varepsilon, \nu) = \sup_{\mu \in \mathcal{M}_n(B)} \min_{\mathscr{A} \in \mathbf{A}_\varepsilon} \max_{\alpha \in \Lambda} I(\nu^\alpha \mu).$$

Finally, the capacity $C(\nu)$ of the channel ν defined as the limit $C(\nu) = \lim_{\varepsilon \to 0} c(\varepsilon, \nu)$ may be expressed in the form:

$$C(\nu) = \sup_n \sup_{\mu \in \mathcal{M}_n(B)} \min_{\alpha \in \Lambda} I(\nu^\alpha \mu).$$

Proof. If F_μ is the probability distribution function of V_μ (cf. (2.2)) then the equality

$$F_\mu(t) = \sum\{\xi_\alpha : \alpha \in \Lambda,\ I(\nu^\alpha \mu) \leqq t\}, \quad t \in R$$

implies together with Theorem 2 the assertion stated in the first part of the Corollary. Since, for $\Theta \leqq \min_{\alpha \in \Lambda} \xi_\alpha$, $\mathbf{A}_\Theta = \{\mathscr{A} : \mathscr{A} \subset \Lambda\}$, the second part is an immediate consequence of the first part.

Remark 2. Making use of the expression given for $c(\Theta, \nu)$ in Remark 1, we obtain for Λ finite that

$$C(\nu) = \sup_{\mu \in \mathcal{M}_{\mathrm{erg}}(B)} \min_{\alpha \in \Lambda} I(\nu^\alpha \mu);$$

the latter formula was deduced by the author in [13] (cf. p. 908 loc. cit.) for a more general case of ergodic components. Applying now the preceding corollary to the case that Λ is a one-element set, we find that, for any channel ν with finite memory, the capacity function c_ν is constant and equal to the usual transmission-rate capacity; in symbols:

$$c_\nu(\Theta) = C(\nu) = \sup_{\mu \in \mathcal{M}_{\mathrm{erg}}(B)} I(\nu\mu) .$$

Before stating another theorem on the capacity function, let us point out that both $m(\nu^\alpha)$ and $s(\nu^\alpha)$ as functions of parameter α are measurable on $(\Lambda, \mathbf{A})$; it is because both the conditions (1.2.1) and (1.3.1) may be required to be valid only for input sequences lying in a countable set which is dense in the space B^I with respect to the usual topology having as its open base the class of finite-dimensional cylinders. This topology may be induced by the distance function (3.37) given in [9], p. 238; as to the proof of measurability of $m(\nu^\alpha)$, cf. Sec. 5, p. 249 loc. cit.*.

Now we shall give some other notations which are necessary for stating the theorem. If $\mathscr{A} \in \mathbf{A}$, $\xi(\mathscr{A}) > 0$ (cf. assumptions (2.1)),

$$\xi_{\mathscr{A}}(B) = \frac{\xi(\mathscr{A} \cap \mathscr{B})}{\xi(\mathscr{A})} \quad \text{for} \quad \mathscr{B} \in \mathbf{A} ,$$

and if $m(\nu^\alpha) \leqq m$ for $\alpha \in \mathscr{A}$ where m is a non-negative integer, then the truncated channel $\nu^{\mathscr{A}}$ with the duration of past history $m(\nu^{\mathscr{A}}) \leqq m$ is associated with the channel ν by definition

(2.10) $$\nu^{\mathscr{A}} = \int \nu^\alpha \, \mathrm{d}\xi_{\mathscr{A}}(\alpha) ;$$

cf. (1.3.6) and (1.3.4). Since (cf. (1.3.2))

(2.11) $$\gamma_s = \xi(\mathscr{A}_s) \nearrow 1 \quad \text{where} \quad \mathscr{A}_s = \{\alpha \in \Lambda : s(\nu^\alpha) \leqq s\}$$

for $s = 0, 1, \ldots$, we shall associate with the given channel ν the truncated channels

(2.12) $$\nu_s = \nu^{\mathscr{A}_s} \quad \text{for} \quad \gamma_s > 0 ,$$

i.e. for s sufficiently large. The following theorem establishes the relation between the capacity function of the channel ν and those of truncated channels ν_s.

Theorem 3. *If ε is a continuity point of the capacity function c_ν of a channel ν decomposable into finity-memory components ν^α $(\alpha \in \Lambda)$ by (2.1), then*

$$c(\varepsilon, \nu) = \lim_{s \to \infty} c(\varepsilon, \nu_s) ;$$

---

* In that proof the range of the last operator $\cap$ in the formula for $\{\alpha : m(\gamma^\alpha) \leqq m\}$ is to be $y \in D \cap T_B^m[y]$.

the capacity function $c(\Theta, \nu_s)$ *of the truncated channel* ν_s *defined by* (2.12) *has the property that, for* $0 < \Theta \leqq 1$,

$$\frac{1}{s+n} r_{n,s}(\Theta, \nu_s) \leqq c(\Theta, \nu_s) \leqq \frac{1}{s+n} r_{n,s}(\Theta, \nu_s) + \frac{s}{s+n} \log_2 d_\nu$$

where d_ν *is given by* (1.3.8): *hence*

$$\lim_n \frac{1}{s+n} r_{n,s}(\Theta, \nu_s) = c(\Theta, \nu_s) .$$

Functions $r_{n,s}(\Theta, \nu_s)$ *defined in* (2.7) *are expressible as maxima*

$$r_{n,s}(\Theta, \nu_s) = \max_{\mu \in \mathcal{M}_{s+n}(B)} \inf_{\mathcal{A} \in \mathbf{A}_\Theta} \operatorname{ess.sup}_{\alpha \in \mathcal{A}[\xi_s]} \mathcal{R}_{n,s}(\nu^\alpha \mu) \quad \textit{where} \quad \xi_s = \xi_{\mathcal{A}_s} \quad (\textit{cf.}\ (2.11)) .$$

Remark 3. As an immediate consequence of definition (1.1.8) we obtain the inequality $\mathcal{R}_{n,m}(\omega) \leqq \log_2 (\text{card } B^{m+n})$ valid for any probability measure ω on $\mathbf{F}_{AB}$, which yields the relation

$$c(\Theta, \nu) \leqq \log_2 d_\nu \quad (0 < \Theta \leqq 1)$$

for the capacity function of an arbitrary channel ν.

Corollary. *Under the assumption* (2.1), *if* $s(\nu^\alpha) \leqq s$ *for all* $\alpha \in \Lambda$ a.s. $[\xi]$ *and for some non-negative integer* s *then* (*for* $0 < \Theta \leqq 1$)

$$\frac{1}{s+n} r_{n,s}(\Theta, \nu) \leqq c(\Theta, \nu) \leqq \frac{1}{s+n} r_{n,s}(\Theta, \nu) + \frac{1}{s+n} \log_2 \mathrm{d}_\nu$$

so that

$$c(\Theta, \nu) = \lim_n \frac{1}{s+n} r_{n,s}(\Theta, \nu) .$$

If the index set Λ *is finite, then the preceding statement holds for* $s \geqq \max_{\alpha \in \Lambda} s(\nu^\alpha)$, *and if* $\nu = \sum \xi_\alpha \nu^\alpha$, $\xi_\alpha > 0$ $(\alpha \in \Lambda)$ *then*

$$r_{n,s}(\Theta, \nu) = \max_{\mu \in \mathcal{M}_{s+n}(B)} \min_{\mathcal{A} \in \mathbf{A}_\Theta} \max_{\alpha \in \mathcal{A}} \mathcal{R}_{n,s}(\nu^\alpha \mu) .$$

The corollary immediately follows from Theorem 3 the proof of which will be given in Section 5.

The last theorem we are going to state, is to show that the capacity function of a channel decomposable into memoryless components is, indeed, identical with that given for this case in [12]. Let us remark that for continuity points we may obtain the equality just mentioned directly from Theorem 1 and the theorem on ε-capacity stated in [12].

In what follows we shall associate with any probability B-vector $p \in P(B)$ (cf. (1.1.3)) the probability measure μ^p uniquely determined by

(2.13) $$\mu^p[y]_i = \prod_{j=0}^{n-1} p(y_j), \quad y \in B^n ;$$

measure μ^p is usually called a memoryless input of the channel v (given here by (2.1)). It follows from (2.4) and (2.5) that

(2.14) $$\mathcal{M}_1(B) = \{\mu^p : p \in P(B)\} \subset \mathcal{M}_{\text{erg}}(B) ;$$

in words, an input is memoryless iff it is 1-independent. In the following theorem we have set

(2.15) $$r(\Theta, v) = \sup \{r_p(\Theta, v) : p \in P(B)\} \quad \text{where}$$
$$r_p(\Theta, v) = q(\Theta, \xi; \{\mathcal{R}_1(v^\alpha \mu^p), \alpha \in \Lambda\}, p \in P(B))$$

for $0 < \Theta \leqq 1$; cf. (1.1.1), (1.1.10), and (2.13). Let us point out that the latter notations coincide with those given in [12], Sec. 1, (1.3.4) and (1.3.7) loc. cit. According to (2.14), (2.6) and (2.7) we obtain that

(2.16) $$r_p(\Theta, v) = r_{1,0}(\Theta, v\mu^p), \quad r(\Theta, v) = r_{1,0}(\Theta, v) .$$

Theorem 4. *If v is a channel decomposable into memoryless components v^α $(\alpha \in \Lambda)$, and if $v = \int v^\alpha \, d\xi(\alpha)$, then the capacity function c_v depends only on the class of memoryless inputs according to the equalities*

$$c(\Theta, v) = \sup_{\mu \in \mathcal{M}_1(B)} c(\Theta, v\mu) = r(\Theta, v), \quad 0 < \Theta \leqq 1 ,$$

where $r(\Theta, v)$ is defined in (2.15) and may be expressed in the form

$$r(\Theta, v) = \max_{p \in P(B)} \inf_{\mathcal{A} \in \mathbf{A}_\Theta} \operatorname{ess.\,sup}_{\alpha \in \mathcal{A}[\xi]} \mathcal{R}_1(v^\alpha \mu^p) .$$

Corollary. *If v is a channel decomposable into a finite number of memoryless channels v^α $(\alpha \in \Lambda)$, and if $v = \sum \xi_\alpha v^\alpha$, $\xi_\alpha > 0$ $(\alpha \in \Lambda)$ then*

$$c(\Theta, v) = r(\Theta, v) = \max_{p \in P(B)} \min_{\mathcal{A} \in \mathbf{A}_\Theta} \max_{\alpha \in \mathcal{A}} \mathcal{R}_1(v^\alpha \mu^p) .$$

The assertion of the corollary follows from Corollary to Theorem 3 and from the relations (2.14) and (2.16). The proof of Theorem 4 will be postponed to Section 5.

Remark 4. We shall see in Sec. 5 that, under the assumptions of Theorem 4,

$$r_{n,m}(\Theta, v) = r_{n,0}(\Theta, v) = n\, r(\Theta, v)$$

for $0 < \Theta \leqq 1$ $(m = 1, 2, \ldots)$.

Example. Given a positive integer k, we shall consider the channel v defined symbolically by

$$v = \sum_{\alpha=1}^{k} \xi_\alpha v^\alpha , \quad \xi_\alpha = \frac{1}{k} \quad (\alpha = 1, \ldots, k)$$

with alphabets $A = \Lambda = \{1, 2, \ldots, k\}$ and $B = \Lambda \times \Lambda$, where, for $\alpha \in \Lambda$, v^α is the memoryless channel uniquely determined according to (1.3.3) by

$$v^\alpha[i \mid (\beta, j)] = \frac{1}{k} \quad \text{for} \quad \beta \neq \alpha , \quad v^\alpha[i \mid (\alpha, i)] = 1$$

for i, j, α, β in the index set Λ. The formula in the preceding Corollary to Theorem 4 enables to find an explicit expression for the capacity function of the given channel, the capacity of which (i.e. $\lim_{\Theta \to 0} c_v(\Theta)$) was first calculated by J. Nedoma in [5], Example 1, pp. 385–389. To the author's suggestion the formula for c_v of the latter channel v was first derived by J. Kadlec in an unpublished thesis (cf. [3]).

We shall associate with any $p \in P(B)$ and $\alpha \in \Lambda$ a probability A-vector $\pi_p^\alpha \in P(A)$ defined by

$$\pi_p^\alpha(i) = p(\alpha, i) + \frac{1}{k}(1 - p_\alpha) \quad \text{where} \quad p_\alpha = \sum_{i=1}^{k} p(\alpha, i) .$$

An easy calculation yields the expression

$$\mathcal{R}_1(v^\alpha \mu^p) = \mathcal{H}(\pi_p^\alpha) - (1 - p_\alpha) \log_2 k ,$$

where $\mathcal{H}$ is the entropy operator given by

$$\mathcal{H}(\pi) = - \sum_{i=1}^{k} \pi(i) \log_2 \pi(i) , \quad \pi \in P(A) .$$

Let f map $P(B)$ into itself by definition

$$(fp)(\alpha, i) = \frac{1}{k} p_\alpha ;$$

then $\mathcal{R}_1(v^\alpha \mu^p) \leqq \mathcal{R}_1(v^\alpha \mu^{fp}) = p_\alpha \log_2 k$, as follows from the well-known properties of entropy $\mathcal{H}$. It is immediately seen that $\mathcal{A} \in \mathbf{A}_\Theta$ if card $(\mathcal{A}) \geqq k\Theta$. From the above considerations we conclude that

$$r(\Theta, v) = \max_{p \in P(B)} \min_{\text{card}(\mathcal{A}) \geqq k\Theta} \max_{\alpha \in \mathcal{A}} p_\alpha \log_2 k .$$

Assertion. For $s = 1, 2, \ldots, k$ and for every $p \in P(B)$ there is $\mathcal{A} \subset \Lambda$ such that

$$\text{card}(\mathcal{A}) \geqq s , \quad \max_{\alpha \in \mathcal{A}} p_\alpha \leqq \frac{1}{k - s + 1} .$$

Proof. Suppose the contrary that, for some s and p, card $(\mathscr{A}) \geqq s$ implies $\max_{\alpha \in \mathscr{A}} p_\alpha > t^{-1}$ where $t = k - s + 1$. Without any loss of generality we may assume that $p_\alpha \geqq p_{\alpha+1}$; then, for $\mathscr{A} = \{\alpha : t \leqq \alpha \leqq k\}$, $p_t = \max_{\alpha \in \mathscr{A}} p_\alpha > t^{-1}$, which yields the contradiction that $\sum_{\alpha \leqq t} p_\alpha > 1$.

Putting $\langle k\Theta \rangle = s$ if $s - 1 < k\Theta \leqq s$ $(s = 1, 2, \ldots, k)$, we obtain now the expression for the capacity function in the form

$$c(\Theta, v) = r(\Theta, v) = \frac{\log_2 k}{k - \langle k\Theta \rangle + 1}, \quad 0 < \Theta \leqq 1 ;$$

so the capacity $C(v) = k^{-1} \log_2 k$, but $C(v^\alpha) = \log_2 k$ for $\alpha = 1, \ldots, k$ as shown in [5].

3. The upper and the lower bounds

Throughout this section we shall make the assumption that we are given a non-negative integer s and a channel v decomposable into finite-memory components v^α $(\alpha \in \Lambda)$ such that

(3.1) $$v = (\mathbf{A}) \int v^\alpha \, d\xi(\alpha), \quad s(v^\alpha) \leqq s \quad \text{for every} \quad \alpha \in \Lambda ;$$

cf. (1.3.4), (1.3.5), and (1.3.2). In other words, the duration of memory of the components of the channel is bounded from above by s. Let us notice that the latter assumption has the consequence that

$$\xi(\mathscr{A}_s) = 1, \quad \text{i.e.} \quad \xi_{\mathscr{A}_s} = \xi, \quad v_s = v,$$

as follows from definitions (2.11) and (2.12). In what follows we shall set (cf. (1.3.8))

(3.2) $$d = d_v, \quad \text{i.e.} \quad d = \max(\text{card}(A), \text{card}(B)).$$

We shall see in this section that in order to find the upper bound for the length of n-dimensional simple codes (with error less than ε) it is necessary first to derive the upper bound for the length of multiple codes in case that $s = 0$, i.e. in case that the components v^α are all memoryless. The latter will be done in the following lemma.

Lemma 1. *If $s = 0$, and if ε and Θ are positive and such that $\varepsilon < \Theta \leqq 1$, then there is a constant $K = K(\varepsilon, \Theta; v)$ for which it holds the following*

Assertion on upper bound: If Q is a k-dimensional multiple code of multiplicity $\leqq \sigma$, and if $e(Q, v) < \varepsilon$ (cf. (1.2.3)) then (cf. (1.2.4), (2.15))

$$l(Q) < \sigma \exp_2(k\, r(\Theta, v) + k^{1/2} K) ; \quad k = 1, 2, \ldots .$$

For the multiplicity $\sigma = 1$, i.e. for simple codes, the assertion of the lemma coincides with Theorem 2 in [12], Sec. 3. For an arbitrary $\sigma \geqq 1$, the proof of Theorem 2 loc. cit. is immediately modified by making use of the following lemma for truncated channels instead of Lemma 2 from [12]. In the lemma $F_n(p)$ is the set of all n-dimensional p-sequences as defined by (2.2) in Sec. 2 of [11] (the definition is due to Wolfowitz; cf. [14], Chapter 2, Sec. 2.1, p. 6), as to the notations $P(B)$ and $v^{\mathscr{A}}$, cf. (1.1.3) and (2.10) in the preceding sections.

Lemma for truncated channels. *If $s = 0$ then for ε' lying in the open interval $(0, 1)$ there is a constant K_0 such that if $\mathscr{A}(\mathscr{A} \in \mathbf{A}$; cf. (3.1)) is a nonempty set with $\xi(-1) > 0$, if $p \in P(B)$, and if $0 < \varepsilon < 1 - \varepsilon'$, then it holds the following*

Assertion on upper bound: If $Q = \{Q(y), y \in Y\}$ is an n-dimensional multiple code of multiplicity $\leqq \sigma$ having the property that $Y \subset F_n(p)$, and if $e(Q, v^{\mathscr{A}}) < \varepsilon$ then (cf. (2.13), (1.2.7), (1.1.10))

$$l(Q) < \sigma(1 - \varepsilon - \varepsilon')^{-1} \exp_2 (n\mathscr{R} + n^{1/2} K_0)$$

$$\textit{where} \quad \mathscr{R} = \operatorname*{ess.\,sup}_{\alpha \in \mathscr{A}[\xi]} \mathscr{R}_1(v^{\alpha}\mu^{p}) .$$

Proof. As in the proof of Lemma 2 in [12] we deduce from the assumptions that the desired inequality is valid by making in the course of the proof the following changes: add $n^{-1} \log_2\sigma$ to the expression for $\mathbf{H}p_n$ and make use of a modified inequality for $\pi(Y_{\alpha})$ by multiplying the right-hand side of the original one by σ. Q.E.D.

The lower bound is given for a channel decomposable into memoryless components in the following

Lemma 2. *If $s = 0$, and if ε and Θ' are positive and such that $\Theta' < \varepsilon < 1$, then there is a constant $K' = K'(\varepsilon, \Theta'; v)$ with the property that, for $k = 1, 2, \ldots$, there is a k-dimensional code Q for which (cf. (1.2.3), (1.2.4), (2.15))*

$$e(Q, v) < \varepsilon , \quad l(Q) > \exp_2 (k\, r(\Theta, v) - k^{1/2} K') .$$

The assertion of the lemma is equivalent to that of Theorem 3 stated in Sec. 3 of [12] (for $\Theta' > 0$; cf. definitions (1.3.7) loc. cit.).

Now we shall proceed to the general case assuming that the given channel v satisfies conditions (3.1) for some fixed integer $s \geqq 0$. For $t = 1, 2, \ldots$, let $(v^{\alpha})_t$ be the memoryless channel with alphabets A^t and B^{s+t} which is uniquely determined by the requirement that (cf. (1.3.3))

$$(v^{\alpha})_t [\bar{a} \mid \bar{b}] = v^{\alpha}[\bar{a} \mid \bar{b}] \quad \text{for} \quad \bar{a} \in \bar{A} = A^t , \quad \bar{b} \in \bar{B} = B^{s+t}(\alpha \in A) .$$

The sense of the definition is guaranteed by the fact that $v^{\alpha}[\bar{a} \mid \bar{b}]$ is defined according to (1.2.2) because $m(v^{\alpha}) \leqq s(v^{\alpha}) \leqq s$; this auxiliary construction of memoryless channels approximating a channel with finite memory is due to Wolfowitz (cf. [14], Chapter 5).

The family $\{(\nu^\alpha)_t, \alpha \in \Lambda\}$ of memoryless channels is measurable on $(\Lambda, \mathbf{A})$ so that the relation

$$(3.3) \qquad (\nu)_t = \int (\nu^\alpha)_t \, d\xi(\alpha)$$

defines a channel with alphabets A^t and B^{s+t} and with the duration of past history $m((\nu)_t) = 0$; cf. (1.3.6) and (1.3.4). In other words, for $t = 1, 2, \ldots, (\nu_t)$ is the channel with zero past history decomposable into memoryless components $(\nu^\alpha)_t$ $(\alpha \in \Lambda)$ with respect to the probability field $(\mathbf{A}, \xi)$ in Λ.

Since there is a one-to-one correspondence between the class $\mathcal{M}_{s+t}(B)$ of $(s + t)$-independent inputs of ν (cf. (2.5)) and the class $\mathcal{M}_1(B^{s+t})$ of memoryless inputs of the channel (ν_t) (cf. (2.14)) given by associating with an $(s + t)$-independent input μ of ν the memoryless input $\bar{\mu}$ of $(\nu)_t$ determined by

$$\bar{\mu}[\bar{b}] = \mu[\bar{b}] \, , \quad \bar{b} \in B^{s+t} \quad (\text{cf. } (2.4)) \, ,$$

it is immediately seen from (1.1.8) and (1.2.7) that

$$(3.4) \qquad \mathcal{R}_{t,s}(\nu^\alpha \mu) = \mathcal{R}_1((\nu^\alpha)_t \, \bar{\mu}) \, , \quad \alpha \in \Lambda \, .$$

Setting in this section for the sake of simplicity

$$(3.5) \qquad r_t(\Theta, \nu) = r_{t,s}(\Theta, (\nu) \, , \quad 0 < \Theta \leqq 1 \, ,$$

we obtain from (3.4) taking into account definitions (2.6), (2.7), and (2.15) that

$$(3.6) \qquad r_t(\Theta, \nu) = r(\Theta, (\nu)_t) \, , \quad 0 < \Theta \leqq 1 \, .$$

In the remainder of this section we shall assume that t $(t = 1, 2, \ldots)$ is kept fixed, and set

$$(3.7) \qquad \bar{\nu} = (\nu)_t \, , \quad \bar{A} = A^t \, , \quad \bar{B} = B^{s+t} \, ;$$

this means that $\bar{\nu}$ is defined by (3.3), and that $\bar{A}$ is its output alphabet and $\bar{B}$ its input alphabet.

For $k = 1, 2, \ldots$ and for integers $m \geqq m(\nu)$, we define the transformations

$$\varphi_k : A^{k(s+t)} \to \bar{A}^k \, , \quad \psi_{k,m} : B^{m+k(s+t)} \to \bar{B}^k$$

by relations

$$(3.8) \qquad (\varphi_k x)_i = \{x_j, \, i(s + t) + s \leqq j < (i + 1)(s + t)\} \, , \quad x \in A^{k(s+t)} \, ,$$
$$(\psi_{k,m} y)_i = \{y_j, \, i(s + t) \leqq j - m < (i + 1)(s + t)\} \, ,$$
$$y \in B^{m+k(s+t)} \, ; \quad i = 0, 1, \ldots, k - 1 \, .$$

It follows from the construction of channel $\bar{\nu}$ that

(3.9) $$\bar{\nu}[\bar{E} \mid \bar{y}] = \nu[\varphi^{-1}(\bar{E}) \mid y] \quad \text{for} \quad y \in \psi^{-1}(\{\bar{y}\}),$$

where $\bar{E} \subset \bar{A}^k$, $\bar{y} \in \bar{B}^k$, $\varphi = \varphi_k$, $\psi = \psi_{k,m}$ $(k = 1, 2, \ldots, m \geqq m(\nu))$.

Proof. Given $y \in \psi^{-1}(\{\bar{y}\})$, $\eta \in [y]_{-m}$, and $\bar{x} \in \bar{A}^k$, we conclude from (3.8), (1.3.1), (1.2.2), (1.2.1), and from the definition of $(\nu^\alpha)_t$ by (1.3.3) that

$$\nu^\alpha_\eta[\varphi^{-1}(\{\bar{x}\})] = \nu^\alpha_\eta(\bigcap_{i=0}^{k-1} [\bar{x}_i]_{s+i(s+t)}) = \prod_{i=0}^{k-1} \nu^\alpha[\bar{x}_i \mid \bar{y}_i] = (\nu^\alpha)_t [\bar{x} \mid \bar{y}]\,;$$

it is because $k(\nu^\alpha) \leqq s(\nu^\alpha) \leqq s$, $m(\nu^\alpha) \leqq s(\nu^\alpha) \leqq s$ according to (3.1) and (1.3.2). Summing up for $\bar{x}$ running over $\bar{E}$, and integrating for α running over Λ with respect to ξ, we obtain the equality (cf. (3.1), (1.3.4))

$$\nu_\eta[\varphi^{-1}(\bar{E})] = \bar{\nu}[\bar{E} \mid \bar{y}]\,,$$

which gives the desired result because of $m \geqq m(\nu)$. (It is easy to see that, moreover, $\nu^\alpha_\eta[\varphi^{-1}(\bar{E})] = \nu^\alpha[\varphi^{-1}(\bar{E}) \mid y]$ since $\varphi^{-1}(\bar{E})$ is a cylinder in coordinates $\geqq s$.)

Lemma 3. *For ε and Θ such that $0 < \varepsilon < \Theta \leqq 1$, there is a constant $K_t = K_t(\varepsilon, \Theta; \nu)$ for which it holds*
Assertion on upper bound: If $n = k(s + t)$, and if Q is an n-dimensional code of past history m where $m \geqq m(\nu)$, then the inequality $e(Q, \nu) < \varepsilon$ (cf. (1.2.3)) *implies that*

$$l(Q) < d^m \exp_2 (n[W_t(\Theta) + k^{-1/2} K_t])$$

(cf. (1.2.4) *and* (3.2)) *where* (cf. (3.5))

$$W_t(\Theta) = \frac{r_t(\Theta, \nu)}{s + t} + \frac{s}{s + t} \log_2 d\,; \quad k = 1, 2, \ldots.$$

Proof. For ε, Θ, and for the channel $\bar{\nu} = (\nu)_t$ associated with ν by (3.3), let $K = K(\varepsilon, \Theta; \bar{\nu})$ be a constant such that the assertion on the upper bound of Lemma 1 holds for the channel $\bar{\nu}$ with K; set

$$K_t = \frac{K}{s + t}\,.$$

Given k and $n = k(s + t)$, let $Q = \{Q(y), y \in Y\}$ be an n-dimensional code of past history $m \geqq m(\nu)$ for the channel ν (given by (3.1)) with the property that $e(Q, \nu) < \varepsilon$. Associate with Q the k-dimensional multiple code $\bar{Q} = \{\bar{Q}(\bar{y}), \bar{y} \in \bar{Y}\}$ for the channel $\bar{\nu}$ defined by

$$\bar{Y} = \psi(Y)\,, \quad \bar{Q}(\bar{y}) = \varphi(Q'(\bar{y})) \quad \text{for} \quad \bar{y} \in \bar{Y}\,,$$

where

$$Q'(\bar{y}) = \bigcup\{Q(y) : y \in \psi^{-1}(\{\bar{y}\}) \cap Y\},$$

and where $\varphi = \varphi_k$ and $\psi = \psi_{k,m}$ (cf. (3.8)). It immediately follows from (3.9) that

$$\bar{v}[\bar{Q}(\bar{y}) \mid \bar{y}] \geqq v[Q(y) \mid y] \quad \text{for} \quad \bar{y} = \psi(y), \quad y \in Y$$

so that $e(\bar{Q}, \bar{v}) < \varepsilon$. Clearly, the multiple code $\bar{Q}$ is of multiplicity $\leqq \operatorname{card}(A^{ks})$, and

$$l(Q) \leqq \operatorname{card}(B^m) \,.\, l(\bar{Q}),$$

as follows from the construction of $\bar{Q}$. Applying Lemma 1, we obtain according to (3.6) that

$$l(\bar{Q}) < d^{ks} \exp_2\left(n\left[\frac{r_t(\Theta, v)}{s + t} + k^{-1/2} K_t\right]\right) = \exp_2\left(n[W_t(\Theta) + k^{-1/2} K_t]\right);$$

since $l(Q) \leqq d^m\, l(\bar{Q})$, we get from here the desired result.

Lemma 4. *If ε and Θ' are positive numbers such that $\Theta' < \varepsilon < 1$, then there is a constant $K'_t = K'_t(\varepsilon, \Theta'; v)$ with the property that, for any integer $m \geqq m(v)$ and for $k = 1, 2, \ldots$, if $n = k(s + t)$ then there exists an n-dimensional code Q of past history m such that (cf.* (1.2.3), (1.2.4), *and* (3.5))

$$e(Q, v) < \varepsilon, \quad l(Q) > \exp_2\left(n\left[\frac{r_t(\Theta, v)}{s + t} - k^{-1/2} K'_t\right]\right).$$

Proof. According to Lemma 2 applied to the channel $\bar{v} = (v)_t$, associated with v by (3.3), and according to (3.5), if k is given, then there is a k-dimensional code $\bar{Q} = \{\bar{Q}(\bar{y}), \bar{y} \in \bar{Y}\}$ for the channel $\bar{v}$ such that $e(\bar{Q}, \bar{v}) < \varepsilon$ and

$$l(\bar{Q}) > \exp_2\left(n\left[\frac{r_t(\Theta', v)}{s + t} - \frac{K'}{s + t} k^{-1/2}\right)\right]$$

for $n = k(s + t)$, where K' is a constant depending on ε, Θ', and $\bar{v}$. Given an integer $m \geqq m(v)$, associate with each $\bar{y} \in \bar{Y}$ the element $y = f(\bar{y})$ in B^{m+n} for which $\psi y = \bar{y}$, $y_j = b_0$ $(0 \leqq j < m)$, where b_0 is a fixed element in B, and where $\psi = \psi_{k,m}$. Setting

$$Y = f(\bar{Y}), \quad Q(y) = \varphi^{-1}(\bar{Q}(\psi y)) \quad \text{for} \quad y \in Y,$$

where $\varphi = \varphi_k$, we obtain an n-dimensional code $Q = \{Q(y), y \in Y\}$ of past history m (for the channel v) having the property that $e(Q, v) < \varepsilon$, as follows from (3.9). Since $l(Q) = l(\bar{Q})$, we get the desired inequality for $l(Q)$ from that for $l(\bar{Q})$ by putting $K'_t = (s + t)^{-1} K'$.

4. Some properties of the capacity function

As in Section 2, we shall assume throughout the entire section that we are given a channel ν decomposable into finite-memory components ν^α ($\alpha \in \Lambda$) by (2.1). Since the equality $\omega(R_{AB}) = 1$ holds for any 1-invariant measure ω on $\mathbf{F}_{AB}$ (cf. [8], Lemma 2, p. 138; R_{AB} is the set of regular points), and since, owing to (1.2.7), $\nu\mu$ is a 1-invariant measure for any 1-invariant input μ, we may conclude that, according to (1.1.5), $\nu\mu(R_{AB}) = 1$ for any $\mu \in \mathcal{M}_{\mathrm{erg}}(B)$; hence

$$V_{AB}(z) = I(\omega_z) \quad \text{a.s.} \quad [\nu\mu] \quad \text{for any} \quad \mu \in \mathcal{M}_{\mathrm{erg}}(B)\,, \tag{4.1}$$

as follows from definition (1.1.12); cf. (1.1.11).

Making use of a well-known result of Khinchin on ergodicity of finite-memory channels (cf. [4], Sec. 11), we obtain that

$$\nu^\alpha\mu \in \mathcal{M}_{\mathrm{erg}}(A \times B) \quad \text{for} \quad \mu \in \mathcal{M}_{\mathrm{erg}}(B)\,, \quad \alpha \in \Lambda\,,$$

because we have assumed that each of the channels ν^α has finite memory. The latter fact enables us to apply equality (5.8) given in [9], p. 251 (cf. also the proof of Lemma 2.1, p. 119 loc. cit.) to the given channel ν, which yields the relation

$$\nu\mu\{z \in R_{AB} : I(\omega_z) \leqq t\} = \xi\{\alpha \in \Lambda : I(\nu^\alpha\mu) \leqq t\} \quad \text{for} \quad \mu \in \mathcal{M}_{\mathrm{erg}}(B)\,, \quad t \in R\,.$$

From here and from (4.1) we conclude that, for $\mu \in \mathcal{M}_{\mathrm{erg}}(B)$, the random variable V_{AB} has (with respect to $\nu\mu$) the same probability distribution as the random variable V_μ (taken with respect to ξ) defined by (2.2); consequently,

$$c(\Theta, \nu\mu) = q(\Theta, \nu\mu) \quad \text{for any} \quad \mu \in \mathcal{M}_{\mathrm{erg}}(B) \tag{4.2}$$

and for $0 < \Theta \leqq 1$, as follows from definitions (1.2.8) and (2.3).

Let us associate with any n-ergodic input μ, i.e. $\mu \in \mathcal{M}_{n-\mathrm{erg}}(B)$, the probability measure $\tilde{\mu}$ on $\mathbf{F}_B$ uniquely determined by

$$\tilde{\mu}[z]_j = \frac{1}{n}\sum_{i=0}^{n-1} \mu[z]_{i+j}\,, \quad z \in A^k \quad (k = 1, 2, \ldots)\,. \tag{4.3}$$

It is easy to see that $\tilde{\mu} \in \mathcal{M}_{\mathrm{erg}}(B)$ provided that $\mu \in \mathcal{M}_{n-\mathrm{erg}}(B)$; as to the ergodicity of $\tilde{\mu}$ in this case, cf. Theorem 7.1, p. 732 in [7]. On the other hand, it follows from equality (3.29) in [9], p. 237 that

$$I(\nu^\alpha\tilde{\mu}) = I(\nu^\alpha\mu) \quad \text{for} \quad \mu \in \mathcal{M}_{n-\mathrm{erg}}(B)\,, \quad \alpha \in \Lambda\,;$$

hence we obtain according to (2.2) that

$$q(\Theta, \nu\,\tilde{\mu}) = q(\Theta, \nu\mu)\,, \quad 0 < \Theta \leqq 1\,, \tag{4.4}$$

for any $\mu \in \mathcal{M}_{n-\mathrm{erg}}(B)$.

Now we can show that, for the supremal quantiles $c(\Theta, v)$ and $\bar{c}_n(\Theta, v)$ defined by (1.2.8) and (2.7), respectively, it holds

Lemma 5. *For* $0 < \Theta \leqq 1$, $c(\Theta, v) = \bar{c}_n(\Theta, v)$.

Proof. Relations (4.4) and (4.2) imply that $\bar{c}_n(\Theta, v) \leqq c(\Theta, v)$. As shown in [6], Satz 1, for any $\mu' \in \mathscr{M}_{\text{erg}}(B)$ there is $\mu \in \mathscr{M}_{n-\text{erg}}(B)$ such that $\mu' = \tilde{\mu}$ where $\tilde{\mu}$ is defined by (4.3). The latter fact implies that $c(\Theta, v) \leqq \bar{c}_n(\Theta, v)$. Q.E.D.

Making use of the inequality given in [1], Sec. 2, p. 30 (cf. also [11], Sec. 2, (2.14), p. 121), we deduce from (1.1.8) (cf. (1.2.7)) that if m and s are non-negative integers such that $s \geqq m$ then

$$\mathscr{R}_{n,s}(v^\alpha\mu) \leqq \mathscr{R}_{n+(s-m),m}(v^\alpha\mu), \quad \alpha \in \Lambda \tag{4.5}$$

for any input μ. In Section 5 we shall make use of the inequality

$$r_{n,s}(\Theta, v) \leqq r_{n+(s-m),m}(\Theta, v), \quad 0 < \Theta \leqq 1. \tag{4.6}$$

which follows (for $s \geqq m$) from (4.5) and (1.1.2), owing to definitions (2.6) and (2.7), by taking the supremum of the corresponding quantiles for μ running over $\mathscr{M}_{s+n}(B)$; cf. (2.5).

According to (1.1.10) we have as a special case of (4.5) the inequality

$$\mathscr{R}_{n,m}(v^\alpha\mu) \leqq \mathscr{R}_{m+n}(v^\alpha\mu), \quad \alpha \in \Lambda \quad (m = 0, 1, \ldots).$$

From Lemma 3.6 in [9], p. 238, we obtain that if μ is $(m+n)$-independent then

$$k\,\mathscr{R}_{m+n}(v^\alpha\mu) \leqq \mathscr{R}_{k(m+n)}(v^\alpha\mu) \quad \text{for} \quad k = 1, 2, \ldots;$$

from here and from (1.1.9) we conclude that

$$\frac{1}{2^t(m+n)}\,\mathscr{R}_{2^t(m+n)}(v^\alpha\mu) \nearrow I(v^\alpha\mu) \quad \text{for} \quad t \to \infty$$

so that

$$\frac{1}{m+n}\,\mathscr{R}_{n,m}(v^\alpha\mu) \leqq I(v^\alpha\mu) \quad \text{for} \quad \mu \in \mathscr{M}_{m+n}(B)$$

where $\alpha \in \Lambda$. The latter relation implies similarly as above according to (1.1.2) and (2.3) that

$$\frac{1}{m+n}\,r_{n,m}(\Theta, v\mu) \leqq q(\Theta, v\mu) \quad \text{for} \quad \mu \in \mathscr{M}_{m+n}(B), \tag{4.7}$$

$m = 0, 1, \ldots, 0 < \Theta \leqq 1$. Taking in (4.7) the supremum for μ running over $\mathscr{M}_{m+n}(B)$, and then making use of Lemma 5 and of (2.8), we get in accordance with (2.7) the inequalities

$$(4.8) \quad \frac{1}{m+n} r_{n,m}(\Theta, \nu) \leqq c_n(\Theta, \nu) \leqq c(\Theta, \nu), \quad 0 < \Theta \leqq 1, \quad m = 0, 1, \ldots .$$

Lemma 6. *If* $m(\nu^\alpha) \leqq m$ *for all* $\alpha \in \Lambda$ *and for some non-negative integer* m, *then, for* $0 < \Theta \leqq 1$,

$$\sup_n \frac{1}{m+n} r_{n,m}(\Theta, \nu) = c(\Theta, \nu) .$$

Proof. Taking into account relation (4.8), assume on the contrary that, for some $\delta > 0$, there is an ergodic input μ such that (cf. (1.2.8))

$$(1) \qquad \sup_n \frac{1}{m+n} r_{n,m}(\Theta, \nu) \leqq c(\Theta, \nu\mu) - 2\delta .$$

From the left-continuity of $c(\Theta, \nu\mu)$ at the point Θ it follows that $c(\Theta, \nu\mu) - \delta < c(\Theta', \nu\mu)$ for some positive $\Theta' < \Theta$. According to (1.1.9).

$$\frac{1}{m+n} \mathscr{R}_{n,m}(\nu^\alpha\mu) \to I(\nu^\alpha\mu) \quad \text{for all} \quad \alpha \in \Lambda$$

so that, owing to Egorov's theorem, there is a set $\mathscr{B} \in \mathbf{A}$ with the properties that $\xi(\mathscr{B}) > 1 - (\Theta - \Theta')$, and that, for n sufficiently large,

$$(2) \qquad \frac{1}{m+n} \mathscr{R}_{n,m}(\nu^\alpha\mu) > I(\nu^\alpha\mu) - \delta \quad \text{for all} \quad \alpha \in \mathscr{B} .$$

Fixing n for which relation holds, construct the $(m + n)$-independent input μ' associated with μ that is uniquely determined by

$$\mu'[y] = \mu[y] \quad \text{for} \quad y \in B^{m+n} ;$$

cf. (2.4) and (1.1.4). Since $m(\nu^\alpha) \leqq m$, it follows from (1.2.7) and (1.1.8) that

$$(3) \qquad \mathscr{R}_{n,m}(\nu^\alpha\mu') = \mathscr{R}_{n,m}(\nu^\alpha\mu) ; \quad \alpha \in \Lambda .$$

Now we obtain from (1) for the constructed input μ' that

$$\xi\{\alpha \in \Lambda : \mathscr{R}_{n,m}(\nu^\alpha\mu') \leqq (m+n)\left[c(\Theta, \nu\mu) - 2\delta\right]\} \geqq \Theta ;$$

since $\xi(\Lambda - \mathscr{B}) < \Theta - \Theta'$, it follows from (2), (3), and (4.2) that

$$\Theta' \leqq \xi\{\alpha \in \mathscr{B} : \mathscr{R}_{n,m}(\nu^\alpha\mu) \leqq (m+n)\left[c(\Theta, \nu\mu) - 2\delta\right]\} \leqq$$
$$\leqq \xi\{\alpha \in \mathscr{B} : I(\nu^\alpha\mu) \leqq c(\Theta, \nu\mu) - \delta < c(\Theta', \nu\mu) = q(\Theta', \nu\mu)\} < \Theta' ,$$

which is the desired contradiction.

Remark. It may be shown that, under the assumptions on $m(\nu^\alpha)$ given in Lemma 6,

$$r_{n,m}(\Theta, \nu) \leqq r_{n+1,m}(\Theta, \nu), \quad k\, r_{n,m}(\Theta, \nu) \leqq r_{k(m+n),m}(\Theta, \nu)$$

for $0 < \Theta \leqq 1$. It is easy to prove that if $\{t_n\}$ is a sequence of real numbers such that $t_n \leqq t_{n+1}$, $kt_n \leqq t_{k(m+n)}$, then

$$\lim_n \frac{t_n}{m+n} = \sup_n \frac{t_n}{m+n}.$$

From here it follows that the assertion of Lemma 6 may be completed with the relation

$$\lim_n \frac{1}{m+n} r_{n,m}(\Theta, \nu) = c(\Theta, \nu)$$

the latter will be shown to hold without any restriction in Lemma 9 of Sec. 5.

In Section 5 we shall need some more facts about the capacity function, which we are going to state here. Rewriting relations (3.1) in [10], Sec. 3, p. 322, for the families $\{V_\mu,\ \mu \in \mathcal{M}_{\text{erg}}(B)\}$ and $\{\{\mathcal{R}_{n,m}(\nu^\alpha\mu),\ \alpha \in \Lambda\},\ \mu \in \mathcal{M}_{m+n}(B)\}$ (cf. (2.2), (2.5), (1.2.7), and (1.1.8), and employing the equality $c(\Theta, \nu) = \bar{c}_1(\Theta, \nu)$ stated in Lemma 5, we obtain for the truncated channel ν_s defined in (2.12), where $\gamma_s = \xi(\mathcal{A}_s) > 0$, the inequalities

$$c(\gamma_s\Theta, \nu) \leqq c(\Theta, \nu_s) \leqq c([1 - \gamma_s + \gamma_s\Theta] + 0, \nu), \tag{4.9}$$
$$r_{n,m}(\gamma_s\Theta, \nu) \leqq r_{n,m}(\Theta, \nu_s) \leqq r_{n,m}([1 - \gamma_s + \gamma_s\Theta] + 0, \nu)$$

valid for $0 < \Theta < 1$ $(m = 0, 1, \ldots)$. The latter relations yield the inequalities

$$c(\Theta_s, \nu_s) \leqq c(\Theta, \nu), \quad r_{n,m}(\Theta_s, \nu_s) \leqq r_{n,m}(\Theta, \nu) \tag{4.10}$$

for $\Theta_s = \Theta - (1 - \gamma_s)$ provided that $\Theta > 1 - \gamma_s$.

5. Proofs of the theorems

Let us begin with stating an assertion which is valid for any channel ν having past history of duration $m(\nu)$: for any non-negative integers m and q, if $m \geqq m(\nu)$, $E \subset A^n$, $y \in B^{m+n}$, $E' \subset A^{n+q}$, $y' \in B^{m+(n+q)}$, and if

$$E' = \{x \in A^{n+q} : \{x_j,\ 0 \leqq j < n\} \in E\} \tag{5.1}$$

and $\{y'_j,\ 0 \leqq j < m + n\} = y$, then

$$\nu[E' \mid y'] = \nu[E \mid y]. \tag{5.2}$$

The latter equality is an immediate consequence of definition (1.2.2) and of the construction (5.1) of E'.

If $Q = \{Q(y), y \in Y\}$ is an n-dimensional code of past history $m \geqq m(v)$ for a channel v, and if q is a non-negative integer, let us associate with each $y \in Y$ the input sequence $y' \in B^{m+(n+q)}$ which is determined by

$$y'_j = y_j \quad \text{for} \quad 0 \leqq j < m + n\,, \quad y'_j = b_0 \quad \text{for} \quad j \geqq m + n\,,$$

where b_0 is a fixed element in the input alphabet B, and define the $(n + q)$-dimensional code $Q' = \{Q'(y'), y' \in Y'\}$ of past history m associated with Q by

$$(5.3) \qquad Y' = \{y' : y \in Y\}\,, \quad Q'(y') = (Q(y))' \quad \text{for} \quad y \in Y\,,$$

where we have set $(E)' = E' \subset A^{n+q}$ for $E \subset A^n$ according to (5.1). It follows from the construction of Q' and from (5.2) that

$$(5.4) \qquad l(Q') = l(Q)\,, \quad e(Q', v) = e(Q, v)\,;$$

cf. (1.2.3) and (1.2.4).

As in the preceding sections, we shall assume in the following lemmas that we are given a channel v having the property

$$v = \int v^\alpha \, d\xi(\alpha)\,, \quad v^\alpha \text{ f. mem., whereas } \quad d = d_v \quad (\text{cf. } (3.2))\,.$$

Lemma 7. *If* $0 < \Theta < \varepsilon < 1$ *then for any* $\lambda > 0$ *there is an index* n'_0 *such that, for* $n' \geqq n'_0$ *and for any integer* $m \geqq m(v)$, *there exists an* n'*-dimensional code* Q' *of past history* m *with the properties that* (*cf.* (1.2.3), (1.2.4) *and* (1.2.8))

$$e(Q', v) < \varepsilon\,, \quad l(Q') > \exp_2\left(n'[c(\Theta, v) - \lambda]\right).$$

Proof. Choose first Θ' arbitrarily but such that $\Theta < \Theta' < \varepsilon$. Choose then s sufficiently large so that $\Theta \leqq \gamma_s \Theta'$; this is possible because of (2.11), and the first inequality in (4.9) yields the relation

$$(1) \qquad c(\Theta, v) \leqq c(\Theta', v_s)$$

for the truncated channel v_s given by (2.12). Finally choose t such that

$$(2) \qquad c(\Theta', v_s) < \frac{1}{s + t}\, r_{t,s}(\Theta', v_s) + \tfrac{1}{4}\lambda\,;$$

the possibility of such a choice is guaranteed by Lemma 6. Let K'_t be a constant associated with ε, Θ', and v_s such that the assertion of Lemma 4 holds for v_s with the constant K'_t. Let k_0 be a positive integer having the property that

$$(3) \qquad 4k^{-1/2} K'_t < \lambda\,, \quad 2 \log_2 d < (k + 1)\,\lambda \quad \text{for} \quad k \geqq k_0\,.$$

Given $n' \geqq n'_0 = k_0(s + t)$, let k be the positive integer for which $k(s + t) \leqq \leqq n' < (k + 1)(s + t)$; hence $k \geqq k_0$. Applying Lemma 4 to $n = k(s + t)$ and taking into account (1), (2), and (3), we shall find that there is an n-dimensional code Q of past history m such that (cf. (3.5))

$$e(Q, v) < \varepsilon, \quad l(Q) > \exp_2(n[c(\Theta, v) - \tfrac{1}{2}\lambda]) .$$

For $q = n' - n$, let Q' be the n'-dimensional code of past history m associated with Q by definition (5.3). Since $c(\Theta, v) \leqq \log_2 d$ according to Remark 3 in Sec. 2, we obtain from (5.4) that $e(Q', v) < \varepsilon$ and

$$\frac{1}{n'} \log_2 l(Q') > \frac{k}{k + 1}\left(\frac{1}{n} \log_2 l(Q)\right) > c(\Theta, v) - \tfrac{1}{2}\lambda - (k + 1)^{-1} \log_2 d ,$$

which together with (3) yields the desired result.

Lemma 8. *If* $\gamma_s = \xi(\mathscr{A}_s) > 0$ $(s = 0, 1, \ldots$; *cf.* (2.11)) *so that the truncated channel* v_s *is defined by* (2.12), *then, for* $0 < \Theta \leqq 1$,

$$(5.5) \qquad \frac{1}{s + n} r_{n,s}(\Theta, v_s) \leqq c(\Theta, v_s) \leqq \frac{1}{s + n} r_{n,s}(\Theta, v_s) + \frac{s}{s + n} \log_2 d ;$$

hence

$$(5.6) \qquad \lim_n \frac{1}{s + n} r_{n,s}(\Theta, v_s) = c(\Theta, v_s) .$$

Proof. Given $\lambda > 0$ and $\Theta(0 < \Theta \leqq 1)$, choose first Θ' such that $0 < \Theta' < \Theta$, $c(\Theta', v_s) > c(\Theta, v_s) - \lambda$; the existence of Θ' is guaranteed by the left-continuity of the capacity function. Then choose ε in the open interval (Θ', Θ). Given n, conclude from Lemma 7 that, for k sufficiently large, there is $k(s + n)$-dimensional code Q_k for v_s (i.e. past history $m(v_s)$: cf. § 1.2) with the properties that

$$e(Q_k, v_s) < \varepsilon , \quad \frac{1}{k(s + n)} \log_2 l(Q_k) > c(\Theta', v_s) - \lambda .$$

It follows from Lemma 3 that, for a suitable constant $K_n = K_n(\varepsilon, \Theta; v_s)$,

$$\frac{1}{k(s + n)} \log_2 l(Q_k) < W_n(\Theta) + k^{-1/2} K_n + \frac{s}{k(s + n)} \log_2 d ,$$

where we have made use of the inequality $m(v_s) \leqq s$. Summarizing the above relations, we obtain for $k \to \infty$ the inequality (cf. (3.5))

$$c(\Theta, v_s) - 2\lambda \leqq W_n(\Theta) = \frac{r_{n,s}(\Theta, v_s)}{s + n} + \frac{s}{s + n} \log_2 d .$$

From the arbitrariness of λ we conclude that the second inequality in (5.5) must hold; the first follows from Lemma 6. Q.E.D.

The equality (5.6) immediately follows from the relation given in Remark to Lemma 6 in Sec. 4 because of $m(v^\alpha) \leqq s$ for $\alpha \in \mathscr{A}_s$; this means that (5.6) may be shown to hold independently of Lemma 3 and Lemma 7. By using (5.6) only together with facts established in Sec. 4 we can show the validity of

Lemma 9. *For* $m = 0, 1, \ldots,$ *and* $0 < \Theta \leqq 1$,

$$c(\Theta, v) = \lim_n (m + n)^{-1} r_{n,m}(\Theta, v).$$

Proof. Given Θ, m, and $\delta > 0$, choose s sufficiently large so that $s \geqq m$, $\Theta_s = \Theta - (1 - \gamma_s) > 0$, and $c(\Theta, v) - \delta < c(\gamma_s \Theta_s, v)$; cf. (2.1) and take into account the left-continuity of c_v. With s fixed, let (cf. (5.6))

$$\frac{1}{s + n} r_{n,s}(\Theta_s, v_s) > c(\Theta_s, v_s) - \delta \quad \text{for} \quad n \geqq n_s .$$

Now it follows from (4.9) and (4.10) that

$$c(\gamma_s \Theta_s, v) - \delta \leqq c(\Theta_s, v_s) - \delta < \frac{1}{s + n} r_{n,s}(\Theta_s, v_s) \leqq \frac{1}{s + n} r_{n,s}(\Theta, v)$$

for $n \geqq n_s$; so according to (4.6)

$$c(\Theta, v) - 2\delta < \frac{1}{m + [n + (s - m)]} r_{n+(s-m),m}(\Theta, v)$$

for $n \geqq n_s$, i.e. $c(\Theta, v) - 2\delta < (m + n)^{-1} r_{n,m}(\Theta, v)$ for $n \geqq n_s + (s - m)$. The arbitrariness of δ implies the desired result.

Lemma 10. *If ε and Θ are positive and such that $\varepsilon < \Theta \leqq 1$ then, for any $\lambda > 0$, there is an index n_0 such that, for $n \geqq n_0$ and for any integer $m \geqq m(v)$, it holds*

Assertion on Upper Bound. If Q is an n-dimensional code of past history m such that $e(Q, v) < \varepsilon$ (cf. (1.2.3), (1.2.4)) then

$$l(Q) < d^{m - m(v)} \exp_2 (n[c(\Theta, v) + \lambda]) .$$

Proof. Choose s sufficiently large that $\varepsilon < \Theta_s = \Theta - (1 - \gamma_s)$; cf. (2.11). Choose t such that $s + t \geqq m(v)$ and $s(s + t)^{-1} \log_2 d < \frac{1}{4}\lambda$. Let K_t be a constant associated with ε, Θ_s, and v_s (cf. (2.12)) for which the assertion of Lemma 3 is valid, and let k_0 be an integer with the properties that $k_0 > 3$, and that

$$K_t k^{-1/2} < \tfrac{1}{4}\lambda , \quad \frac{1}{k - 3} \log_2 d < \tfrac{1}{4}\lambda \quad \text{for} \quad k \geqq k_0 .$$

Given $n \geqq n_0 = k_0(s + t)$, let k be the positive integer for which $(k - 1)(s + t) < < n \leqq k(s + t)$; consequently, $k \geqq k_0$. Let Q be an n-dimensional code of past history $m \geqq m(v)$ having the property that $e(Q, v) < \varepsilon$. For $q = k(s + t) - n$, let Q' be the $k(s + t)$-dimensional code of past history m associated with Q by definition (5.3). Since $e(Q', v) < \varepsilon$ by (5.4), we may apply Lemma 3 so that

$$l(Q') < d^{m - m(v)} \exp_2 (k(s + t) [c(\Theta_s, v_s) + \tfrac{1}{2}\lambda] + m(v) \log_2 d)$$

as follows from above and from Lemma 6. Since $c(\Theta_s, v_s) \leqq c(\Theta, v) \leqq \log_2 d$ by (4.10) and by Remark 3 in Sec. 2, since $m(v) \leqq s + t$, and since $l(Q) = l(Q')$ by (5.4), an easy calculation yields that

$$\frac{1}{n} \log_2 (l(Q)\, d^{m(v) - m}) < \left(1 + \frac{1}{k - 1}\right)(c(\Theta, v) + \tfrac{1}{2}\lambda) + \frac{\log_2 d}{k - 1} < c(\Theta, v) + \lambda$$

because of $k \geqq k_0$; the latter inequalities show the validity of the lemma.

Lemma 11. *If* $s(v^\alpha) = 0$ *for all* $\alpha \in A$, *i.e.* v^α *have zero memory, then, for* $m = 1, 2, \ldots$, $0 < \Theta \leqq 1$,

$$r_{n,m}(\Theta, v) = r_{n,0}(\Theta, v) = n\, r(\Theta, v) \quad [\text{cf. } (2.15)] .$$

Proof. It follows from Sec. 3.2 of [2] that

$$\mathcal{R}_n(v^\alpha \mu') \leqq \sum_{i=0}^{n-1} \mathcal{R}_1(v^\alpha \mu^{p_i}) \quad \text{for} \quad \mu' \in \mathcal{M}_n(B) \tag{1}$$

where $p_i \in P(B)$, $p_i(b) = \sum\{\mu'[y] : y \in B^n,\ y_i = b\}$ (cf. (2.13), (1.1.3)), and that

$$\mathcal{R}_n(v^\alpha \mu^p) = n\, \mathcal{R}_1(v^\alpha \mu^p), \quad p \in P(B) . \tag{2}$$

Since (cf. Theorem 7, Sec. 2 in [7], p. 700)

$$\frac{1}{n} \sum_{i=0}^{n-1} \mathcal{R}_1(v^\alpha \mu^{p_i}) \leqq \mathcal{R}_1(v^\alpha \mu^{(1/n)\Sigma p_i}) ,$$

it follows from (1) and from the latter inequality that

$$r_{n,0}(\Theta, v) \leqq n\, r(\Theta, v) ;$$

cf. (1.1.2). Since $\mathcal{M}_1(B) \subset \mathcal{M}_n(B)$ (cf. (2.5), (2.4)), we obtain from (2) and from (1.1.2) that

$$n\, r(\Theta, v) \leqq r_{n,0}(\Theta, v) .$$

It is easily seen that $\mathcal{R}_{n,m}(v^\alpha \mu) = \mathcal{R}_n(v^\alpha \mu)$ because of (1.3.3), where μ is an arbitrary input; hence it is immediately deduced that $r_{n,m}(\Theta, v) = r_{n,0}(\Theta, v)$ by associating $\mathcal{M}_n(B)$ and $\mathcal{M}_{m+n}(B)$ for this special case, which proves the lemma.

Proof of Theorem 1. Let $\varepsilon \notin \mathscr{D}_\nu$, $0 < \varepsilon < 1$ so that $c(\varepsilon, \nu) - \frac{1}{2}\lambda < c(\Theta, \nu) < c(\varepsilon, \nu) + \frac{1}{2}\lambda$ for $\varepsilon - \delta < \Theta < \varepsilon + \delta$ for a suitably chosen $\delta > 0$. Given $n \geqq \max(n_0, n_0')$, we obtain under the assumption that n_0' is an index associated with $\frac{1}{2}\lambda$ for which the assertion of Lemma 7 holds, that the coding theorem is valid for n. Similarly, if n_0 is an index associated with $\frac{1}{2}\lambda$ for which the assertion of Lemma 10 holds, then the strong converse of the coding theorem must be valid for n.

Proof of Theorem 2. The first part of Theorem 2 coincides with Lemma 5. The second and the third part follow from (4.8) and Lemma 9 (cf. Remark 1 to Theorem 2).

Proof of Theorem 3. The first part of the theorem is an immediate consequence of relations (4.9). The second part coincides with Lemma 8, and the assertion concerning the existence of maxima follows from (3.4) and from Theorem 1 given in Sec. 2 of [12] because of the one-one correspondence between $P(B^{s+t})$ and $\mathscr{M}_{s+t}(B)$.

Proof of Theorem 4. The theorem is an immediate consequence of Lemma 11 and Theorem 3.

Remark. According to Remark 2 in Sec. 2, Theorem 1 together with Theorem 3 yield Wolfowitz's Coding Theorem and its Strong Converse for finite-memory channels; cf. [14], Chapter 5.

References

[1] Dobrushin, R. L. (1959). A general formulation of Shannon's fundamental theorem in information theory (in Russian). *Usp. matem. nauk*, **14**, 3—104.

[2] Feinstein, A. (1953). "Foundations of information theory". *New York*.

[3] Kadlec, J. (1972). On decomposable information channels (in Czech). Unpublished Thesis, *Charles University*, Prague.

[4] Khinchin, A. I. (1956). On the fundamental theorems of information theory (in Russian). *Usp. Matem. nauk*, **11**, 17—25.

[5] Nedoma, J. (1960). On non-ergodic channels. In: *Transact. of the 2nd Prague Conf. on Inform. Theory etc.*, Prague 1959. Academia, Prague, 363—395.

[6] Nedoma J. (1963). Über die Ergodizität und r-Ergodizität stationärer Wahrscheinlichkeitsmasse. *Z. Wahrscheinlichkeitstheorie*, **2**, 90—97.

[7] Winkelbauer, K. (1960). Channels with finite past history. In: *Transact. of the 2nd Prague Conf. on Inform. Theory etc.*, Prague 1959. Academia, Prague, 685—831.

[8] Winkelbauer, K. (1970). On the asymptotic rate of nonergodic information sources. *Kybernetika* **6**, 127—148.

[9] Winkelbauer, K. (1971). On the coding theorem for decomposable discrete information channels. *Kybernetika* **7**, 109—123 (Part I), 230—255 (Part II).

[10] Winkelbauer, K. (1971). On the regularity condition for decomposable communication channels. *Kybernetika* **7**, 314—327.

[11] Winkelbauer, K. (1972). On discrete channels decomposable into memoryless components. *Kybernetika* **8**, 114—132.

[12] WINKELBAUER, K. (1974). Information channels with memoryless components. In: *Transact. of the 7th Prague Conf. on Inform. Theory etc.*, Prague, 559—576.
[13] WINKELBAUER, K. (1973). On the capacity of decomposable channels. In: *Transact. of the 6th Prague Conf. on Inform. Theory etc.*, Prague 1971. Academia, Prague, 903—914.
[14] WOLFOWITZ, J. (1964). "Coding theorems of information theory". 2nd edition, *Berlin*.

INSTITUTE OF INFORMATION THEORY AND AUTOMATION OF CZECHOSLOVAK ACADEMY OF SCIENCES, PRAGUE, CZECHOSLOVAKIA

Received March 1976

ABSTRACTS

J. ANDĚL: *On Interpolation of Multiple Autoregressive Processes*

Let $\mathbf{X}_1, \ldots, \mathbf{X}_N$ be a p-dimensional autoregressive process of the order n. Suppose that $\mathbf{X}_1, \ldots$ $\ldots, \mathbf{X}_{i-1}$ and $\mathbf{X}_{i+q}, \ldots, \mathbf{X}_N$ are known and we want to estimate $\mathbf{X}_j$ $(i-1 < j < i+q)$. If each of those known parts has at least n members then the best linear interpolation depends only on n previous and n subsequent known vectors, i.e. only on $\mathbf{X}_{i-n}, \ldots, \mathbf{X}_{i-1}$ and $\mathbf{X}_{i+q}, \ldots, \mathbf{X}_{i+q+n-1}$. A very simple interpolation formula is obtained in the case $q = 1, j = i$.

R. R. BAHADUR: *A Note on UMV Estimates and Ancillary Statistics*

In a given statistical framework let t be a UMV estimate, i.e., a square integrable function which is the uniformly minimum variance unbiased estimate of its own expected value, and let u be an ancillary statistic. This note describes certain general conditions which imply that t and u are independent, and also gives an example where t and u are not independent.

R. H. BERK and I. R. SAVAGE: *Dirichlet Processes Produce Discrete Measures: An Elementary Proof*

Let $\mathscr{X}$ be a space and $\mathscr{A}$ a σ-field of subsets, and let α be a finite non-null measure on $(\mathscr{X}, \mathscr{A})$. Then a stochastic process P indexed by elements A of $\mathscr{A}$ is said to be a Dirichlet process on $(\mathscr{X}, \mathscr{A})$ with parameter α if for any measurable partition $(A_1, \ldots, A_k)$ of $\mathscr{X}$, the random vector $(PA_1, \ldots$ $\ldots, PA_k)$ has a Dirichlet distribution with parameter $(\alpha A_1, \ldots, \alpha A_k)$. The paper follows up with results by Ferguson and Blackwell and gives an elementary proof of Ferguson's main theorem stating that $Pr(P \text{ is discrete}) = 1$.

P. J. BICKEL and E. L. LEHMANN: *Descriptive Statistics for Nonparametric Models IV. Spread*

The spread of a probability distribution is studied as an alternative to the dispersion which does not require the assumption of symmetry. Measures of the spread are defined as functionals satisfying certain equivariance and ordering conditions. The different measures are compared in terms of the asymptotic relative efficiencies of their consistent estimates.

T. DALENIUS and R. A. VITALE: *A New Randomized Response Design for Estimating the Mean of a Distribution*

The mean μ of a characteristic X in a population of individuals is to be estimated by an interview sample survey based on simple random sampling with replacement. The problem is to design a survey which protects the anonymity of the individual respondent. The paper presents a new randomized response design which is simple to use and which affords a large measure of anonymity. The individual selected for the survey is asked to spin the spinner and report whether the value S he sees is at least the value of his characteristic X. The observations provide an unbiased estimate of μ; its variance may be reduced by requiring each respondent to spin the spinner s times.

T. DALENIUS: *Heuristics for Survey Sampling Design: A Case Study*

A general problem of the present survey sampling methods and theory is that of constructing an efficient survey sampling design. By means of a case study, the author shows how this problem can be tackled by way of a heuristic approach in the case that the design problem is not amenable to an analytic approach. The heuristic method uses a large-scale computer as a basis for generating data which are then used to reflect the performance of some "trial design"; i.e., it analyzes the realizations of a certain scheme in order to detect possibilities for improvements.

V. DUPAČ: *Asymptotic Normality of the Continuous Robbins-Monro Stochastic Approximation Procedure*

The continuous time Robbins-Monro process described by the Itô stochastic differential equation

$$\mathrm{d}\mathbf{X} = a(t)\,(R(\mathbf{X})\,\mathrm{d}t + \sum_{r=1}^{k} \sigma_r(t, \mathbf{X})\,\mathrm{d}\xi_r(t))$$

is considered. Under conditions analogous as in the monograph by Nevelson and Hasminskii, where the asymptotic normality is proved for $a(t) = a/t$, the result is extended to the case of $a(t) = a/t^{\alpha}$ with $\frac{1}{2} < \alpha < 1$.

J. DUPAČOVÁ: *A Note on Rejective Sampling*

In connection wih rejective sampling, defined by (1) of the paper, the following question was raised: Can the inclusion probabilities π_j, $1 \leq j \leq N$, be arbitrarily prescribed? The question is answered affirmatively in the main theorem of the paper (Section 2).

L. G. GVANCELADZE and D. M. CHIBISOV: *On Tests of Fit Based on Grouped Data*

The problem of testing a simple hypothesis H_0 specifying a density function $p_0(x)$ of a real-valued random variable is considered. The sample of size n is grouped into k class intervals, equiprobable under H_0, with the respective frequencies $\nu_1, \ldots, \nu_k$ ($\nu_1 + \ldots + \nu_k = n$). It is proved that, under $k \to \infty$ and $n \to \infty$, any sequence of tests symmetric in $\nu_1, \ldots, \nu_k$ asymptotically does not distinguish H_0 from a contiguous alternative of the form

$$H_{1n}: p_n(x) = p_0(x)\,(1 + n^{-1/2}\,h(x) + r_n(x))\,;$$

i.e. the difference of the size and of the power of such tests tends to zero as $k \to \infty$ and $n \to \infty$. It is then proved with the aid of this result that the values of k maximizing the power of the Pearson's chi-square test against H_{1n} tend to a finite limit.

R. Z. HASMINSKII: *Lower Bounds for the Risks of Nonparametric Estimates of the Mode*

The main goal followed by the paper is to show that Hájek's ideas concerning the asymptotically minimax lower bounds for the risks of the estimates can be successfully applied to the type of nonparametric estimates with degenerate analogue of Fisher's information. Utilizing Hájek's ideas, the author derives the asymptotically minimax lower bounds for the estimates of the mode of a probability distribution.

M. HUŠKOVÁ: *The Rate of Convergence of Simple Linear Rank Statistics under Alternatives*

The rate of convergence of distribution functions of the simple linear rank statistics under general alternatives is treated. The conditions are analogous to those considered by Hájek (1968) in his Theorems 2.1 and 2.2. The main result is proved with the aid of the projection method suggested by Hájek (1968) and with the method for treating the difference of characteristic functions developed by Bjerve (1973) (see also Bickel (1974) and Hušková (1975)).

J. JUREČKOVÁ: *Nuissance Medians in Rank Testing Scale*

Let $X_1, \ldots, X_m$ and $X_{m+1}, \ldots, X_{m+n}$ be two samples with the respective densities $\sigma_1^{-1} f(\sigma_1^{-1}(x - \mu))$ and $\sigma_2^{-1} f(\sigma_2^{-1}(x - \nu))$, $\sigma_1, \sigma_2 > 0$. Rank testing $\sigma_1 = \sigma_2$ against $\sigma_1 \neq \sigma_2$ (or $\sigma_1 > \sigma_2$, or $\sigma_1 < \sigma_2$) is considered under the conditions that $\mu \neq \nu$ are unknown nuissance parameters and are replaced by the proper estimates. If $\mu = \nu$ then proper rank tests for testing $\sigma_1 = \sigma_2$ are based on the statistics $S_N = s_N(X_1, \ldots, X_N)$, $N = m + n$, where $s_N(x_1, \ldots, x_N)$. is the simple linear rank statistic corresponding to a score-generating function $\varphi(t)$, $0 < t < 1$. The adjusted statistics $S_N^* = s_N(X_1 - \bar{\mu}_N, \ldots, X_m - \bar{\mu}_N, X_{m+1} - \bar{\nu}_N, \ldots, X_N - \bar{\nu}_N)$ are shown to be asymptotically normally distributed under $\sigma_1 = \sigma_2$ as well as under contiguous scale alternatives. The result includes Hájek's earlier result concerning symmetric f and φ.

L. LE CAM: *On a Theorem of J. Hájek*

The paper is devoted to a generalization of Theorem 4.1 of Hájek's fundamental paper "Local Asymptotic Minimax and Admissibility in Estimation". Hájek's results are extended to abstract versions which are applicable to a variety of situations which are not necessarily related to the local asymptotic normality assumed by Hájek. The use of the generalized results is then exemplified by considering the sequence of experiments satisfying the assumptions closely related to Hájek's local asymptotic normality.

P. MANDL: *On Aggregating Controlled Markov Chains*

The problem of optimal control of a controlled Markov chain corresponding to an aggregate consisting of independent units A', A'', ... checked at times $0, \Delta, 2\Delta, \ldots$ is considered. The criterion is of the form $\mathsf{E}e^{-Rh\Delta}$, $h > 0$, and it keeps the sequence of total rewards $\{V_n, n = 0, 1, \ldots\}$ within prescribed bounds l_0, l_1; $R = \inf\{n : V_n \notin (l_0, l_1)\}$. In general the control minimizing $\mathsf{E}e^{-Rh\Delta}$ will be a function of the state J_n of the aggregate as well as of V_n, and therefore it will be rather awkward to compute. The paper deals with the controls of the form $\{u(V_n), n = 0, 1, \ldots\}$. A sufficient condition for the convergence of a continuous random function, associated with $\{V_n, n = 0, 1, \ldots\}$ to a controlled diffusion process as $\Delta \to 0$ is given. To the processes satisfying the condition, the theory of one-dimensional Markovian optimization problems with boundaries, developed earlier by the author, is applicable; this enables to obtain approximately optimal controls.

J. Oosterhoff and W. R. van Zwet: *A Note on Contiguity and Hellinger Distance*

Two sequences of product probability measures $\prod_{i=1}^{n} P_{ni}$ and $\prod_{i=1}^{n} Q_{ni}$ are considered. Necessary and sufficient conditions in terms of the marginal distributions P_{ni} and Q_{ni} are derived for the contiguity of the sequences. Boundedness for $n \to \infty$ of the sum of squares of the Hellinger distances of the marginals is one of the conditions. By strengthening these conditions one obtains sufficient and (almost) necessary conditions for the asymptotic normality of the log likelihood ratio statistic.

J. Pfanzagl: *First Order Efficiency Implies Second Order Efficiency*

It is shown that for any test obtained from a sufficiently regular test statistic by asymptotic studentization the coincidence of the power with the envelope power up to $o(n^0)$ implies coincidence up to $o(n^{-1/2})$. This holds true even if the estimators of the nuissance parameters (applied for asymptotic studentization) are inefficient. These results on tests are used to show that for any asymptotically efficient estimator of a vector parameter which is componentwise median unbiased $o(n^{-1/2})$, the term of order $n^{-1/2}$ in the Edgeworth-expansion of its distribution is uniquely determined.

M. L. Puri, J. S. Rao and Y. Yoon: *A Simple Test for Goodness-of-Fit Based on Spacings with some Efficiency Comparisons*

The problem is to test whether the common distribution of independent random variables $X_1, \ldots, X_{n-1}$ is the uniform one on (0, 1). A simple class of tests $R_n' = R_n'(n\delta_n)$ based on the spacings $T_i = X_i' - X_{i-1}'$, $i = 1, \ldots, n$ where $X_1' \leqq X_2' \leqq \ldots \leqq X_{n-1}'$ are the order statistics $(X_0' = 0, X_n' = 1)$ is proposed. The exact null distribution and the asymptotic null distribution of the test statistic are obtained. The asymptotic relative efficiency and Bahadur efficiency of R_n relative U_n are discussed, where U_n is another spacings test considered by Rao (1969). The statistic R_n^* which has the maximum limiting efficiency in the class $R_n(n\delta_n)$ is considered. A table that can be used to obtain the critical values of R_n^* is also provided.

P. K. Sen: *Nonparametric Tests for Interchangeability under Competing Risks*

Some nonparametric tests for the hypothesis of stochastic interchangeability of the elements of a random bivariate vector under competing risk model are proposed and studied. Both fixed sample and sequential procedures are considered. The case of progressively censored non-parametric procedures is also presented. The theory of these procedures rests on a stochastic process approach based on the Brownian motion approximations for some allied rank statistics. The choice of optimal score functions is discussed.

J. SETHURAMAN: *Characterization of Distributions by Large Deviation Rates*

The exponential rate of convergence to zero of large deviation probabilities of the mean of independent and identically distributed random variables is well known. The present paper shows that this large deviation rate completely determines the common distribution. This feature becomes noteworthy when contrasted with the fact that the limits of the probabilities of ordinary and moderate deviations of the mean depend very little on the common distribution.

Z. ŠIDÁK and S. HOJEK: *Monte Carlo Comparisons of Some Rank Tests Optimal for Uniform Distributions*

Power functions of the E-test (proposed by J. Hájek), Haga test and Wilcoxon test are compared by Monte Carlo method for the case of the two-sample location problem with uniform distributions. Motivation of the choice of uniform distributions is that the former two tests are locally most powerful rank tests in this case. Some tentative practical conclusions are drawn from the comparisons.

J. ŠTĚPÁN: *Simplicial Measures*

A metrizable compact convex set X in a locally convex topological vector space is considered. A Borel probability measure μ on X is called simplicial if it is an extreme point in the set of all probability measures with the same barycenter as μ. A necessary and sufficient condition for a probability measure to be simplicial is found. The result generalizes that of E. M. Alfsen (1971) from finite to infinite-dimensional spaces.

I. VINCZE: *On the Cramér-Fréchet-Rao Inequality in the Non-Regular Case*

Let

$$f = \{f(x, \vartheta) = f(x_1, \ldots, x_n; \vartheta) : \vartheta \in \Theta\}$$

be a system of densities with respect to a σ-finite measure μ on $\mathscr{B}_n$, let $g(\vartheta)$ be an identifiable function of ϑ and let $t(X)$ be its unbiased estimate. For fixed $\vartheta, \vartheta' \in \Theta$, let

$$f_\alpha = \{f_\alpha(x, \vartheta, \vartheta') = (1 - \alpha) f(x, \vartheta) + \alpha f(x, \vartheta'),\ 0 < \alpha < 1\}.$$

The paper provides a survey of variance inequalities for $t(X)$ with X being distributed according to $f_\alpha(x, \vartheta, \vartheta')$ which hold in the non-regular case, namely in the case that the support of $f(x, \vartheta)$ depends on ϑ. Some points differ from earlier investigations of the problem.

J. Á. VÍŠEK: *Asymptotic Distribution of Simple Estimate for Rejective, Sampford and Successive Sampling*

Asymptotic normality of the simple estimate for the rejective, Sampford and successive sampling ic established. The results concerning Sampford and successive sampling are new; the part conserning the rejective sampling gives a simpler proof of the result proved by Hájek in 1964.

K. WINKELBAUER: *Discrete Communication Channels Decomposable into Finite-Memory Components*

The paper is devoted to the study of the asymptotic behaviour of the maximum length of n-dimensional error-correcting codes having their maximum probability of errors less than a prescribed level. The investigation is performed for the case of discrete communication channels which are decomposable into a finite or infinite number of component channels having each finite memory. The result is a generalization of the author's earlier result concerning the channel decomposable into memoryless components.

AUTHOR INDEX

SUBJECT INDEX